Gut Feelings

publication supported by a grant from
The Community Foundation for Greater New Haven
as part of the **Urban Haven Project**

Gut Feelings

The Microbiome and Our Health

Alessio Fasano and Susie Flaherty

The MIT Press
Cambridge, Massachusetts
London, England

MIT Press first paperback edition, 2022

This book was set in Stone Serif and Stone Sans by Westchester Publishing Services. Printed and bound in the United States of America.

Library of Congress Cataloging-in-Publication Data

Names: Fasano, Alessio, author. | Flaherty, Susie, author.
Title: Gut feelings : the microbiome and our health / Alessio Fasano and
    Susie Flaherty.
Description: Cambridge, Massachusetts : The MIT Press, [2021] |
    Includes bibliographical references and index.
Identifiers: LCCN 2020012663 | ISBN 9780262044271 (hardcover)
    ISBN 9780262543835 (paperback)
Subjects: LCSH: Gastrointestinal system--Microbiology.
Classification: LCC QR171.G29 F37 2021 | DDC 612.3/2--dc23
LC record available at https://lccn.loc.gov/2020012663

10 9 8 7 6 5 4

Contents

Preface

To better appreciate the potential role of the human microbiome in health and disease, let's take a look at where we've been, where we are now, and where we are going on our journey into the microbial world. Not so long ago, we focused our studies on host-microbe interactions, convinced that microorganisms are exclusively responsible for infectious disease.

The germ theory of infectious disease suggesting that some diseases are caused by microorganisms was first formulated by Girolamo Fracastoro in 1546. His theory challenged the much more popular and almost universally accepted miasma theory from Galen (and Greek pollution) suggesting that endemic diseases such as cholera or the "Black Death" were due to a noxious form of "bad air." The Black Death, a bubonic plague pandemic caused by the bacterium *Yersinia pestis*, killed between seventy-five million and two hundred million people throughout Eurasia and North Africa in the mid-fourteenth century.

The pioneering work of Francesco Redi, Agostino Bassi, Ignaz Semmelweis, and John Snow led to the seminal discoveries of Louis Pasteur and Robert Koch, which not only confirmed the germ theory but also established the new scientific fields of bacteriology and infectious diseases. Ever since, we have been engaged in warfare with microorganisms to eradicate them to efficiently treat infectious diseases. The concept has been (and in many minds still is) that microorganisms are indiscriminately nasty enemies that need to be defeated at all costs by developing and deploying weapons of mass destruction (antibiotics).

Now the entire field of disease pathogenesis and possible prevention has been revolutionized by the appreciation that the microbial ecosystem is instrumental in maintaining a state of health, and when disturbed, may instigate the onset of a series of diseases in genetically predisposed individuals. With the mapping of the human genome followed by the appreciation of the complexity of the human microbiome, we realize that we are immersed in an extremely rich environment of microorganisms.

Previously, we have only been able to cultivate 5 percent of the members of this microbial environment that live parallel lives to our human ones. Now we have the tools to take us far beyond anywhere we even dreamed of just five or ten years ago in our quest to uncover the components of not only the microbiome, but also the "virome," the "parasitome," the "fungome," and more.

We have realized that, just like there are belligerent people, there are belligerent microorganisms that can make us sick. The SARS-CoV-2 virus responsible for the recent COVID-19 pandemic provides a devastating example. But there also are friendly bugs that live a peaceful and symbiotic life with their hosts. The more we learn about this cross-talk between ourselves and our microbiome, the closer we will be to predicting disease as well as understanding its development.

Before we delve more deeply into this brave new world of microorganisms and their habitats, let's take a look at our own species. We pose this question to the reader: Throughout human history, what is the most abiding trait of a researcher? Think about researchers in any endeavor: philosophy, geography, science, aeronautics, or the history of rock 'n' roll. Think about any area in which humans are looking for answers to specific questions. What is the common denominator in all these areas of exploration? The answer is intellectual curiosity.

From the time that humans moved beyond the continent of Africa to inhabit the far reaches of the globe, curiosity—sparking our compulsion to explore—has been our motivating force. Christopher Columbus was compelled to explore the shores of the American continent in the same way unnamed Polynesian boatmen traveled vast distances to settle in new lands.

Humans have always been explorers. From Magellan to Columbus, from traveling the Silk Road to traversing the Amazon basin, exploration is a deeply rich part of our history. Once we uncovered all the reaches of the continents, we explored the oceans—think of Jacques Cousteau's *Calypso*, a British minesweeper transformed into a research vessel. Along with sea exploration came our insatiable desire to explore space, and now we are exploring Mars.

We are driven by curiosity and a deep need to find new frontiers and new extraterrestrial worlds. What drives us to explore? Are we trying to see if we are truly alone and if we are the only inhabited planet? We know we are a small part of a huge space, and it seems we can't be here on our own. So, what are we looking for? What do we hope to find when we pursue the possibility of a new civilization?

Do we think these extraterrestrials might look like us, use the same tools or even idioms that we do? Think about all the fantastic space literature and movies with bizarre creatures and cybermonsters. In our imaginations, we have dreamed up many implausible creatures as we search for that elusive frontier beyond our borders.

This incredible ecosystem that we call the human microbiome is the latest frontier. We have been looking for new civilizations far, far away, while the most fascinating, complex, and sophisticated civilization ever discovered has been living within us all this time; we just didn't appreciate it. Our body is the world the microorganisms inhabit, and these microscopic species can move from us to another universe by moving to another body or environment. They grow like we do, they interact like we do, and they speak different languages like we do. Their genetic language is similar to ours in many senses. They also speak a metabolic language that shows us what is going on in their world.

During their millions of years of evolution, they have studied the human host very carefully and found a way to communicate with us. They understand very clearly our anatomy and physiology, our strengths and our weaknesses, and our biological necessities and goals. The bottom line is that they have developed a full array of lines of communication to understand who we are and how we function.

Conversely, we know very little about our tenants. Sometimes we understand them; sometimes we do not. Our challenge is to close this knowledge gap. Now that we have discovered this new universe, we have the opportunity to learn more about our cohabitants and to establish a friendlier line of communication.

We can change our attitude from the myopic view of microorganisms as enemies to an informed perspective on a civilization that needs to be respected and engaged with for our own benefit. The goal of this book is to help the reader understand that we are at the dawn of another scientific revolution—one that will lead to a paradigm shift in science and medicine, opening up new ways to treat and prevent diseases as we have never been able to do before.

Authors' Note: We completed writing this book before a novel coronavirus initiated a global pandemic, severely testing the international public health response and world economies. We have added some material on COVID-19 to certain sections, but we recognize that the public health and research landscape will change rapidly in the coming year. Our best hope is that a highly collaborative and rapid response from the international research community, aided by industry and supported by thoughtful world leaders, will continue to develop effective treatments and vaccine candidates by the time of this book's publication and beyond.

I The Wisdom of a Microscopic Species

1 Evolutionary Biology Explains Bacterial Adaptability

First Living Microorganisms on Planet Earth

After emerging as the first inhabitants of our planet nearly four billion years ago, microbes rapidly spread to all parts of the world. From a melting patch of snow in Greenland to volcanic dust from Western Australia, scientists have observed the ubiquitous and tenacious nature of early microbial life. They have long confirmed the enormous effect one minuscule organism can have on our physical environment. In one example, a single species of cyanobacteria recently identified in western Greenland seems to be accelerating the darkening of the Greenland Ice Sheet. The microbe causes dust and soot to clump together, leading to a more rapid melting of the largest ice body in the Northern Hemisphere.[1]

Another group of extremophiles, as microbes that can survive harsh conditions have been named, has been studied in the far reaches of the Kamchatka Peninsula. In this volcanic region of the Pacific Rim of Fire, organisms survive conditions unthinkable for most forms of life.[2] Archaea, single-celled microorganisms with no nucleus, are a distinct domain of life along with bacteria and eukaryotes (see below). They are well represented in the acidic and extremely hot temperatures found in the volcanoes, geysers, and hot springs marking this spectacular landscape.

Archaea, Greek for "ancient things," thrive on sulfur instead of oxygen and are thought to be some of the oldest organisms on the planet. Scientists are scrambling to see how high temperatures can go and still sustain life for these primordial creatures. One group found a community of archaea deep in a thermal vent in the Pacific Ocean that can survive for two hours at 130 degrees Celsius.[3] And in 2012 scientists from the University of Colorado Boulder found a handful of newly identified bacteria, fungi, and archaea in South American volcanoes that survive at a temperature range of –10 to 56 degrees Celsius amid high ultraviolet radiation.[4]

Researchers have classified microbial fossils from nearly four billion years ago, not long after planet Earth was formed, according to current scientific thought. Steve Schmidt, founder of the Alpine Microbial Observatory, is among the new generation of microbe hunters tracking desolate and inhospitable places on planet Earth and beyond. Schmidt is collaborating with other scientists from around the globe to crack open core ice samples from Antarctica to study cryoconites, pockets of microbial life, which might provide some clues to community settlement in an extreme environment. From their characterization of single-strand DNA viruses from three sites in Taylor Valley, Antarctica, they determined the presence of a "unique regional virome" in freshwater hosts in the region of the McMurdo Dry Valleys.[5]

Given the rapid, continual changes in atmosphere and climate that occurred so long ago, we can predict that adjusting to these sudden changes required a great deal of genetic flexibility. It is no surprise that microorganisms, which are known for their exhaustive ability to adapt quickly, were most likely the first living creatures to colonize our planet. And because of the lack of oxygen in Earth's early atmosphere, the vast majority of microorganisms were anaerobic, meaning they could survive without oxygen.

Evolution of Eukaryotes

To more fully appreciate the great deal of adaptivity that characterizes the prokaryotic lifestyle, we need to examine the evolutionary aspect of prokaryotes and eukaryotes. Prokaryotes, which include the form of life we call bacteria, are single-celled organisms that lack a nucleus. In the evolutionary chain, they are thought to precede the development of eukaryotes, which are multicellular organisms that include fungi, plants, and animals. Evolutionary history has been written by mainly focusing on eukaryotic organisms, overlooking the microbes that represent the most ancient and diverse assemblage of organismal diversity. This limits our overall understanding of evolutionary biology.

The evolutionary path of most eukaryotes is defined as anagenesis, a gradual process in which a species slowly accumulates adaptive genetic changes under the pressure of environmental modifications. Eventually, as in thousands or millions of years, the originating species becomes extinct in favor of the new species.

A classic example of anagenesis is the evolution of the horse, which began its evolutionary journey as a small mammal called *Eohippus*, meaning "dawn horse."[6] A short animal twelve to twenty inches tall, it had an arched back, a short neck and snout, and a long tail. A long chain of species changes over tens of millions of years led the evolution from *Eohippus* to the modern horse, *Equus*. The most visible evolutionary changes were the overall size of the animal and dramatic changes to its lower extremities.

About fifty-six million years ago *Eohippus* lived mainly in forests and jungle environments and fed primarily on leaves. But over time, the preferred habitat of the Equidae (evolutionary family of the modern horse) shifted from forests to plains. Roaming the wide, open spaces of the plains forced the Equidae to become bigger, so they could discourage smaller predatory animals and outrun larger ones. Ancestors of the modern horse gradually lost most of their toes and padding, developing a single hoof as they slowly adapted from forests to plains.

Evolution of Prokaryotes

Conversely, cladogenesis, an evolutionary path more typical but not exclusive of prokaryotes, is the evolutionary splitting of a parent species into two distinct species, forming a "clade." This event usually occurs when a few organisms end up in new, often distant areas, or when environmental changes cause several extinctions, opening up new ecological niches for the survivors.

Analysis of the phylogenetic trees for microbes suggests that macroevolution (major evolutionary change) may differ between prokaryotes and both micro- and macroeukaryotes. Phylogenetic trees, diagrams that represent evolutionary relationships between organisms, have suggested that prokaryotes and some microbial eukaryotes conform to expectations that assume a constant rate of cladogenesis over time and among lineages.

However, the key difference found in macroevolutionary dynamics between prokaryotes and many eukaryotes is due, at least in part, to differences in the prevalence of lateral gene transfer mainly mediated by bacteriophages between the two groups. Inheritance is predominantly, if not wholly, vertical within eukaryotes, a feature that allows for the emergence and maintenance of heritable variation among lineages over long periods of time, which is not ideal for rapid adaptation.

By contrast, the frequency of lateral gene transfer in prokaryotes may increase the rate of cladogenesis through heritable variation, which results from the emergence of rapid environmental changes. Taking all this into account, the inferred difference in macroevolution might reflect the relative exclusivity of key innovations in eukaryotes and their more unrestrained nature in prokaryotes.

Humans Benefit from a Happy Accident of Fate

With the appearance of oceans and oxygen in the atmosphere, a symbiotic relationship between some of the early anaerobic microorganisms and other unicellular organisms known as mitochondria occurred. Mitochondria, specialized structures within cells, are the cellular "engines" that produce most of the energy that cells need to survive.

The interaction was between so-called "simple," one-celled or unicellular organisms; this phenomenon can be viewed as the first example of how developing a mutually beneficial relationship may have given an evolutionary edge to both participants. By engaging in an intercellular lifestyle, mitochondria gained access to the metabolic pathways of their early microbial hosts, which became instrumental for the development of the mitochondria.

At the same time, bacteria gained the tremendous advantage of maximizing energy production through glycolysis, the breakdown of glucose by enzymes. The oxidation of one single glucose molecule under anaerobic conditions produces approximately two moles of adenosine 5'-triphosphate (ATP), a molecule that stores and transfers energy in a cell.

Under aerobic conditions, however, ATP production from one glucose molecule skyrockets to thirty-two moles. The addition of oxygen makes ATP production sixteen times more efficient. Even in the earliest phases of the development of life on our planet, joining forces and working together gave a tremendous advantage to both parties, something that holds true today in examples of symbiotic bacteria and other microbes, as well as in animal and human communities.

Fighting a War—or Learning to Coexist?

The next evolutionary step occurred almost two billion years ago with the development of single-celled organisms into pluricellular organisms. How that happened remained a mystery until the recent past; it looks like Japanese researchers have now found the missing link.[7] Scientists have long known that there must have been predecessors along the evolutionary road. But to judge from the fossil record, complex multicellular species simply appeared out of nowhere. The new species discovered by the Japanese scientists on the floor of the Pacific Ocean, called *Candidatus* Prometheoarchaeum syntrophicum strain MK-D1, turns out to be just such a transitional form, helping to explain the origins of all animals, plants, fungi—and, of course, humans.

One of the most eloquent examples of early multicellular organisms is the worm *Caenorhabditis elegans*, an ancient roundworm made up of only a handful of cells. It has no respiratory or nervous system, but this multicellular eukaryotic organism exemplifies the major advantage that is the paramount goal of any living organism: preservation of the species. With a pluricel-lular lifestyle, and with only eight cells making up its gastro-intestinal system, *C. elegans* can bring the efficiency of energy harvesting to the next level.

But even in such a basic, rudimentary pluricellular organ-ism, the optimization of energy harvesting and reproduction is highly dependent on another example of symbiosis. Indeed, the intestines of these animals cannot function at their best unless they have been colonized with bacteria. Whether we are talk-ing about worms, dinosaurs, or human beings, there is no plu-ricellular life on Earth that can occur without coevolution with microorganisms. Just as one microbe can affect the largest ice ecosystem in the Northern Hemisphere, one microbe can have an enormous effect on the human ecosystem.

Of course, our current understanding of the different ecosys-tems to which we are exposed is still in the early stages. Nev-ertheless, it is obvious and intuitive that if we study microbes found in different habitats—in the soil, the ocean, springs and rivers, volcanoes, the air, and plants and animals—as distinct ecosystems, these are not compartmentalized and totally segre-gated environments. Instead, there is a continuous exchange of microorganisms from the soil to human beings, from humans to water to air, and so on. To understand the human microbiome and its implications for health and disease, we need to consider the whole of the earth's ecosystem as a continuous circle of life.

"Missing Microbes" Hypothesis

If we want to appreciate the health consequences in departing from this evolutionary co-adaptation, we need to understand the evolutionary path of this human-microbiome symbiotic relationship. We know that the composition and function of the human microbiome, particularly the gut microbiome, is influenced by a variety of factors including habitat, lifestyle, and diet. Although every person has a different microbiome, researchers have identified certain trends by studying the human microbiome across the globe.

For example, the gut microbiome in people from industrialized countries seems to contain 15 to 30 percent fewer species when compared with the gut microbiome of people from non-Western nations. Aside from the obvious differences in lifestyle, there are several theories that attempt to explain these differences. One proposal is the "missing microbes" hypothesis put forth in 2014 by Martin Blaser. He theorized that losses of particular bacterial species of our ancestral microbiota, due in large part to antibiotic treatments, have altered the context in which immunological, metabolic, and cognitive development occur in early life, resulting in increased chronic inflammatory disease.[8] But on what premise is this theory of missing microbes based?

We will revisit Blaser's theory in more detail and examine the comparative analysis of ancient versus modern microbiomes in chapter 2. Assuming that the extinct microbiome theory is correct, by definition no living population today carries an "ancestral" microbiome.

In trying to close this knowledge gap, scientists have completed analyses of the microbiome composition of human mummified remains and coprolites (fossilized fecal matter). The results show an evolution of the human microbiome that reveals major shifts that coincide with radical changes in lifestyle, like changing from forager-gatherer (typical of apes) to hunter-gatherer (typical of hominids) or the advent of agriculture.

Gut Microbiome of Nonhuman versus Human Primates

Comparative analyses of human and nonhuman primate gut microbiomes by Rob Knight, Emily Davenport, and others show that humans have lower gut microbial diversity, increased relative abundances of *Bacteroides* (a genus of Gram-negative, anaerobic bacteria), and reduced relative abundances of the genera *Methanobrevibacter* (anaerobic archaea) and *Fibrobacter* (anaerobic, ruminal bacteria). These changes seem to reflect the shift toward a carnivorous diet. Other major historical and evolutionary changes identified by comparing nonhuman primate and human microbiomes also provide an indication of how quickly the human microbiome is changing. Human gut microbiome composition appears to have diverged from the ancestral state at an accelerated pace when compared to that of the great apes.[9]

Some hallmarks of human evolution and history potentially responsible for this accelerated shift in gut microbiome composition include the advent of agriculture, cooked food, urbanization, and increased population density. The human interpopulation differences seem to be driven by diet, while in nonhuman primates the key driving force seems to be related to host habitat and season.

To summarize, it seems likely that there is a strong interrelationship between the host and the host's gut microbiome with a potentially mutual influence on evolutionary traits; what remains to be established are the health consequences. More specifically, it will be important to determine when changes in microbiome composition and function are necessary for evolutionary adaptations, as opposed to detrimental, unplanned changes that may lead to disease onset.

Based on the evidence presented earlier, it is likely that humans and their microbiome share a coevolutionary destiny that seems to be species-specific. For example, "humanized mice," in which the rodent's microbiome is replaced with a human microbiome, show impairment of a fully mature intestinal immune system and are less protected from infection when compared to mice with native microbiota. This evidence leads to the introduction of another concept of human-microbiome coevolution, namely the hologenome.

Proponents of this theory of evolution argue that there is a strong and mutual influence between phenotypes induced in the human host by its genome and those induced by microbes in a combined host-microbe system, or holobiont. These evolutionary theorists conceptualize this link as so intertwined that they consider the host-microbe system an extension of a unitary organism on the same evolutionary journey.

Nine Months versus Twenty Minutes

It should come as no surprise that a microorganism adapted to live at 56 degrees Celsius in a volcano or at −67 degrees Celsius in the Arctic can quickly acclimatize to a 37 degrees Celsius environment if it needs to cohabit with a human being. This exquisite ability to adapt is the consequence of the rapid life cycle of a microorganism that can reproduce itself every fifteen to twenty minutes. During this process, random genetic mutations occur. The ones beneficial to a new lifestyle will be conserved, while other adaptations that are useless or even detrimental will not survive this evolutionary process.

While we are not currently observing the tremendous changes that occurred early in Earth's genesis, the climate changes we are now witnessing demonstrate, once again, that adaptation is the winning attribute needed to survive. Some multicellular species are on the verge of extinction because of changes in climate. While nimble bacteria can take advantage of these climatic and environmental changes, human beings in industrialized countries are experiencing varying epidemics of inflammatory diseases, including the recent COVID-19 pandemic, that are potentially related to these changes.

If human beings could reproduce every twenty minutes, it is certain that we would enjoy the same plasticity that characterizes bacteria and other microorganisms. But our biological clock requires nine months for reproduction. Random genetic mutation in the human, whether beneficial or detrimental, occurs too rarely to adapt to sudden changes in our environment.

As single-celled organisms, bacteria could be considered "bit" players compared to more complex multicellular organisms like *Sequoiadendron giganteum* (giant sequoia) or *Homo sapiens*. As we use new technologies to drill down into the microbial domain of life to study the microorganisms in and around us, we are expanding our concept of molecular and cellular biology to include evolutionary biology and ecology as essential in the role of the human microbiome.

Cholera Teaches Us a Lesson

An exquisite example of bacterial flexibility from the microbial world is *Vibrio cholerae*, an enteric pathogen responsible for cholera, the most rapidly fatal diarrheal disease affecting humankind. Cholera remains a major cause of morbidity and mortality in many parts of the developing world. And when natural or human-made calamities occur, such as earthquakes or wars, cholera can spread rapidly through refugee camps in suboptimal sanitary conditions, presenting a significant worldwide threat.

Vibrio (the Greek word for "comma") are single, sharply curved Gram-negative rods with a long polar flagellum. With this comma-like "tail," *V. cholerae* employs its characteristic, rapid motility in a watery environment. These lethal bacteria are great swimmers.

Typically transmitted through the oral-fecal route, the deadly pathogen is spread primarily through contaminated food and water. During London's cholera epidemic in 1854, John Snow observed that the disease was spread through contaminated water drawn from a water pump on Broad Street.[10] After he convinced public authorities to remove the pump handle, new cases declined substantially. Using autopsy samples from hospital patients and washerwomen during the same cholera pandemic, Filippo Pacini isolated the comma-shaped bacillus that he called "vibrion" in intestinal mucosa in 1854.[11]

But it is Robert Koch, a physician-scientist from Baden-Baden, Germany, who is most remembered for isolating *V. cholerae* as the cause of cholera in India in 1884.[12] His pioneering discoveries in anthrax and tuberculosis led him to develop four postulates for laboratory research in microbiology that still hold true today. Using his rigorous research methods to identify bacterial agents that cause disease, Koch firmly upended the theory of spontaneous generation and replaced it with the germ theory of disease.

During the past two centuries, seven cholera pandemics have spread across the world. The first recorded cholera epidemic began in the Ganges River Delta in 1817 before traveling to Asia, the Middle East, and the Mediterranean. Five of the six subsequent cholera outbreaks have originated in India. The seventh outbreak occurred in Indonesia in 1961 and traveled to Asia and the Middle East, reaching Italy, Japan, and the South Pacific in the 1970s.

One hundred years after cholera was vanquished in South America, an outbreak in Peru spread across the continent in 1991, killing ten thousand people. It was a similar strain to the seventh pandemic, which had faded out more than a decade earlier. In 1994, an outbreak in Rwandan refugee camps in the Democratic Republic of the Congo killed tens of thousands.

The earthquake that devastated Haiti in 2010 was followed by a cholera epidemic from the same cholera species, killing approximately two hundred thousand Haitians. A new cholera species was discovered in Bangladesh in 1992 and reemerged in Calcutta in 1996.[13] Epidemiologists and researchers have been monitoring the new species and the coexistence of the earlier classical biotype.

No Zero-Sum Game for *Vibrio cholerae*

There are more than thirty *V. cholerae* species identified to date that can infect humans. What we don't know is where *V. cholerae* resides in between pandemics, when it's not affecting humans. Remember that *V. cholerae* is a great swimmer, thanks to its comma-like tail flagellum. Since water is the main vehicle of disease transmission, it has been hypothesized that its natural habitat is an aquatic environment.

But how can *V. cholerae* suddenly move from a stream or river to a totally different environment like the human intestine? This puzzling question was resolved in 1998 by Jim Kaper and coworkers at the University of Maryland Center for Vaccine Development (CVD). While Alessio Fasano (coauthor of this book) was working in Kaper's lab trying to develop a cholera vaccine, Michele Trucksis and other colleagues determined that the genetic material of *V. cholerae* is arranged in two different chromosomes: a small one with genes needed to adapt to the aquatic environment, and a larger one with genes needed to adapt to the environment of the human intestine.[14]

This second chromosome includes genes that encode for cholera's colonization factors and the elaboration of its toxin, both of which are acquired by lateral transfer through a bacteriophage, also called "phage," a type of virus that infects bacteria. After attaching to the bacteria, the phage inserts itself and stably integrates its genome in the host genome. The insertion of the *V. cholerae* pathogenic island phage allows the elaboration of a pilus, a rigid, hairlike strand used by the bacterium to colonize the human intestine. The cholera-toxin phage that elaborates the powerful cholera toxin that causes the purging diarrhea typical of cholera is an exquisite example of symbiosis between viruses and bacteria.

Vibrio constantly switches from one environment to another, representing a superb example of sustained plastic adaptability. This bacterium does not need to constantly look to mutation, but instead keeps both lifestyles in mind when deciding which genes to express and which ones to suppress. This microbe also teaches us another lesson, that the host-microbe interaction, which is the cornerstone of this book, is a two-way cross-talk. The outcome of the communications from microorganisms to the host and the host to the microorganisms dictates not only our overall well-being but also, quite possibly, our susceptibility to a variety of pathological conditions.

How did we come to appreciate this new paradigm of science? As typically happens, this was the result of reshaping scientific knowledge by challenging incorrect or only partially correct concepts; the old paradigm is overturned by the new paradigm.

Bacterial Not-So Basics

Although microorganisms are almost four billion years old, we have become acquainted with their existence only in the past few centuries, thanks to the work of pioneers like Pacini, Koch, Pasteur, Alexander Fleming, and many, many others. This interest in bacteriology has been fueled by the knowledge that, until the recent past, infectious diseases were the main reasons for the morbidity and mortality of humankind.

Developing tools to observe bacteria under the microscope or cultivate them on a petri dish and eventually kill them with antibiotics has been at the forefront of cutting-edge scientific achievements as we wage war with these antagonists of humankind. Until the very recent past, mainstream thinking in the field of microbiology and immunology has regarded an invasion of microorganisms as the major cause of human death and disease. In the last thirty years, however, this landscape has changed completely, thanks to two major accomplishments. The first was the completion of the Human Genome Project in April 2003.

Humans Are Really Genetically Rudimental

When the Human Genome Project was launched in 1990 as a worldwide effort, there was much skepticism about whether the project's ambitious goals could be achieved. The original projections were a cost of $3 billion and fifteen years to achieve the goal of mapping the human genome. The project was completed three years ahead of schedule with a $2.7 billion investment by the International Human Genome Sequencing Consortium, led in the United States by the National Human Genome Research Institute and the Department of Energy.[15]

At the same time, a parallel private effort with a very bold approach led by J. Craig Venter's group at The Institute for Genomic Research (TIGR) in Rockville, Maryland, also completed sequencing the human genome. Venter's group attained similar results, which were presented in a joint press conference with Francis Collins, director of the National Human Genome Research Institute, and President Bill Clinton in 2000.

The magnitude of this accomplishment is best appreciated through the words of Nobel laureate James Watson, who together with Francis Crick first described the DNA double helix in 1953: "Never would I have dreamed in 1953 that my scientific life would encompass the path from DNA's double helix to the three billion steps of the human genome. But when the opportunity arose to sequence the human genome, I knew it was something that could be done and that must be done. The completion of the Human Genome Project is a truly momentous occasion for every human being around the globe."[16]

The excitement that transpired from Dr. Watson's statement was based on the expectation that sequencing the human genome would lead to the solution of all diseases affecting humankind. At that time, the paradigm of one gene encoding for one protein causing one disease was the driving force of our biological knowledge. Nevertheless, the subsequent appreciation that the human being is made up of only twenty-five thousand encoding genes, since downgraded to fewer than twenty-three thousand, has left the entire world of science with a puzzling conundrum.

As president of TIGR, Claire Fraser helped to usher in the field of comparative genomics and establish the field of microbial genomics. Now director of the Institute for Genome Sciences at the University of Maryland School of Medicine, she noted that early estimates of the number of human genes were seventy-five thousand to one hundred thousand. When the human genome was initially reported to contain twenty-five thousand genes, researchers were very surprised.[17] Along with the assumed high number of genes, the concept of one gene as the cause of one disease was also discarded. Scientists were compelled to go deeper to explain the role of genetics in disease development.

How do we explain that one of the most rudimentary genetic beings, *Homo sapiens*, is complex and sophisticated enough to become the dominant species on our planet? This question has forced both basic and translational researchers back to their lab benches and clinical trials to try to make sense of what appeared to be inexplicable results. This led to the second major scientific accomplishment in recent times, the Human Microbiome Project, established through the National Institutes of Health's Common Fund in 2008.

Our Parallel Universe of Microbes

This effort to identify and classify microorganisms found in the human microbiota capitalized on technologies developed during the Human Genome Project. The classical approach to studying microorganisms, that is, viewing them under the microscope and creating cultures in an aerobic atmosphere, can only identify a small percentage of the entire microbial ecosystem present in our body.

Before the genomic sequencing revolution in the 1990s, some of the remaining microbes were identified through various techniques, including culturing anaerobic bacteria in Square-Pak flasks as described in 1982.[18] This and other techniques help us appreciate the historic complexity of the study of the microorganisms of the human microbiome, given its continual changes over time, even within the same individual.

With a limited number of genes, humans enjoy a stable human genome. We inherit it from our parents, and it can give us a genetic predisposition for a variety of biological and pathological traits. Conversely, the human microbiota expresses one hundred times more genes than humans, is extremely plastic, and can change from individual to individual and within the same individual over time, all as the consequence of a variety of environmental factors that can shape its composition and function. Our human genome has coevolved with the trillions of constantly changing microorganisms found in and on the human body.

Given these facts, it is logical and, at the same time, quite tantalizing to hypothesize that we are actually the products of these two interacting holobiontic genomes. Studying one—namely, the human genome—without analyzing the other—namely, the microbiome—will not provide the answers as to why we develop diseases given a specific genetic background.

Limiting study to only one system also will not explain why we are experiencing unprecedented surges of autoimmune and other chronic diseases, and most important, what we can do to stop these epidemics. With a better understanding of the interaction of these two systems and the potential to manipulate the microbiome to our advantage, we move closer to the promise of precision and preventive medicine, a topic addressed in part III of this book. But first, we will examine the role of the human microbiome in human evolution and human health and disease, and in particular, noninfectious, chronic inflammatory diseases.

2 The Ancestral Microbiome

Our Evolving Microbiome

Before we came to more fully appreciate the complexity of the human microbiome and its potential role in human health and disease, some scientists proposed a different reason for why we have experienced such a sharp rise in the rates of noninfectious, chronic inflammatory diseases during the past fifty years. The "hygiene hypothesis" was first postulated by epidemiologist David Strachan in 1989.[1] He proposed that increased sanitation through hand washing and water and sewage management, along with social changes like increasingly urbanized lifestyles and smaller households, led to a lower incidence of infection in early childhood that was linked to the rise in pediatric allergic disease.

Scientists have subsequently expanded the hygiene hypothesis. They argue that rising incidences of asthma, inflammatory bowel disease (IBD), multiple sclerosis, type 1 diabetes (T1D), celiac disease, and other chronic inflammatory diseases may, at least in part, be due to lifestyle and environmental changes that have made us too clean for our own good health. Evidence from some equatorial countries, where hygiene remains suboptimal and the rate of parasitic infestations remains high, further supports the hygiene hypothesis. In these areas, there has been no corresponding rise in chronic inflammatory conditions.[2]

However, when people migrate from these developing areas to industrialized countries and undertake a new lifestyle, they are subjected to the same increased risk of chronic inflammatory diseases. This seems to support the hypothesis that a "too-clean" Western lifestyle might strip us of some of the environmental stimuli still present in less developed areas of the world that keep our immune systems working properly to protect us rather than inflict damage. Perhaps some of these environmental factors that dictate the composition and function of our symbiotic microbiome play a role in boosting our natural immune protection against noninfectious, chronic inflammation.

More recently, the hygiene hypothesis has been challenged by the observation that despite increased hygiene in some developing countries, inhabitants have not experienced the same sharp rise in chronic diseases recorded in the Western world. As we learn more about the complexity of the microbiome and its potential role in dictating the balance between tolerance and the immune response that can lead to disease, the "microbiome hypothesis" has been presented as an alternative explanation.

If we view the recent changes we have experienced as a species over the last fifty years, the changes are much more complex than just increased hygiene. Diseases associated with chronic inflammation, a prolonged response that can result in changes in cell types at the site of inflammation, are expected to continue to rise during the next thirty years in the United States. In 2014, the RAND Corporation estimated that nearly 60 percent of people in the United States had at least one chronic condition;

42 percent had more than one; and 12 percent of adults had five or more chronic conditions.[3] A more in-depth analysis of the social, economic, and political changes experienced since the beginning of the sharp rise of these noninfectious, chronic inflammatory diseases reveals many other factors in play that can more comprehensively explain these epidemics.

Dental Calculus from Neanderthals

Between four and six million years ago, as early hominids evolved on the African continent, our ancestors made the shift from four legs to two. Paleoanthropologists tell us that as humans progressed, we developed a larger, more complex brain, the ability to make and use tools, and the capacity for language.

Like any field of science, paleoanthropology is not without controversy, and which human species lived when and how they might have overlapped have been subjects of debate for decades. We do know that about two million years ago, several species of the genus *Homo* made the migration out of Africa into Europe. *Homo neanderthalensis*, whose remains have been found in present-day Europe, is considered by many scientists to be our closest extinct human relative.

When *H. sapiens* made our evolutionary debut around two hundred thousand years ago, we accelerated the development of specialized tools to more easily procure food. Multiple interactions between Neanderthals and modern human species have been confirmed through genomic data, but controversy continues about whether these earlier hominids were assimilated into *H. sapiens* or died out from competition with them.

In 2017, scientists expanded the human-centered evolutionary discussion to the holobiont, which is the coevolutionary concept of the human microbiome described in chapter 1. They performed shotgun sequencing of the oral microbiota of five Neanderthal samples and compared the results with present-day subjects from Belgium and Spain. From dental analysis, scientists were able to determine very different dietary preferences between the two groups. Nevertheless, results showed that the dominant bacterial phyla in our modern mouths, Actinobacteria, Firmicutes, Bacteroidetes, Fusobacteria, Proteobacteria, and Spirochetes, were also dominant in Neanderthals.[4]

What is even more provocative are sampling results for potential pathogens as a sign of disease. The Neanderthals' oral microbiota contained fewer potentially pathogenic, Gram-negative species and were more similar to the sample from the historic chimpanzee also included in the study than to modern humans. As the authors state, "The increased diversity of Gram-negative immunostimulatory taxa in modern humans is strongly linked to a wide-range [sic] of Western diseases."[5] As well as providing data about the diet, behavior, and health of early hominids, this line of research could provide interesting insights into the evolution of microbial species among members of the genus *Homo*.

As *H. sapiens* multiplied, domesticating animals and developing agriculture, humans made the transition from hunter-nomads to farmer-settlers around 12,500 BC. And as trade expanded and settlements grew, human density gradually increased. Although Rome has been named by some scholars as the first city to reach one million inhabitants, this is still open to debate by historians and others. Whichever ancient city first hit that milestone, with the expansion of maritime trade and the rise of cities, exchanges of diverse microbial communities became more frequent. But these social and economic changes did not come without a price.

Not surprisingly, these increased microbial exchanges led to pandemics of infectious diseases, including bubonic plague, influenza, smallpox, and, more recently, COVID-19. Along with ensuring a more reliable supply of meat and milk from domesticated animals, humans also increased the risk of sickness through zoonosis (see chapter 5). From the Greek words "zoon" for animal and "nosos" for disease, zoonosis occurs when a pathogen jumps from its natural non-human animal host to a human. This can be due to mutations that allow the microorganism to adapt and infect humans. Viruses, bacteria, fungi, and parasites can all be disease trajectories through zoonosis.

Chasing the Origins of COVID-19

The word "zoonosis" took on greater global significance in late 2019, after the emergence of a novel coronavirus that was thought to originate from a market selling live wild animals in Wuhan, the capital city of Hubei Province in southern China. The causative agent of COVID-19, the disease responsible for a disruptive, global pandemic that took on epic proportions, is severe acute respiratory syndrome (SARS) coronavirus 2 (CoV-2). How SARS-CoV-2 moved from an animal or animals to enter its human host has yet to be fully determined. But the first outbreak of SARS-CoV in 2002–2003 provides some early clues into the nature of the transmission of novel coronaviruses.

In November 2002, SARS was first recognized as a human disease in Guangdong Province, China, before spreading to more than thirty countries, infecting approximately eight thousand people and causing more than seven hundred fatalities.[6] In 2003, researchers isolated SCoV (SARS coronavirus–like viruses) from Himalayan palm civets in a live-animal market in Guangdong, along with other animals and humans working at the market.[7] The coronavirus isolated in the civet shares 99.8 percent of its genome with the initial SARS coronavirus, leading to the identification of the civet as the reservoir host. In 2006 two research groups identified several horseshoe bat species as the reservoir host for viruses sharing a close genetic relationship with the SARS coronavirus.[8]

Fast-forward seventeen years from the initial SARS-CoV epidemic to enter a world vastly changed from the devastating effects of the SARS-CoV-2 virus. This pathogen is closely related not only to the initial novel coronavirus transmitted to humans but also to coronaviruses found in horseshoe bats and pangolins. The only mammal with scales, the pangolin is a small anteater prized for its meat and its scales; the latter are used in traditional Chinese medicine.

Metagenomic sequencing of pangolin samples indicates that the mammal is a likely intermediate host. But some scientists caution that the genomic similarity of approximately 91 percent between the pangolin coronavirus and SARS-CoV-2 is not high enough to confirm an evolutionary relationship, and that other animals could be identified as intermediate hosts.[9] No matter what the course of transmission is from animal to human, the COVID-19 outbreak is a savage reminder of how deeply human history is intertwined with the microorganisms that share our planet.

Devastation from Deadly Microbes

The first recorded outbreak of bubonic plague, transmitted through fleas from small mammals, is estimated to have eliminated a quarter to half of the human population from 540 to 750 AD.

The bacillus *Y. pestis*, carried by rats, returned to Europe from Asia centuries later when twelve "death ships" from the Black Sea made port in Messina, Italy, in 1347. In the Black Death that followed, an estimated twenty-five million people, almost a third of the continent's population, perished throughout Europe.[10]

Influenza, also first transmitted through animals, caused the "Spanish flu" during World War I, killing approximately fifty million people worldwide. From the U.S. Navy hospital in Great Lakes, Illinois, nurse Josie Brown wrote, "The morgues were packed almost to the ceiling with bodies stacked one on top of another. ... We didn't have the time to treat them. ... We would give them a little hot whisky toddy; that's about all we had time to do."[11] Unlike most deadly pathogens, the influenza virus demonstrates an ingenious ability to reinvent itself by changing its surface proteins, making it more difficult for the immune system to detect the "new" intruder; hence the need for yearly vaccinations against the ever-evolving pathogen.

Along with warfare, influenza, smallpox, and measles all helped to devastate Indigenous populations during the spread of colonialism around the globe. Smallpox aided European conquerors in Mexico as large numbers succumbed to the disease, and measles killed approximately two million Indigenous Peoples in Mexico in the seventeenth century. From 1618 to 1619, smallpox also eliminated 90 percent of Indigenous natives in the area that would become the Massachusetts Bay Colony,[12] where the inhabitants were already weakened through encounters with Europeans decades before. On the other side of the globe, smallpox killed more than half of the Indigenous population of Australia several decades later.[13]

A more recent, deadly microbial event is the HIV/AIDS epidemic, with the first five cases of a rare lung infection identified in June 1981.[14] A new pandemic emerged in late 2019, causing COVID-19, a serious and potentially life-threatening disease affecting millions of individuals in more than two hundred countries and territories around the world and causing hundreds of thousands of deaths.[15] Patients with severe cases of COVID-19 experience life-threatening viral pneumonia that can progress to acute lung injury (ALI), acute respiratory distress syndrome (ARDS), and possible systemic inflammation with subsequent multiorgan failure and death.

These are the obvious, visible testimonials of the sometimes devastating consequences of microbes spreading from one species to another and among individuals from the same species. But these events represent the tip of the iceberg in the exchange of microbial communities that occurs all the time between individuals.

Modern Modes of Microbial "Transport"

Another key element fueling the exchange in microbiome communities is the dramatic change in our modes of transportation. Seventy years ago, a trip from Europe to the United States would take three weeks by sea. Now you can fly from any city in Europe to the United States in a matter of hours and, as has occurred with COVID-19 infections, a disease can spread from one continent to another in a few days. This implies that microbial exchanges between different communities with different lifestyles can occur in real time, rather than through gradual adaptation.

These considerations clearly emphasize the concept that increased hygiene is only one of the many factors in play that can influence the host microbiome in a detrimental way, leading to the development of diseases given a specific genetic background.

Just as the spread of infectious diseases has been highly influenced by the modernization of our lifestyle over centuries, epidemics of chronic inflammatory diseases have been influenced by similar changes.

Infectious disease epidemics caused by a single pathogen can ravage a community within weeks or even days, as SARS-CoV-2 has so brutally reminded us. But the phenomenon of rising chronic inflammatory diseases has occurred over decades, fueled by a more complex change in microbial communities. Thus, it seems more logical to consider the microbiome hypothesis as the driving force responsible for the epidemics of noninfectious, chronic inflammatory diseases, a concept that will be covered in more detail in later chapters.

Put simply, our modern lifestyle has placed the microbiome off-balance by changing the terms of holobiontic coevolution. Some scientists think that communications between the dynamic microbial communities and the complex human immune system, an interaction that plays a key role in determining the state of health or disease, have been disrupted. The human microbiome is no longer simply evolving as planned by evolutionary biology but is reacting and adapting to massive and sudden changes in the human environment that have occurred in the past few decades. These compositional and functional microbiome modifications cause epigenetic changes that increase the shift from genetic predisposition to clinical outcomes, which explains the accelerating rate of noninfectious, chronic inflammatory diseases in the Western world.

Dissecting the factors that influence microbiome composition and function may hold the key to understanding how a potential genetic predisposition to a particular disease may become an actual clinical outcome in any given person. If this is the case, studying the factors that influence microbiome composition and, most important, function may hold the key to possible targeted interventions for personalized medicine and ideally, primary prevention.

Tracking the Ancestral Microbiome

As we move toward a new scientific paradigm in this field, our current knowledge of the human microbiome is made up of very few facts and many, many uncertainties. Claire Fraser, first introduced in chapter 1, is a pioneer in the field of metagenomics and a leader in the Human Microbiome Project. She compares the early days of microbiome research, trying to describe the microbiome with a limited set of tools, to the parable of the six blind men and the elephant.[16] Each man determined the nature of the beast according only to what he could feel with his hands, leading to six wildly differing descriptions.

One point of consensus among researchers is that just as human behavior has altered our physical environment—resulting in air and water pollution, radiation exposure, and other climate-change factors—so too lifestyle modifications, particularly during the post-industrialization era, have had a major impact on the composition of the human microbiome. Our overuse of antibiotics, our increased intake of processed foods, and changes in lifestyle and environmental factors all have a direct impact on our microbiome.

As discussed in chapter 1, some proponents suggest that these recent changes have caused the extinction of certain ancient microorganisms that coevolved with the human species, putting us at a higher risk of developing certain diseases. Without an ideal human model to challenge this hypothesis, the best proxy is the evaluation of studies focused on comparative biology with nonhuman primates and with populations living in remote parts of the globe that still embrace a hunter-gatherer lifestyle.

The vast majority of our current studies on the microbiome, particularly the gut microbiome, have been performed on people from industrialized countries. To redress this imbalance, investigators interested in holobiontic evolutionary biology have focused their research efforts on studying the microbiome composition in hunter-gatherers living in Africa, South America, and Asia. Combined, these studies show that the gut microbiomes in these groups are composed of different microbes than those found in Western gut microbiomes, and, in general, their microbiomes appear to be more diversified.

Microbial Surprises from the Amazon

Modern-day microbe hunter Maria Dominguez-Bello studies microbial function in animals and humans across the globe. Integrating data from many fields, including ecology, architecture, and environmental engineering, Dominguez-Bello has analyzed microbial signatures from the electronic teller pads of ATM machines in New York City and from hunter-gatherers in South America and southern Africa. With a broad-based collaborative team of leading microbiome scientists including Jeff Gordon and Rob Knight, she is trying to characterize the impact that Westernization has had on the human microbiome. As part of a small group that is racing to collect microbial signatures from far-flung regions of the globe, she believes that Indigenous populations have much to teach us about human health.[17]

Working with the Yanomami, a group of seminomadic hunter-gatherers living in the Amazon jungle in Venezuela, Dominguez-Bello and collaborators provided data in 2015 showing that this group exhibited "unprecedented levels of bacterial diversity."[18] Microbial diversity from skin and fecal samples was significantly higher in the Amerindian group than in U.S. subjects. Contrary to U.S. subjects, the Yanomami showed high rates of *Prevotella* and low rates of *Bacteroides*.[19]

This is perhaps not surprising, given the Yanomami's high-fiber diet of wild bananas, seasonal fruits, plantain, palm hearts, and cassava, along with birds, small mammals and crabs, frogs and small fish, and meat from peccaries, monkeys, and tapirs. *Prevotella* have been associated with the fermentation of dietary fibers into short-chain fatty acids (SCFAs), whose beneficial impact on many gastrointestinal functions has been amply reported in the literature.

But one of the most compelling findings of this study conducted in the High Orinoco region is the characterization of the group's resistome, an individual's collection of antibiotic resistance (AR) genes. Because the Yanomami have remained relatively isolated for more than eleven thousand years, according to Dominguez-Bello, genomic divergence from the Western samples is not surprising. However, it is intriguing that without known exposure to pharmacological antibiotics, AR genes are present in the microbiome of the Amerindian group. Dominguez-Bello and her colleagues posit that AR genes could have evolved in the soil-dwelling ancestor of a human commensal or arrived through antibiotic-producing soil microbes, or possibly through traded objects or a chain of human contact.[20]

Even if these are plausible hypotheses, a possible alternative explanation is that some resistome components have been acquired by lateral gene transfer through bacteriophages that integrate their genome into one of their bacterial hosts. These viruses bring in AR genes to more efficiently create a niche for their host, which by successfully eliminating antibiotic-susceptible bacteria competitors will assure bacteriophage survival and proliferation. This is a sublime example of symbiotic advantage achieved by lateral gene transfer.

Findings from the Dominguez-Bello group show that these AR genes identified in antibiotic-naïve microbiota might help explain how antibiotic resistance to new classes of antibiotics can arise so rapidly. The group asserts that the "discovery of functional AR genes encoded by antibiotic-sensitive *Escherichia coli* severely underestimates the potential community resistome."[21]

Threatened by illegal gold miners, along with killing diseases such as malaria and measles, the Yanomami tribe members' ability to maintain their way of life, with or without a rich microbiome and resistome, is uncertain. Working in a culturally sensitive manner with Indigenous Peoples like the Yanomami to gather saliva, blood, tissue, and stool samples is a race against the clock. As Dominguez-Bello and her collaborators state, characterizing the commensal resistome of diverse peoples "will inform the design and prudent deployment of antibiotics to minimize enrichment for preexisting resistance."[22]

One possible limitation of these kinds of studies is the impact the researchers will have on microbiome samples from the native Yanomami during the time they spend with the tribe. Even a short coexistence with individuals coming from the Western world and a different lifestyle may result in microbiome exchange and therefore alter the original composition of the Yanomami's microbiome.

When Microbiomes Don't Match

Dominguez-Bello also has collaborated on studies that demonstrate a lack of Bifidobacteria in the microbiome of hunter-gatherer groups from Tanzania and Peru. In clinical and translational research into the Western gut microbiome, this phylum of gut bacteria has been determined to be a sign of healthy gut microbial composition.

By contrast, the hunter-gatherer groups include *Treponema pallidum*, a spirochete bacterium that can cause syphilis. In what sense can we determine if one of these microbiomes is "healthier" than the other? Is it simply a matter of adapting to the local food supply and combating local pathogens and environmental threats, or is it more complex?

These differences illustrate another challenge that we face when we try to define a healthy human microbiome. It is highly unlikely that we will ultimately identify a microbiome composition that could be healthy for everyone. Rather, the composition of the microbiome that can keep us in a state of health would strongly depend on the genetic composition of the host and the environment in which the host lives. No doubt it also depends on other factors yet to be determined, and perhaps on microbial adaptations specific to that individual.

As for the hunter-gatherers, their specific gut microbial composition most likely reflects the food supplies, water, climate, seasonal changes, social and animal groups, and other environmental variables that characterize their particular corner of the world. For the complete bacterial domain of so-called beneficial and harmful bacteria, even at the species level in bacteria such as *Treponema*, there are some species that can cause disease and others that can be useful symbiotic partners. One example is the useful *Treponema* species that helps to digest foodstuffs like carbohydrates, as shown by Cecil Lewis and colleagues in their work on Peruvian hunter-gatherers.[23]

Interestingly, similar *Treponema* strains were also found in Hadza hunter-gatherers from Tanzania and in nonhuman primates, while they are totally absent from the gut ecosystem of industrialized populations. These findings led the researchers who performed these studies to suggest that this strain "may represent a part of the human ancestral gut microbiome that has been lost to the adoption of industrial agriculture and/or other lifestyle changes."[24]

Missing Microbes

Their conclusions are in line with the provocative missing-microbes hypothesis introduced in chapter 1. Blaser suggests that changes in modern lifestyles, particularly in terms of diet, sanitation, and the overuse of antibiotics, have led to the depletion of some components of the "ancient microbiome," which is now extinct. He posits that these changes may therefore be responsible for the rise of chronic inflammatory diseases, including autoimmune diseases, food allergies, obesity, and neuroinflammatory diseases found throughout industrialized countries during the past fifty years.[25]

Even if Blaser's theory is intriguing and plausible, it assumes that the remote geographical groups present in today's world are a repository of these ancient bacteria, which are otherwise extinct in other parts of the world. An alternative explanation is that microbiome differences can simply be reflective of the practices of a traditional lifestyle that allow current microorganisms to find an ideal niche in which to flourish. In other words, rather than *Treponema* being one of those ancient bacteria that have been preserved in this population, they may simply represent more modern species that flourish in the gut of current hunter-gatherers because of their lifestyle.

It is also possible that gut microbial composition can be influenced by other factors, including the host genome, the presence of other microorganisms in the gut such as viruses, fungi, and parasites, or other unidentified variables. Investigators have shown that the presence of specific parasites, particularly *Entamoeba*, can highly influence the composition of gut microbiota that favors a signature found in hunter-gatherers, namely increased diversity and representation of the *Treponema* species. We still have to determine how parasites can affect microbiota composition. Several nonmutually exclusive possibilities are that the parasites eat a particular bacterial species and that they consume specific metabolites, thus limiting the bacteria's food supply or directly eliminating some components of the microbiome, or both.

We might never settle the question of the extinction of ancient microbes versus lifestyle adaptation. But it becomes fairly obvious that the gut microbiome of rural populations that embrace a sustainable lifestyle, meaning that they live only on products from their immediate natural environment, is substantially different from the gut microbiome composition of people living in industrialized countries. Determining whether these differences are the cause of the current epidemic of chronic inflammatory diseases, or the consequence of these diseases, or merely an epiphenomenon, remains one of the most thought-provoking challenges faced by scientists interested in the role of the human microbiome in health and disease.

Hunters, Gatherers, and the Mediterranean Diet

If you study human anatomy and physiology, it quickly becomes apparent that we are an omnivorous species. In a simple-minded view of human evolutionary biology, we can assume that our hunter-gatherer ancestors typically devoted 90 percent of their time to food procurement and 10 percent to reproduction. Why was so much time needed to find food? The simple answer is the unpredictability of the food supply. In this milieu, we can assume that fruits and vegetables and tubers and nuts were the "easy food" that was always available to the gatherers and in abundant supply.

Conversely, the hunter role, which requires skill and patience to catch animals such as buffalo, antelope, or even fish, would have a much higher level of unpredictability and a high rate of failure. So evolutionarily speaking, let us assume that our species has relied much more on the gathering segment of the food chain than on hunting large animals.

As a corollary of this concept, a high daily intake of fruits, vegetables, tubers, nuts, and oil and limited consumption of an animal-based diet would more closely reflect our evolutionary plan, thus representing the ideal balanced diet to maintain health. This probably holds true, at least in theory, in the vast majority of circumstances. This concept also reflects many current dietary recommendations similar to the "Mediterranean diet," which was formally espoused by Ancel Keys as a healthy way of eating to reduce cardiovascular risk.

Through his interest in human physiology and nutrition, Keys developed the "K-ration," pocket-sized daily rations for mobile U.S. forces in World War II. After noticing that the Italians he met in small towns in Southern Italy enjoyed better health than their New York City relatives, Keys focused on the intersection of diet, lifestyle, and cardiovascular health. He launched his decades-long "Seven Countries Study" in 1958, evaluating the health and nutrition of twelve thousand healthy, middle-aged men from Italy, the Greek islands, Yugoslavia, the United States, the Netherlands, Finland, and Japan.[26]

Keys concluded that a high consumption of saturated fats was related to increased cardiovascular disease. His findings at the time, and in the decades since, have not been without controversy, but the concept of the Mediterranean diet (see figure 2.1) as a healthy nutritional intervention is now widely accepted and remains one of Keys's main legacies.

A native of Southern Italy, coauthor Alessio Fasano can personally attest to the superiority of the Mediterranean diet in most aspects of nutrition and life. However, diving more deeply into the relationship between diet and our gut microbiota, we find exceptions to this nutritional lifestyle that are strictly related to specific environmental characteristics. An exquisite example of this exception is the case of the Hadza hunter-gatherers living around the shores of Lake Eyasi in Northern Tanzania.

A Seasonal Microbiome

The region inhabited by the Hadza is characterized by extreme swings between wet and dry seasons, which are related to the unique weather pattern in Eastern Africa. The rainy season lasts from November to April and is followed by extremely dry weather from May to October. Although the total population of the Hadza is estimated at 1,000, approximately 150 people live predominantly as hunter-gatherers, eating mainly wild

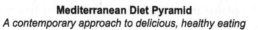

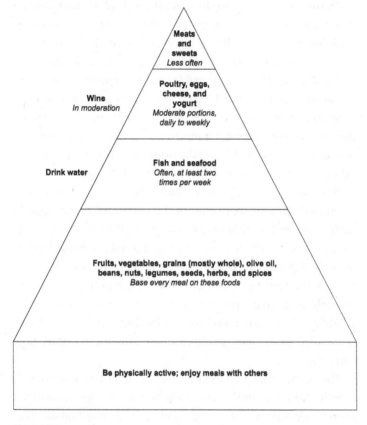

Figure 2.1
Food pyramid based on the Mediterranean diet and lifestyle. Adapted from https://oldwayspt.org.

game and plants in a subsistence lifestyle that has endured for millennia.

During the dry season, the Hadza can forecast that animals will gather where rare sources of water are available, making a ready supply of large game more predictable in dry weather. The scarcity of foliage in the dry season also makes the animals more visible and much more vulnerable to Hadza hunting skills. With no means to preserve their catch, the Hadza must consume the meat in a short period. Along with the scarcity of vegetation, this consumption of meat during a six-month period creates an increase in protein and fat intake from animals and a drop-off in calorie intake from other sources for the hunter-gatherers.

But during the wet season, Tanzania becomes a luxuriant, green land with an abundance of flowers, fruits, berries high in natural sugars, and wild honey. And when the rains come, large animals scatter to become camouflaged in foliage and harder to catch, which leads the Hadza to rely on a diet based on their gathering skills. The Hadza also eat fibrous baobab fruit and tubers throughout the year. As parents in the Western world cajole their children to consume more fiber in the form of fruits and vegetables, the Hadza families enjoy a large amount of dietary fiber every day.

The Hadza group and their researchers have provided us with insights into the impact of seasonality on the interconnected microbial environment, namely soil, animal, and human. This leads to the key question of what a normal, healthy microbiome might have looked like before humankind created artificial conditions to modify its composition. Our challenge is to learn about the functional contributions of those species that might now be missing or scarce in our modern-day microbiome.

The Circadian Microbiome

Making the story even more complicated and, at the same time, more intriguing, is the recent observation that there is a circadian modification of the microbiome. Several gastrointestinal functions, including gastric secretions, blood flow, stem cell regeneration, digestion, and immunity, are known to be influenced by molecular circadian rhythms that oscillate every twenty-four hours.

Similarly, it has been recently demonstrated that the circadian clock of the intestine, which is regulated not only by light but also by other stimuli, such as the sense of smell, routine feeding practices, the vision or thought of appealing food, and the anticipation of food ingestion, also influences gut microbiome composition and function. Therefore, anything that disrupts the circadian rhythms of gastrointestinal functions, such as jet lag or poor sleeping patterns in teenagers, could induce dysbiosis, as has been demonstrated in both mice and humans.

Using circadian-arrhythmic, $Per1^{-/-}$, $Per2^{-/-}$ double-knockout mice, Christoph Thaiss and colleagues have elegantly demonstrated that the abundance, composition, and function of gut microbiota are dramatically altered in these mice. Also intriguing is the evidence that an unbalanced microbiome may cause changes to the diurnal circadian gastrointestinal clock, further indicating a bidirectional relationship between gastrointestinal circadian rhythms and the gut microbiome.[27] Given the major impact of the gut microbiome on intestinal immunity, it is perhaps not surprising that the circadian microbiome rhythms are intimately intertwined with the circadian rhythms of mucosal immunity.

The circadian clock regulates the diurnal release of defensins, which are released in the small intestine in response to exposure to bacterial endotoxins such as lipopolysaccharides (LPS), suggesting the presence of a mechanism of mucosal defense against bacteria ingested during feeding. Furthermore, the expression of toll-like receptors and interleukin 6 (IL-6), key elements of the mucosal innate immune response against enteric pathogens, is subject to circadian control.

Other key elements of the mucosal immune response, like macrophages and the maturation of immature T cells into the Th17 cells that function in the adaptive immune response against autoimmunity, are most frequently generated during the diurnal part of the circadian gastrointestinal clock. This once again points out the potentially negative consequences of a disturbed circadian rhythm on gut immune and microbiome functions, which is possibly linked to chronic inflammatory diseases.

A Developing Microbiome in an Industrialized World

With this premise in mind, what would be the best strategy to regain our peaceful cohabitation with our microbiome? Who comprises our most beneficial microbial companions? Will eating only organic food or embracing any specific diet, along with supplements of prebiotics or probiotics, do the trick?

If results from the study of the Hadza are relevant to the goal of regaining the proper microbiome composition defined as "normal," maybe we should focus on the seasonal and daily circadian rhythm of life rather than a one-size-fits-all approach. In other words, would the hunter-gatherer microbiome of an Inuit in the Arctic, an Aborigine in Australia, and a Hadza from Tanzania be similar? And how would the microbiome from these groups compare to the gut ecosystem of a native New Yorker, a Parisian tourist, or a stockbroker from Singapore?

These are clear examples of how, over time, we become the result of this intricate interplay of the environment we inhabit. In other words, not only are we what we eat, but we are also "where we live." An effort to identify the so-called normal microbiome is a purely theoretical concept that may have very little impact as we seek out the optimal interaction for proper equilibrium with our microbial ecosystem to maintain health.

Nevertheless, if we consider Eastern Africa as the cradle of human evolution, we could view the Hadza tribes as a possible example of early human evolution, with the clear understanding that they cannot be considered living fossils of what human life was like hundreds of thousands of years ago. Keeping this limitation in mind, the Hadza people, with their hunter-gatherer lifestyle, natural birth practices, extended breastfeeding time, limited access to Western medications and culture, and intimate interaction with the natural microbial circle of life in water, soil, animals, and vegetables, may be the best proxy for us to appreciate what our ancient microbial ecosystem might have looked like.

Metabolic Function: The Big Microbial Picture

Researchers studying the Hadza tribes learned that drastic changes in feeding habits during the two distinct seasons also led to remarkable changes in the gut microbiota composition of the hunter-gatherer group members. Nevertheless, those changes do not necessarily imply changes in function. It is possible that despite the drastic changes in microbial communities, the metabolic functions exerted by this microbial ecosystem could remain the same over time.

If the ideal symbiotic lifestyle between ourselves and our intestinal microbiome implies stability in the metabolic pathways that the gut ecosystem influences, a dramatic shift in the composition of the gut microbiome has to be programmed to maintain metabolic stability. Indeed, different microbiota can influence metabolic pathways similarly, thus preserving function stability. In other words, nature always has a backup plan.

If this concept holds true, then our goal to target the microbiome to ameliorate or reverse the chronic inflammatory diseases prevalent in the Western world should not be aimed at changing its composition by adding specific probiotics or prebiotics. We should instead look at changing metabolic functions that have been negatively affected by the Western lifestyle, such as the multiple abnormalities seen in insulin secretion in people with type 2 diabetes (T2D).

Learning more about manipulating microbiota composition to modulate metabolic functions is a key step toward capitalizing on the potential of the entire field of microbiome study. Jumping in too early to try to achieve this goal, as we are currently doing with the multi-billion-dollar probiotic industry, may jeopardize the tremendous potential of targeting the microbiome as a possible strategy to treat chronic inflammatory diseases. Thus, we could make the same mistake that we made with the promiscuous use of penicillin when it was applied as a remedy for any kind of infection.

We now know that penicillin affects only Gram-positive bacteria, and its indiscriminate use has led to the emergence of antibiotic-resistant bacterial strains. There is no disputing the heroic role of penicillin in fighting deadly infections and saving millions of lives. However, its widespread use has not come without a cost, which includes a decreased capability to use penicillin as an extremely effective treatment against fatal infections.

Differences in Dietary Intake

Similar to researchers conducting studies in Tanzania, Peru, and the Amazon, researchers studying the intestinal microbial composition of the Hadza found a scarcity of the genus *Bacteroides*, which conversely is the most abundant genus in the gut microbiome of populations in the Western world. While we have no clear understanding of the reason for these striking differences, the most likely explanation may be related to the contrast between our Western diet and the Hadza diet.

While the overall quality and quantity of nutrient intake could be relatively similar, the most remarkable difference is the extremely low fiber intake in our current standard American and Western diet (less than twenty grams of fiber per day), compared to the many-times-higher fiber intake in the Hadza population. Could our poor fiber intake translate into a different microbial environment? This could be a setting that favors *Bacteroides* as a dominant species in our gut at the expense of other components that may be important in maintaining a balanced ecosystem.

In another striking finding, researchers showed that Hadza people carry *Oxalobacter formigenes*, an oxylate-degrading gut microbe that has almost disappeared in the Western world. *O. formigenes* has recently gained attention for its ability to prevent the formulation of calcium oxalate kidney stones. People with lower levels of this Gram-negative anaerobic bacterium run a higher risk of developing hyperoxaluria and recurrent kidney disease.

Researchers also found that the Hadza microbiome is extremely rich in the genus *Prevotella*, containing more than a dozen species of this microorganism, compared to only two that have been identified in the human gut of Western subjects.[28] Because

Prevotella has been associated with some chronic inflammatory diseases, particularly arthritis, it is tantalizing to hypothesize that a diversified *Prevotella* species composition could be beneficial in preventing chronic inflammation when compared to a decreased diversity within the genus *Prevotella*. Preliminary research findings from the microbiome of an Indigenous group at the opposite end of the globe could provide some additional clues to the *Prevotella* puzzle.

A Cautionary Tale from the Arctic

In 1999, after decades of activism from Indigenous tribes, Nunavut was established as the largest and northernmost territory in Canada. It encompasses a total area of more than two million square kilometers with approximately thirty-eight thousand people of mostly Inuit heritage. A nomadic group for thousands of years, some Inuit people still follow a traditional diet from hunting, fishing, and gathering local food.

Narwhal, beluga, and bowhead whale, along with ringed and bearded seal, caribou, polar bear and other land mammals, and saltwater and fresh fish are a large part of the traditional diet. Meat is eaten raw, frozen, or fermented, as well as cooked, and seal blood is an important part of the traditional Inuit diet and culture. Inuit people also gather clams, tubers, roots, berries, grasses and seaweed, and some herbaceous plants.

Using stool samples, researchers from Montreal mapped the gut microbiome of people indigenous to the Canadian Arctic.[29] Comparing the microbiomes of people from Montreal who follow a Western diet to those who follow a mostly traditional Inuit diet, the researchers expected to see a marked divergence in the two samples. Instead, they found that at the overall microbial community level, the gut microbiomes of the urban dwellers and the Inuit people were indistinguishable and had similar diversity levels.[30] Given the similarities between the low-fiber, high-fat animal protein diets of the Inuit people and the Qallunaat (non-Inuit people), this might not be so startling, especially as many Inuit move away from traditional sources of food and embrace the Western diet. But once again, the microbiome holds additional surprises once we look more deeply.

Using oligotyping, the researchers found significant differences in the relative abundance of certain microbial taxa down to the subgenus level. *Prevotella* were enriched in both groups, but the genetic diversity was less significant in the Inuit group. The researchers hypothesize that a low-fiber diet selects against *Prevotella* and also reduces its diversity. In a very small sample of Inuit people who consumed a Western diet, *Prevotella* was even more abundant overall than in the larger group of Inuit people, and the genus was "more diverse in its richness and evenness of oligotypes."[31]

We know that *Prevotella* strains vary from person to person. We agree with the suggestion from the Montreal researchers that certain strains might have contrasting associations with diet and a state of health. They also suggest that some strains may correlate with fiber intake while others may not, and they assert that factors other than fiber may affect the diversity of *Prevotella* subspecies. Finally, in a follow-up study, the researchers confirmed that differences within individuals are usually less pronounced than differences among individuals. They also confirmed a trend noted from other studies that highly diverse gut microbiomes tend to be more stable over time.[32]

Building the Best Microbial Village

The few examples already outlined further exemplify what seems to be a recurring and dominant theme concerning the microbiome and its role in health. It is likely that the emergence of noninfectious, chronic inflammatory disease in the Western world has been the consequence of the loss of the extraordinary diversity of microbes that are meant to be the most beneficial colonizers of our guts. This collection of commensal microbial communities may have provided the key signature for proper functioning of many of our metabolic pathways.

It seems quite clear, and we will revisit this topic again, that the rather rudimental genomic assets of humankind, made up of approximately twenty-three thousand genes, cannot fully explain the extraordinary complexity of our species. Rather, we are the product of the coevolution of our genome with the metagenome (the gene array from our microbiome), which contains from 100 to 150 times more genes than we do, a figure that continues to shift with the developing field of microbiome research.

Losing one-third or possibly one-half of the diversity of our microbiome, as some researchers claim, is like losing a huge component of our genetic identity. It's no surprise that this loss has come with a price. That price could mean that our immune system, through its array of commensal bacteria, is no longer capable of performing at peak capacity if the cross-talk between our genes and what is left of the microbiome genes takes the wrong turn.

It's obvious that with less genetic redundancy, we have fewer alternatives with which to play the game of maintaining the balance between health and disease. And there is no time that is more important in the development of an individual's microbiome than the first one thousand days of a person's life, as we will see in chapter 3.

3   Early Factors Influencing the Human Microbiome

Development of Our Microbial "Organ"

As mentioned elsewhere in this book, the balance between health and disease is primarily the result of gene-environment interactions. What is still not entirely clear is how the environment may epigenetically affect the expression of our genes by shifting genetic predisposition to a clinical outcome. It is becoming increasingly clear, however, that the role of the intestinal microbiome is critical to healthy development in early life. The role of environmental influences on child health outcomes, and most likely on health outcomes across the life span, is closely related to a variety of prenatal, perinatal, and postnatal events (see figure 3.1).

No matter which of these three temporal periods we analyze, an argument can be made that all of these environmental factors can affect microbiome composition and function, making the microbiome a possible transducer of every environmental factor affecting the onset of noninfectious, chronic inflammatory diseases in genetically susceptible individuals. In this chapter, we will analyze current knowledge about prenatal, perinatal,

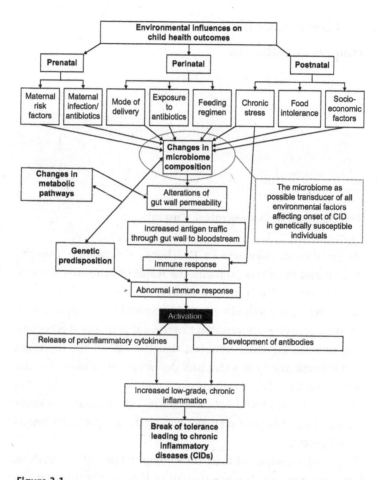

Figure 3.1
Prenatal, perinatal, and postnatal factors influence microbiome composition and function. Adapted from V. J. Martin, M. M. Leonard, L. Fiechtner, and A. Fasano, "Transitioning from Descriptive to Mechanistic Understanding of the Microbiome: The Need for a Prospective Longitudinal Approach to Predicting Disease," *Journal of Pediatrics* 179 (December 1, 2016): 240–248, https://doi.org/10.1016/j.jpeds.2016.08.049.

and postnatal environmental factors affecting the microbiome's composition and most likely the individual's long-term health.

The Maternal Microbiome

Since the infant microbiome is highly influenced by the maternal-offspring exchange of microbiota, it follows that the lifestyle of the mother-to-be may have an impact on this exchange. If the future mother is healthy or has a chronic inflammatory disease, has an active or sedentary lifestyle, is a smoker or a nonsmoker, is a recreational user of drugs, consumes alcohol, undergoes high levels of stress, lives in an urban or rural setting, belongs to a low or high socioeconomic group, is obese or not—these are some of the many variables that might affect the composition of the microbiota that is transmitted to the infant.

Many of these variables will be discussed in greater detail in different sections of this book. It is intuitive, however, that the more the maternal lifestyle departs from the evolutionary standards of hunter-gatherers described earlier, the more likely the microbiota composition and function may depart from the biological plan to transmit the ideal microbiome to the baby. Another important but mainly unexamined variable is the impact of maternal health disparities on the microbiome of mother and baby. There is growing evidence in the literature that stressors related to socioeconomic factors including poverty, malnutrition, and education level can negatively impact the human microbiome composition.[1]

With the appreciation that the human genome is rather rudimental, comprising approximately twenty-three thousand genes, the most likely interpretation to explain the complexity of human biology is that we are the product of two coevolving genomes. The first is the human genome, which is influenced by parental inheritance and does not change over time. It contains genes that in their complexity dictate the risk of developing diseases. The second is the human microbiome, which is extremely dynamic and can be influenced by many factors, including health disparities and poor access to health care.

Health Disparities Related to the Microbiome

Does it follow that an increased risk of developing disease may merely be the consequence of living in a disadvantaged social and economic environment? If a compromised microbial composition can increase our risk of developing disease, then will health disparity experienced by pregnant women affect the destiny of their offspring? If so, this puts their infants at a disadvantage compared to babies born to women who do not experience these social and economic challenges.

Unfortunately, we see tangible examples of this, such as the U.S. epidemics of childhood obesity, infant mortality, and infections, including the current COVID-19 pandemic, which are experienced at higher rates in African American and Hispanic populations than in the general population.[2] Pointing the finger at gut microbial composition as the only cause of these negative outcomes would be an oversimplification of what is most likely a health outcome influenced by multifactorial etiology. However, it will be extremely helpful to expand our current knowledge of the role of microbiota composition in increasing the risk for negative clinical outcomes in childhood.

So, what happens if the microbial signature is sequential to maternal health disparity and is determined to play a role in negative outcomes for these children? If this proves to be the case, then political and economic interventions to mitigate these disparities represent a moral obligation toward families living in poverty. It also presents a formidable opportunity to prevent negative clinical outcomes in children born to mothers with low socioeconomic status.

If a pregnant woman is living in poverty without a social support system, these factors could affect the health of her offspring not only through environmental and external factors, but possibly also through the effect of these factors on the composition of the microbiota she passes on to her child. Conversely, if a more beneficial microbiota composition is passed on to a child, this will give that child an advantage in mitigating the risk of developing disease.

On the one hand, if a pregnant woman is disadvantaged by genetic evolution, for example, and she passes along genes related to colon cancer or T1D to her offspring, there is not much that can be done to avoid the transmission of these genes. On the other hand, if a pregnant woman is at a disadvantage due to environmental and social factors, with an increased risk of transmitting a microbiome that raises her offspring's risk of chronic inflammatory disease, then theoretically at least, those external factors can be modified to mitigate the risk of epigenetically shifting her child's genetic predisposition into clinical outcomes. In other words, whether the fixed human genetic code becomes translated into disease depends on the plastic, genetic code of the microbiome, which can be modified.

Scientists have documented evidence that maternal lifestyle affects the composition and function of the microbiome passed along to offspring, sometimes leading to pathological consequences. University of Maryland School of Medicine (UMSOM) researcher Tracy Bale hit the headlines in 2015 with mice studies conducted at the University of Pennsylvania that showed that inducing stress in the mother leads to an altered gut microbiota in her newborn pups. Bale asserts that stress in the first trimester of the mouse moms can influence brain development in mice offspring through changes in the vaginal microbiomes of the mother mice.[3]

Research groups led by Rebecca Brotman and Jacques Ravel at UMSOM have published results showing that smoking can lead to a reduced number of beneficial microbes in the vaginal microbiome, which would be imprinted on the baby through natural delivery.[4] And in an interesting and unexpected twist in one of Ravel's studies on diet and pregnancy outcomes, providing pregnant women with round-the-clock access to clinicians to answer questions and provide care seemed to positively affect outcomes, possibly through reducing the level of stress and affecting changes in the composition of the microbiome.[5] From these growing lines of evidence related to the human microbiome, we've learned that teasing out the ever-changing and intricate nature of its interactions is indeed a formidable task, but also a formidable opportunity to change the clinical outcome once the intricate human genome-microbiome interactions are fully dissected.

Prenatal Microbiome Imprinting

Henry Tissier of the Pasteur Institute in Paris first isolated *Bifidobacterium* from an infant's gut in 1906. The French pediatrician also introduced the "sterile womb hypothesis," a long-standing doctrine that the fetus develops in a sterile environment. This paradigm is based on past studies using culture methodology suggesting that prior to labor, the amniotic fluid is completely sterile.[6] But our advancing ability to identify noncultivable microbes through sequencing has led to a fierce and far from settled scientific debate on the topic. As this book goes to press, the most recent research (see below) is leaning toward the sterile womb hypothesis; the notion of a placental microbiome remains highly controversial, according to Ravel.[7]

In rebuttal of the sterile womb hypothesis, recent culture-independent polymerase chain reaction (PCR) analysis seems to show that microbial colonization of the amniotic cavity could be a rather frequent event involving a variety of microbes, including *Lactobacillus*, *Prevotella*, and *Bacteroides* genera. Some scientists hypothesize that the baby could receive beneficial microbes through swallowing amniotic fluid in utero. Recent studies also have demonstrated that the placenta may harbor a very diversified microbiome.[8]

Following on from the work of Indira Mysorekar from Washington University, who found bacteria in almost one-third of nearly two hundred placental samples in 2011, Kjersti Aagaard identified bacterial DNA in placental tissue in 2014.[9] Growing evidence suggests that the fetus, through both the placenta and the amniotic fluid, can already be exposed to an imprinting microbiome even before birth.

Cultivating Microbes from Meconium

In other studies, scientists have used sequencing to characterize the microbes in meconium, the first fecal matter a newborn produces. These studies show the presence of a diversity of bacteria that some scientists have interpreted as the strongest argument in favor of a prenatal imprinting and engraftment of the infant's gut microbiome with these mostly beneficial bacteria. Practicing neonatologist Josef Neu of the University of Florida and his colleague Eric Triplett, a microbial ecologist, and their students have analyzed first-pass meconium. These microbes differ from the surrounding skin and thus are unlikely to reflect contamination from the anal skin shortly after birth, according to Neu. Triplett and Neu are part of a growing group of scientists challenging—or at least questioning—the sterile womb hypothesis. Neu thinks there is definitely "something to the narrative that the uterus contains microbes, the placenta contains microbes, and the amniotic fluid contains microbes."[10]

But as with so many areas of microbiome study, we have far more questions than answers. How do we explain the presence of microbes in meconium in healthy infants if not from some form of transmission from the mother? And do studies in insects, vertebrates, and other animals that show direct transmission of microbiota from mother to fetus have any bearing on human studies?[11]

With a group of scientists from the University of Florida and Tbilisi, Georgia, Neu presented evidence in 2010 showing microbial DNA in the meconium of premature infants, suggesting that the microbes were not of postnatal origin.[12] Since then, animal studies and human studies in healthy mothers and infants have provided additional evidence that bacteria are present in first-pass meconium samples.

In another intriguing study, the presence of household pets during pregnancy has been determined by a Finnish group to be associated with a higher microbial diversity in the mothers' microbiome as well as in their offspring's meconium.[13] Results from other studies show that the microbial signature in the meconium of preterm infants differs markedly from that of full-term newborns.

Neu has had many discussions on this topic with many scientists. He thinks the most insightful response might come from his colleague Triplett. According to Neu, Triplett does not view this as a binary question; to Triplett, there is no answer to the argument of sterile or not sterile. Triplett is currently investigating what drives microbial diversity in environments as varied as citrus fruit, agro-ecosystems, and the preterm infant gut, so it is not surprising that he brings a nondichotomous approach to the role of maternal transmission of microbes.

Technical Artifacts as Possible Confounding Factors

But the debate about a prenatal intestinal microbiome is far from settled, and Jacques Ravel, associate director of the Institute for Genome Sciences at the University of Maryland School of Medicine, provides a compelling argument for not discarding the long-standing evidence for a sterile intrauterine environment too quickly. He noted that bacteria in the amniotic fluid or placenta are most likely associated with negative outcomes for the infant.[14] There is general agreement that the placenta and amniotic fluid cannot harbor high concentrations of microorganisms, like those detected in the gastrointestinal tract, without serious clinical consequences to the fetus, including spontaneous abortion.

Another reason Ravel lists for supporting the sterile womb hypothesis is the inherent difficulty of collecting samples, especially during the period of predelivery and postdelivery, and the possibility of contamination in sampling and processing techniques.[15] He is not alone in this viewpoint, which is supported by the difficulty of determining whether the low numbers of bacteria in meconium, amniotic fluid, and the placenta are detected as the consequence of contamination related to experimental methods. Researchers from the University of Pennsylvania compared placental samples from six healthy deliveries to contamination controls and were unable to distinguish between the two groups, suggesting that, for their sample set, they could not distinguish between placental bacteria and contamination introduced during DNA purification.[16]

The term "kitome" has been coined by some skeptical researchers to reflect the very real phenomenon of bacterial DNA contamination in sample collection. Ravel also argues that at this stage in microbiome studies, current technology is too limited to fully resolve the question of the sterile womb. He adds that we need to invest in establishing microscopic evidence of the presence of microbes in these body sites before studying them using molecular methods.[17] In a critical review, Maria Elisa Perez-Munoz and colleagues argue that the evidence in support of the "in utero colonization hypothesis" is weak because it is based on studies that used molecular approaches with an insufficient detection limit to study the "low-biomass" microbial populations detected in the placenta and amniotic fluid samples. The lack of appropriate controls for contamination and limited evidence of bacterial viability increased their skepticism.[18]

Studies specifically designed to rule out potential cross contaminations and involving a sufficiently large sample size to ensure statistical robustness are needed to address this issue. Marcus de Goffau and colleagues in the United Kingdom have recently reported on such a study.[19] The authors analyzed placental samples from 537 women using a thorough DNA-sequencing approach and DNA-extraction toolkits to search for microbial content both in the placental samples and in negative controls, which were supposedly free from biological material. To account for the abundance of possible microbes isolated from the placental samples, they also spiked placental samples with a known amount of *Salmonella bongori* as a positive control. They used both shotgun metagenomics and 16S rRNA gene amplicon sequencing to account for possible technique-specific potential biases.

Their results clearly showed that the placenta does not harbor microbes during healthy pregnancy. The researchers demonstrated that contamination issues were a convincing explanation for the presence of any detected bacteria, with the exception of a rare finding in only 5 percent of samples of a single type of pathogenic bacterium, *Streptococcus agalactiae*. If this pathogen is present in the mother during childbirth, it can be transmitted to the newborn and cause pneumonia, septicemia, and meningitis.

Despite their strong evidence for a sterile placenta, the negative results from de Goffau and colleagues are hard to prove conclusively, since bacteria can overcome many host barriers, and just a handful of microorganisms that are hard to detect with current techniques can reach the gut of the fetus and potentially start in utero colonization. Still, even if the dogma that the womb is free of microbes should be further investigated, it is unlikely that the placenta is a microbial reservoir streaming a complex microbiome to the fetus under healthy conditions.

In addition to the previously mentioned arguments in support of microbial colonization in the womb, there are several other lines of evidence, as Josef Neu reminds us. These include an evolutionary history of maternal transmission of microbes to offspring in various other invertebrates and vertebrates.[20] Also, a lack of significant immune response after delivery, as well as the ability to produce Immunoglobulin M (IgM) and plasma cells, suggests that fetal programming via microbes and development of tolerance may occur with in utero host-microbial interaction. Finally, nonsterility of the female upper reproductive tract and additional studies showing microbes in the brain,[21] placenta,[22] and gastrointestinal tract[23] suggest that a similar prenatal colonization may occur in the uterus as well.

Whether or not the symbiosis of mammalian hosts with their microbiota is established before birth remains an intriguing question. Besides the placenta's route, another theory of fetal-gut microbiota engraftment based on animal evidence suggests that maternal dendritic cells, specialized immune cells capable of sending "periscopes" out into the mother's intestinal lumen, can sample the beneficial bacteria, capture them within their phagosomes, and, through maternal circulation, deliver the bacteria to the placenta, and ultimately to the fetus. Irrespective of the validity of these theories, which are not mutually exclusive, there is growing evidence that the maternal womb, once considered a totally sterile environment, may be the first environment in which the developing fetus is exposed to the complex microbial ecosystem crucial to the infant's survival, assuming that the prenatal microbiome engraftment theory proves to be correct.

Perinatal Microbiome Imprinting

Another critical moment for microbiota engraftment is the perinatal period, which occurs during and immediately after delivery. The specific microorganisms needed to further colonize the baby are intuitively influenced by maternal microbial ecologies related to vaginal and gut microbiota. As babies move through the birth canal and receive microbial imprinting from their mothers, the vaginal microbiome is particularly important.

The complexity of the vaginal microbiome was already known well before the recent human microbiome "revolution." Throughout a woman's life, the composition of the vaginal microbiome undergoes a variety of changes related to specific milestones of a woman's development. The colonies of microbes in the vaginal microbiome have been the object of intense studies for more than a decade, and as mentioned earlier in this chapter, Jacques Ravel is one of the foremost researchers in the field. We spoke with him about recent research into the vaginal microbiome and the role of *Lactobacillus*.

Culture-based studies conducted before the advent of sequencing technology suggested the prevalence of specific components, with particular attention paid to *Lactobacillus* as a component of the vaginal microbiome. A related finding shows that there is a subgroup of women in which the vaginal microbiome is relatively *Lactobacillus*-poor.

Ravel noted that recent studies challenge the idea that only high numbers of *Lactobacillus* constitute a normal or healthy vaginal microbiome.[24] As will be discussed in detail, whether this difference in vaginal microbiome types is associated with a specific clinical outcome of the offspring remains unclear, but it is likely that *Lactobacillus*-poor microbiota, while being normal, are not optimal under certain conditions.[25] Although it has been amply demonstrated that gut microbiota can affect intestinal epithelial cells, research on the impact of the vaginal microbiome on vaginal epithelial cells is just beginning.[26]

Keeping this information in mind, a key question is whether there is a precise, evolutionary program that changes the composition and properties of the vaginal microbiome before, during, and after pregnancy. Several studies in the research literature have attempted to address this issue, but most of these studies are based on cross-sectional observations rather than prospective, longitudinal studies that follow the same woman before, during, and after pregnancy.

Therefore, the dynamic of changes in the vaginal microbiome currently described should be considered with caution, since these changes can merely reflect intersubject differences, rather than true differences in vaginal microbiome composition during pregnancy. However, the recent addition of a few prospective studies may help us mechanistically link a specific vaginal microbiome composition and function to specific clinical outcomes, including premature birth.

Vaginal Microbiome

Limited information is available about the composition of the gut microbiota of women during pregnancy and how any possible changes can be related to both maternal and fetal health needs. It is intuitive that a malnourished mother will conceive babies that have a higher risk for problems throughout childhood than a well-nourished mother. While these poor outcomes have been related to quantitative deficits in nutrients and vitamins, it is possible that changes in microbiota composition and function may impact the baby's clinical outcome as well.

Some reports describe changes in maternal gut microbiota composition during pregnancy in certain groups, while in other populations the microbiome was reported to be more stable throughout pregnancy. This apparent dichotomy suggests that both the maternal microbiota and diet may already have the potential to provide an epigenetic fetal imprinting that could translate into clinical health outcomes for the baby.

If we embrace the microbiome hypothesis as an explanation of the epidemic of noninfectious, chronic inflammatory diseases in the Western world during the past fifty years, we need to understand how the human microbiome, and especially the gut microbiome, is engrafted and then evolves in humans. We know that the mother's nutrition and lifestyle, particularly during the last trimester of pregnancy, may highly influence her own intestinal microbiota composition. This can then affect the baby's fate by impacting fetal nutritional development and lead to possible complications, including premature birth. But along with the gut microbiome, there is another maternal microbiome that plays an essential role at the time of birth, with possible implications throughout the child's lifetime, through early transmission of its microbial imprint: the vaginal microbiome.

Researchers describe the vaginal microbiome as less complex than the gut microbiome. Ravel noted that, contrary to the complexity of the gut microbiome that makes it almost unique for each individual, we have a better understanding of what might be considered "optimal" or "healthy" for the vaginal microbiota.[27] Scientists think that the genus *Lactobacillus*, with more than 170 species, plays a leading role in the healthy microbiome of both the vagina and the gut.

Looking into *Lactobacillus*

The vaginal microbiome, which can fluctuate daily with factors such as stress, alcohol consumption, antibiotic use, sexually transmitted diseases, contraception use, smoking, and other factors, remains more stable in composition in pregnant women than in nonpregnant women. Researchers have stratified microbial vaginal signatures into five "community state types," or CSTs, most of which have a dominant species of *Lactobacillus*.

Lactobacillus is especially dominant in the vaginal microbiomes of women of European descent when compared to African and African American women.[28] Along with converting complex sugars into cellular energy and lactic acid, *Lactobacillus* also provides a protective effect through the mediation of host immune responses in its attack on pathogens. Unlike *Lactobacillus*, which is viewed as a favorable component of the vaginal microbiome, bacteria associated with bacterial vaginosis, including *Prevotella* and *Gardnerella*, are more commonly associated with proinflammatory immune responses.

How *Lactobacillus* came to play such a dominant role in human evolution, at least in the Caucasian vaginal microbiome, remains a mystery, which leaves room for some creative hypothesizing. Ravel has several theories about what might have happened. One premise is that when humans developed agriculture, food was stored for long periods and fermentation occurred. When humans began to eat fermented foods, lactobacilli found their way to the vagina, which Ravel describes as a very hospitable place with availability of nutrients preferred by the vaginal *Lactobacillus*.[29]

According to Ravel and other researchers, in the beneficial relationship of the vaginal microbiome, the human host provides nutrients from exfoliated cells and glandular secretions. Two small glands called Bartholin glands, situated on either side of the vaginal opening, secrete mucus that keeps the vagina moist, also providing nutrients for bacterial growth. In turn, colonies of indigenous bacteria protect the vaginal ecosystem from pathogenic invaders.

Priming the Infant Immune System

A second hypothesis entertained by Ravel is that the migration of *Lactobacillus* to the vaginal microbiome occurred much longer ago and could be linked to pregnancy.[30] When humans began to walk on two legs, the pelvic bone began to rotate inward, and the female pelvic opening began to shrink. But the human brain was also evolving; the head of an infant is proportionately much larger than that of most other primates. These two opposing evolutionary developments gave rise to the "obstetrical dilemma" hypothesis credited to Sherwood Washburn in 1960, in which human females deliver offspring who are much less developed than most primates. Holly Dunsworth is among a group of scientists challenging this hypothesis.

Along with her elegant argument challenging the obstetrical dilemma hypothesis, Dunsworth points out that a larger adult brain size and behavioral complexity are associated with a higher level of dependency in primate offspring, and these are all exaggerated in humans.[31] Ravel noted that human babies are born without many of the skills—the ability to walk or actively seek milk from their mother, for example—that other mammals have, certainly because gestation length in humans is believed to be shorter than in most other primates of comparable size, due to the baby's larger head and the mother's smaller pelvic opening.[32]

He suggests that the coating of *Lactobacillus* that a human infant receives from its mother during vaginal birth could act as a stimulant to the immune system to produce protective antimicrobials as the baby matures. Pregnant women who have *Lactobacillus* consistently present in the vagina tend to have better outcomes (full-term pregnancy) than women who lack

Lactobacillus throughout pregnancy, a condition that can be associated with preterm birth, according to Ravel.[33]

When Baby Comes Early

Evidence from Ravel and others shows that the makeup of the vaginal microbiome can affect the risk of preterm birth, the leading cause of infant mortality worldwide. In several studies, an altered vaginal microbiota has been associated with preterm birth deliveries. Daniel DiGiulio and David Relman from the Stanford University School of Medicine found that CST 4, a vaginal microbial signature characterized by a *Lactobacillus*-poor vaginal community state type, was associated with an increased incidence of preterm birth. They also found that the risk for preterm birth was more pronounced for subjects with CST 4 accompanied by elevated *Gardnerella* or *Ureaplasma* abundances.[34]

Relman's group took weekly samples of a small number of women at risk of premature birth, with results that also highlight the challenge of studying the microbiome, and CST 4 in particular. The authors note that with less frequent sampling they would have missed a "number of excursions to CST 4 and hence would have been hindered in associating this state with preterm birth."[35]

It is well known that prematurity is a complication that occurs more frequently among underserved populations, with higher rates in African American women compared to white women. Based on this premise, Molly Stout and colleagues performed a prospective, longitudinal core study of pregnant African American women to establish whether particular vaginal microbial community characteristics are associated with the risk of subsequent preterm birth.[36] To achieve this goal, seventy-seven pregnant African

American women were enrolled in the study. The composition of their vaginal microbiomes was monitored longitudinally throughout their pregnancies.

Thirty-one percent gave birth prematurely. The authors noted a downward trend in the diversity of the vaginal community in both women with full-term birth and women with preterm birth, but they added, "However, among women who delivered at term, vaginal community richness and Shannon diversity remained stable." Conversely, those mothers with preterm delivery had vaginal microbiomes with "significantly decreased vaginal richness, diversity, and evenness during pregnancy," with the most significant changes occurring between the first and second trimester.[37]

Based on these results, the authors concluded that within a predominantly African American population, a significant decrease of vaginal microbial community richness and diversity was associated with preterm birth.[38] Because this divergent vaginal microbiome composition appeared early in pregnancy, this suggests that the early gestational stage may be an ecologically important time for events that can affect either preterm or full-term birth outcomes.

In a more recent prospective study, David Relman's group analyzed weekly samples of the vaginal microbiome of thirty-nine predominantly Caucasian pregnant women at low risk of premature birth and ninety-six predominantly African American pregnant women at high risk of premature birth. The authors were able to confirm the previously reported association between premature birth and lower *Lactobacillus* and higher *Gardnerella* abundances in the low-risk cohort but not in the high-risk cohort.[39]

When they applied high-resolution bioinformatics to achieve taxonomic assignment to the species and subspecies levels, they found that *Lactobacillus crispatus* was protective against preterm birth in both cohorts, while *Lactobacillus iners* was not, and that a subspecies clade of *Gardnerella vaginalis* explained the genus association with preterm birth. Patterns of co-occurrence between *L. crispatus* and *Gardnerella* were highly exclusive, while *Gardnerella* and *L. iners* often coexisted at high frequencies. Based on their results, the authors argued that the vaginal microbiome is better represented by the quantitative frequencies of these key taxa than by classifying communities into the five CSTs previously described.[40]

Ravel and his colleague Michal Elovitz at the University of Pennsylvania collected samples of the vaginal microbiome from two thousand women throughout their pregnancies, the largest study to date, to identify a subgroup at risk for preterm birth.[41] Using immunological and microbial "signatures" from 120 women who experienced spontaneous preterm birth, the researchers have developed a tool to identify women at risk. They are also developing a "live biotherapeutic" (highly specific mixture of microbial products) to deliver vaginally as an intervention to prevent preterm birth.[42]

As with many aspects of the study of the human microbiome, prospective, long-term studies are needed to help develop effective therapeutic interventions for conditions such as the risk of preterm birth. In the future, these will no doubt be targeted to an individual's microbial makeup and genetic and environmental factors.

Mode of Delivery

It should come as no surprise that recent studies suggest that the clinical fate of babies born by vaginal delivery differs when compared to those born by cesarean (C) section.[43] Some findings show that babies born by C-section have a higher risk of developing a variety of noninfectious, chronic inflammatory diseases, including diabetes, asthma, and celiac disease, to name a few conditions.[44]

It remains unclear why C-section delivery carries a higher risk for these conditions. One explanation is that the microbial components from the mother's microbiome that are imprinted on the infant through vaginal delivery would provide babies with a microbial composition that epigenetically induces more immunological tolerance. This could help infants maintain a state of health irrespective of their genetic predisposition to develop chronic inflammatory diseases.

Babies born by C-section also are mostly imprinted by the maternal skin microbiome. This microbiome has not been as highly selected as the maternal vaginal and intestinal microbiome, but it is instead composed of a mixture of chance microorganisms acquired from the environment along with genetically derived components. Studies show these include microbes from the hospital environment and the personnel present in the delivery room.

Current research into the skin microbiome shows a diverse, loosely organized microbial composition that varies between different body sites and among individuals. In parallel with the work done by gut immunologists, dermatologist Richard Gallo from the University of California, San Diego studies the molecular mechanisms that the skin microbiome employs to benefit the innate immune system.

Research by Gallo, Nina Schommer, and others provides new descriptions of the skin microbiome that have inspired work into the functional significance of the colonies of bacteria, viruses, fungi, and mites found on the skin. This emerging and vital area of research is adding to our understanding of the complexity of the human body as a "mutualistic organism," as Schommer and Gallo describe it.[45]

C-section deliveries, which are sometimes driven by financial or cultural factors rather than medical necessity, are on the rise around the world. We are just starting to tease out what might be the impact on the microbial colonization of vaginally delivered infants versus C-section infants, along with consequences for long-term health and the risk of chronic disease.

Neonatologist Josef Neu, introduced earlier in this chapter, reminds us that the consequences of C-section versus normal delivery can be more complex than simply missing the microbial seeding from the trip through the birth canal. Mothers who deliver via C-section are given antibiotics as a matter of routine preventive intervention, which can affect the mother's breast milk, which in turn can affect the early development of the baby's microbiome, said Neu. Another consequence he noted is that these infants may not be able to feed from their mothers or receive formula as readily as babies who are delivered vaginally. Since C-section hospitalizations are usually longer, the babies are also exposed to hospital microbes for a longer period of time.[46]

Another controversy that has arisen along with our deeper understanding of the human microbiome is the practice of artificially "seeding" infants born by C-section with the microbes from their mother's vaginal microbiome shortly after birth. In this practice, which is only supported by the American College of Obstetricians and Gynecologists (ACOG) under an Institutional Review Board–approved research protocol,[47] vaginal fluids from the mother are collected on a cotton swab and transferred to the mouth, nose, or skin of the newborn.

This practice has created a blistering debate in the literature about its effectiveness and safety. Neu has his doubts about the procedure, pointing out that pathogens could be present, such as herpes simplex or chlamydia, that could negatively affect the infant, a concern echoed by ACOG, which calls for screening of mothers if the procedure is carried out at the mother's request.

In a critical review of current literature, an Australian group concluded that the lack of exposure to vaginal microbiota in C-section newborns is unlikely to be a major contributing factor to neonatal dysbiosis.[48] Conversely, in an investigation of the meconium microbiome in neonates from China, the researchers concluded that "the microbiome and metabolic diversity of vaginally delivered infants were significantly higher than the corresponding factors in the C-section group."[49] Ample evidence shows that C-section infants have lower numbers of *Bacteroidetes* and higher numbers of *Firmicutes* than naturally delivered infants.[50]

Postnatal Microbiome Imprinting

No matter the mode of delivery, once the baby is born and colonized by microorganisms through both prenatal and perinatal exposure, the gut microbiome undergoes a period of highly dynamic and seemingly chaotic changes during the first year of the baby's life. Early colonizers are mainly aerobic bacteria, but these bacteria are almost immediately replaced by anaerobic bacteria, which will represent the largest bacterial subgroup to colonize the gut mucosa during the baby's life span.

Patricio La Rosa and colleagues studied stools from fifty-eight premature infants who weighed less than 1,500 grams at birth, sequencing 922 specimens. They found a patterned progression of the intestinal microbiome that was minimally influenced by external factors, including feeding regimens, antibiotic use, and mode of delivery. In their study, gestational age is shown as the strongest driver of the pace of progression of the microbial communities. La Rosa's group raises the question of whether host biology plays a more important role in driving this microbial progression in the infant gut.[51]

As the baby grows, the gut microbiota will then continue its modification until the milestone of the baby's first birthday, as the microbial inhabitants become more stable. By age one, the baby's gut microbiome has become similar to the composition of an adult's microbiota, a process that continues, albeit at a slower pace, until the age of three.

Unless other environmental and lifestyle factors come into play, this microbiota composition will remain substantially unchanged in any individual until the elderly years, when it seems that gut microbiota again will become more unstable and dynamic. Postnatal factors influencing microbiota composition and function are instrumental in shaping the gut ecosystem in children, and the diversity of the gut ecosystem may have a far-reaching impact on the clinical destiny of babies throughout their entire life span. Feeding regime, infections, and the use of antibiotics are among the environmental factors that have the most impact on the gut microbiota and function, and therefore on the baby's destiny.

It seems intuitive that if a baby is born by vaginal delivery, is fed breast milk, and is not exposed to antibiotics, this infant will be in a biological situation very similar to that of babies born two million years ago. These infants will be following the biological steps with which we evolved as a species. This concept represents more than a theory, since it is supported by biological evidence including the composition of human milk.

Feeding Baby's Gut Microbiome

Along with the many known benefits of exclusive or predominant breastfeeding of infants for the first six months, an increased microbial community resilience can be added to the list. Isabel Carvalho-Ramos and colleagues analyzed stools from eleven infants in an urban area in Brazil during their first year of life. Even within this small sample, the researchers found a fairly stable pattern of microbiome development in children exclusively or predominantly breastfed compared to children with a mixed feeding regime with formula and early food introduction. In breastfed babies, the gut microbiome evolved in an uninterrupted ecological succession, despite external influences including complementary feeding and antibiotic use, a phenomenon not observed in children with mixed feeding and solid food before their fifth month.[52]

One of the most abundant components of breast milk are the human milk oligosaccharides (HMOs). When HMOs were first described, their role puzzled investigators because they are not used as an energy source by the baby. With the discovery and appreciation of the complexity of the human gut microbiome, we subsequently learned that the paramount function of these sugars is to feed specific bacterial species that colonize the infant's intestine.

Recent studies show that the content and composition of HMOs may dictate the infant's growth. Mothers of severely stunted infants show significantly lower concentrations of HMOs, particularly sialylated and mucosylated HMOs.

Furthermore, there is growing evidence suggesting that HMOs are not only important for physical growth but also for brain development and cognition. Other key elements of breast milk that can influence the shape of the microbiome are lactose, the most abundant sugar in milk, and antibacterial molecules. Researchers have recently acknowledged that breast milk, long thought to be sterile, employs its own microbial ecosystem.

Allan Walker has been researching breast milk components for more than forty years. Research from his lab at the Mucosal Immunology and Biology Research Center (MIBRC) at Massachusetts General Hospital (MGH) shows that breast milk stimulates the proliferation of a diverse and well-balanced microbiome and helps to ensure the long-term development of a healthy immune response.[53]

Necrotizing Enterocolitis in Premature Infants

Infant formula is missing the beneficial components of breast milk, including a very rich breast milk microbiome, that help shape a healthy microbiome in babies. This partially explains the reported difference in the gut microbiota of formula-fed versus breast-fed infants. Some research shows that although formula feeding of preterm and low birthweight infants can result in a higher rate of short-term growth, it also carries a higher risk of developing necrotizing enterocolitis (NEC).[54]

This condition, a dangerous intestinal infection often seen in very premature babies, can lead to intestinal perforation and resulting bacterial infection and peritonitis. Canadian researchers estimate that NEC affects 5 to 12 percent of very-low-birthweight infants. NEC, which requires surgery in 20 to 40 percent of cases, is fatal in 25 to 50 percent of these infants.[55] If the mother cannot provide her own human milk, donor milk is often given to the tiniest babies to try to help prevent NEC. Donor milk is pasteurized, which kills off most of the bacteria, noted Neu, whose collaborators at the University of Florida have been able to culture bacteria from human milk.[56]

In a new line of research into NEC, our scientists at the MIBRC are using human intestinal organoids or "mini-guts." Led by Stefania Senger, this collaborative group has shown that there is increased susceptibility to inflammation in the premature intestine when compared to late fetal and adult intestinal organoids.[57] This confirms the importance of breast milk in the maturation of the immature infant gut by providing the protective immunological components listed in table 3.1 from Walker's research group.

Table 3.1
Protective immunological factors found in human milk mitigate characteristics of immature innate immunity

Characteristics of innate immunity in preterm infants	Protective immunological factors found in human milk
Lack of maternal antibodies transferred via the placenta during late pregnancy	**Immunoglobulins** Secretory IgA, IgG
Inadequate extracellular elimination of bacterial infections	**Cytokines** IL-6, IL-8, TNF-α, TGF-β1, and TGF-β2
Reduced pattern-recognition receptor (PRR) and tight junction (TJ) function contributing to inappropriate inflammation	**Growth factors** EGF, TGF-α, and TGF-β
Aberrant intestinal colonization	**Microbiological factors** Lactoferrin, human milk oligosaccharides, probiotic bacteria

Adapted from K. E. Gregory and W. A. Walker, "Immunologic Factors in Human Milk and Disease Prevention in the Preterm Infant," *Current Pediatrics Reports*, 1, no. 4 (September 2013): 222–228, https://doi.org/10.1007/s40124-013-0028-2.

Overuse of Antibiotics in Early Life

As infants progress from the birth experience through their first
year of life, differences in the microbiomes of vaginally delivered
and C-section babies and breastfed versus formula-fed babies
become less pronounced. Several studies support the hypothesis that the introduction of solid food accelerates the infant gut
microbiome as it moves toward a more adult microbiome. One
systematic review showed a significant difference in C-section
versus vaginally delivered babies up until three months, with
this difference disappearing after six months.

Another key element influencing postnatal microbiome
dynamics is the use of antibiotics. This class of drugs used to com-
bat bacterial infections represents a superb tool that positively
impacted infant mortality in the twentieth century. However,
the promiscuous use of antibiotics, typically prescribed throughout the first two years of life—even though the vast majority of
infections are viral in nature—has caused a major shift in the
composition and diversity of the gut microbiome in a multitude
of infants.

How antibiotic use affects vaginal microbiota composition
is also coming under scrutiny. Researchers from the United
Kingdom examined the role of vaginal dysbiosis in the risk of
premature rupture of the fetal membranes. Unlike in full-term
pregnancies, the researchers found that the vaginal microbiome
was lacking in *Lactobacillus* prior to the rupture of the membranes in a third of the cases studied. Additionally, a common
clinical treatment to prevent this condition, prophylactic antibiotics, was found to exacerbate the vaginal dysbiosis. This *Lactobacillus* depletion was also associated with the early onset of
neonatal sepsis.[58]

Microbiome Engraftment and Long-Term Clinical Outcome

In reviewing the complex and sophisticated prenatal, perinatal, and postnatal factors in play in the establishment and dynamic of the gut microbiota in infants, we see that this is not a random process; rather, it is an exquisitely orchestrated program. The progression is dictated by evolutionary needs matured throughout the past two million years of the history of humankind.

The reason this process can dictate the fate of the human host can be better appreciated if we consider that the first one thousand days, starting at conception, are instrumental in shaping the function of the immune system. Furthermore, this shaping of the immune system occurs not only soon after the establishment of the gut microbiota but also throughout the person's life span.

This in turn will control the balance between tolerance (state of health) and immune response (disease) in the context of a person's specific genetic background. We are learning that the unnecessary introduction of antibiotics too early into this delicate interplay can have a deleterious impact on the baby's development.

Training the Immune System

The microbiome, with its diverse composition and balance, trains the immune system about if and when to unleash inflammation to fight danger. Evolutionarily speaking, the immune system has been bioengineered to fight a single enemy, namely microorganisms. Remember that the life expectancy of our Paleolithic ancestors (from 2.6 million years ago to 10,000 BC) was about thirty-three years of age. For our more recent Neolithic ancestors, the average life span was from twenty to thirty-three years of age. Throughout human biological history, the two most frequent causes of mortality were being killed by a predator and being killed by an infection. It is only in recent years that the immune system has encountered a host of new enemies, such as pollution, radiation, exposure to toxic chemicals, and cancer.

When exposed to microorganisms, the immune system, using both innate and adaptive immune branches, creates a state of inflammation to generate a micromilieu that is inhospitable for microorganisms. This environment can become too hot (a temperature of 37 degrees Celsius is optimum for the vast majority of microorganisms) and hostile for microbes. The immune system produces chemicals, including cytokines and chemokines, and it recruits a variety of immune cells that can kill the vast majority of microorganisms.

This inflammatory process results in collateral damage to the host, namely destruction of the tissue where the inflammatory process occurred. But this same sophisticated process saves the host from succumbing to the infection. We can see that the microbiome is the most important trainer and guide for the immune system to determine if and when it should unleash inflammation.

It is likely that a balanced and well-established microbiome, thanks to optimal prenatal, perinatal, and postnatal environmental factors in line with human evolutionary development—including a good state of maternal health, vaginal delivery, breastfeeding, and sparing use of antibiotics—can train the immune system to unleash inflammation only when strictly necessary. These microbial fighters create inflammation only when our systems of surveillance, through pattern-recognition receptors, perceive a potentially dangerous situation (see figure 3.2).

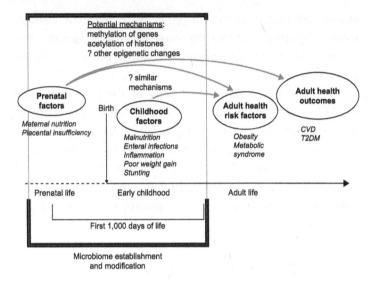

Figure 3.2
Factors and lifestyle during the first one thousand days of life (from conception to two years of age) influence microbiome engraftment and function toward adulthood and adult health outcomes.

An Off-Balance Microbiome

Conversely, if the gut microbiome's composition is influenced by factors that are not aligned with human evolutionary planning, including poor lifestyle of the mother, C-section delivery, formula feeding, and overuse of antibiotics, the establishment of the microbiome will be off-balance (dysbiosis). This dysbiosis can lead to a series of biological consequences, including the alteration of gut permeability, increased antigen trafficking, and an abnormal immune response that instigates the immune system to unleash inflammation even when it is not strictly needed. Ultimately, this chain of events can lead to increased, low-grade chronic inflammation that, together with epigenetic changes induced by the microbiota, can cause a break of tolerance and the onset of chronic inflammatory disease in genetically predisposed individuals.

Since this microbial programming of the baby's intestine would be completed within the first one thousand days of life, we can hypothesize that an inappropriate microbiota composition during this crucial period may have far-reaching consequences on the state of health of this individual. This could result in a continuous, chronic inflammatory process that, for a person with a specific genetic background, could lead to negative clinical outcomes.

4   Cracking the Codes: From the Human Genome to the Human Microbiome

Early Developments in Identifying Bacteria

As our knowledge of the human body has evolved, we have moved from Leonardo da Vinci's clandestine study of cadavers for the basics of anatomy to a more holistic vision of our body's composition and function. With the realization that some tissues and organs can regenerate, thanks to the activity of multipotent or pluripotent stem cells, we broadened our reductionist vision of a human being as a multicellular, static organism.

This expanded concept makes us part of a continuum with our surrounding environment. An appreciation of the symbiotic relationship we share with our microbiome is bringing human physiology to the next level of complexity, one in which epigenetics is part of our plasticity and capability to adapt without genetic mutations.

Molecular microbial ecologist Rita Colwell is among the scientists most qualified to help us appreciate the depth of such a revolutionary and novel view related to the impact of the human microbiome on this evolutionary journey of personal destiny. "There is an integrated fabric and an interwoven connectivity.

We need to understand that we humans are simply another species in this massive evolved environment on planet Earth," said Colwell.[1] She reminds us of the historical foundations of scientific research, stretching back generations, which have brought microbiome science to where we are today.

In the 1920s, scientists examined enzymes and the functions of bacteria by separating and identifying groups based on their physiology. Recalling the history of the 1940s, Colwell described what she defined as a "half moment" for the "unity of biochemistry" when scientists focused on the realization that the Embden-Meyerhof-Parnas pathway of sugar degradation in cows contains the same enzymes as a human being. "That was a big revelation in the 1940s," said Colwell.[2] Now it's an accepted and universal tenet that organisms are strikingly similar at the molecular level.

In the following decades, attention shifted away from commonalities and toward quantitative studies to identify individual organisms. Colwell described how the DNA revolution following the double helix discovery led to the understanding that different species have different base compositions, which led to simple hybridization of organisms, which was followed by "elegant density gradient hybridization."[3]

A self-described "microbial systemicist" as well as a microbial ecologist, Colwell has studied cholera for more than fifty years, written more than eight hundred publications, and received at least sixty honorary degrees. She is a former director of the National Science Foundation and former president of the American Association for the Advancement of Science. With her groundbreaking discoveries in cholera and her expansive understanding of the concepts behind the human microbiome, she could be easily defined as the "Robert Koch" of our times. Her six decades of work span many eras of science and different fields of biology, but her early choice of systematics was related to her gender.

"I was a woman scientist, and women scientists were not welcome in the laboratories in the 1960s," said Colwell. Working in marine microbiology at the University of Washington for her doctorate in microbiology, she recalled that her fieldwork was restricted to daytime cruises only because women were "not welcome on board ship for oceanographic or fishery work, especially overnight."[4]

This gender bias had a silver lining for the scientists now studying any aspect of the collective microbiome of our biosphere called planet Earth. In 1960, using an IBM 650, the first computer employed at the University of Washington, Colwell wrote the first computer program to identify marine bacteria, using a technique called numerical taxonomy. For her PhD thesis, also known as the "little bug" program, Colwell cultured marine bacteria, which she then ran through a battery of tests before calculating similarities. Her early findings, published in influential journals including *Nature* and *Science*, led to the understanding that bacteria could be identified in a quantitative way.[5]

The Friar and the Pea Plants

Before moving on to the more recent—and extremely daunting—task of using advanced technology to study the human microbiome's composition and function, we need to revisit the genetic piece of the puzzle. In 1913, Alfred Sturtevant mapped the relative locations of a series of genes on a drosophila chromosome, and the golden age of quantitative genetics took a quantum leap forward.[6] But the intellectual roots of genetic inquiry stretch back to the mid-nineteenth century with the famous Augustinian friar Gregor Mendel, who crossbred almost thirty thousand pea plants and established the principles of inheritance.

As often happens with a shift in a scientific paradigm, groundbreaking work is initially disparaged or overlooked, which is what happened when Mendel published his 1865 work on the heritability of traits.[7] Mendel's work was rediscovered in the early twentieth century, and this led to the development of quantitative genetics, which greatly influenced many aspects of biology and medicine and many scientists, including Sturtevant and Thomas Hunt Morgan.

The youngest of six children, Sturtevant was encouraged by his older brother Edgar to read a book on Mendel's laws. Alfred applied Mendel's principles to the horses on the family farm in Alabama to explain the inheritance of coat colors. He entered Columbia University in 1908, working under Morgan, who received the Nobel Prize for Physiology or Medicine in 1933 for "his discoveries concerning the role played by the chromosome in heredity."[8] A critical and independent thinker, Morgan was unafraid of challenging scientific orthodoxy.

Working with drosophila around 1909, Morgan turned up a striking number of mutants, including the white-eyed male fruit fly; he subsequently identified white eyes as a sex-linked trait. At that time, chromosomes had yet to be identified as repositories of genetic information. Morgan's studies on drosophila confirmed that genes are contained in chromosomes inside of cell nuclei. Inside those chromosomes, Morgan determined not only that genes are organized in long rows, but also how traits related to each other correspond to genes that are close to one another on the chromosomes. Morgan also discovered the "crossover" phenomenon, in which parts of different chromosomes can trade places with each other.

A Tarnished Double Helix?

Forty years after Sturtevant and Morgan's discoveries, the next major milestone was perhaps the most significant discovery in human science, namely the determination of the double helical structure of the DNA molecule, credited to Francis Crick and James Watson in 1953. Along with Maurice Wilkins, they were awarded the Nobel Prize in Physiology or Medicine in 1962 for "discoveries concerning the molecular structure of nucleic acids and its significance for information transfer in living material."[9] But as often happens, the taking of a great leap forward in science was marred by poor communication, personality conflicts, and controversial accounts of the findings.

Earlier work by Rosalind Franklin and her graduate student Raymond Gosling, which led to the model that Watson and Crick created of the now-familiar double helix image of DNA, has too often been omitted from the story. Franklin, who went on to study the molecular structure of viruses after her DNA studies, died in 1958 and has since received posthumous attention and honors for her pivotal role in unraveling the structural puzzle of DNA.[10]

Through X-ray crystallography studies conducted in the John Randall Biophysics Lab at King's College London, Franklin and Gosling identified two forms (dry and wet) of DNA, which Franklin called Form A and Form B. While Franklin realized Form B was probably helical in structure, with phosphates on the outside of the ribose chains, Franklin's mathematical analysis of

Form A did not show a helical structure. Franklin worked on trying to resolve the differences in the more complex Form A, and by early 1953 she had determined that both forms had helical structures.[11]

Meanwhile, according to some accounts, Wilkins, who also worked on crystallography studies in the Biophysics Research Unit at King's College, shared a crucial photograph taken by Gosling with Watson and Crick, unbeknownst to Franklin. Before Randall assigned Gosling to Franklin as a technician and graduate student, he had worked with X-ray diffraction studies under Wilkins. According to one of Franklin's biographers, Crick's thesis advisor, Max Perutz, also shared unpublished material from Franklin and Gosling's work with Crick and Watson.[12]

The material helped Watson and Crick create their final model of the helical structure, and they subsequently published a brief but groundbreaking article in *Nature* describing the molecular structure of DNA in April 1953.[13] In that same issue, two teams from King's College, one comprised of Wilkins, Alexander Stokes, and Herbert Wilson and the other comprised of Franklin and Gosling, published experimental data that supported evidence of the "likely molecular structure of DNA."[14]

Although the Watson and Crick article of 910 words included no supporting evidence of their findings and did not reference Franklin's findings, it did mention Wilkins and Franklin in its acknowledgments: "We have also been stimulated by a knowledge of the general nature of the unpublished experimental results and ideas of Dr. M. H. F. Wilkins, Dr. R. E. Franklin and their co-workers at King's College, London."[15]

In 2003, Lynne Osman Elkin described their words as "one of the greatest understatements in the writing of scientific history." She adds, "Watson and Crick made one of the most important and impressive scientific discoveries of the 20th century, but their golden helix is tarnished by the way they have treated Franklin and Wilkins."[16] A closer reading of the story would add Raymond Gosling to Elkin's list of pioneers in DNA imaging.

Advances in DNA Sequencing

In the 1970s, Frederick Sanger's work on protein structure led to his successful sequencing of DNA. One of a select few to be awarded two Nobel Prizes in the same category, Sanger won his first Nobel Prize in Chemistry in 1958 and his second Nobel Prize in the same category in 1980, when he shared the prize with Walter Gilbert and Paul Berg.[17] Two years earlier, Hamilton "Ham" Smith, often called the father of modern molecular biology, was one of the winners of the Nobel Prize in Physiology or Medicine for "the discovery of restriction enzymes and their application to problems of molecular genetics."[18] These chemicals are used by bacteria to slice segments of DNA, giving them unparalleled genetic plasticity and adaptability.

Recruited by J. Craig Venter in the early 1990s to The Institute for Genomic Research, Smith was pivotal in the race to sequence the DNA of an entire organism. Before Smith's arrival, Venter had been randomly sampling and sequencing small bits of cDNA. Smith proposed a bolder approach of "shotgunning" the entire genome of an organism.

With an ordinary kitchen blender, the organism's DNA was divided into millions of small fragments. The fragments were run through automated sequencers and then reassembled into the full genome using a high-speed computer and novel software developed at TIGR. In 1996, in collaboration with scientists from the Molecular Infectious Diseases Group at the University of Oxford, the TIGR group published the entire genome of *Hae-mophilus influenzae*, the microorganism that had earlier led Smith to the discovery of restriction enzymes.[19] TIGR soon joined the race to sequence the entire human genome.

Meanwhile, in 1987, in response to the possible threat of radiation causing mutations to the human genome, the U.S. Department of Energy (DOE) established an early genome project. A year later, Congress approved funding for the DOE and the National Institutes of Health (NIH) to explore the feasibility of such a project. James Watson, whom we met earlier in this chapter, was appointed as the first director of the National Center for Human Genome Research, an agency created by the NIH in 1989 and elevated to research institute status as the National Human Genome Research Institute in 1997.[20]

In 1990, the initial planning stage was followed by the publication of a research plan. However, Watson resigned in 1992, and after a brief tenure by Michael Gottesman, Francis Collins was named director, a position he held for fifteen years. The National Human Genome Research Institute spearheaded efforts to complete the Human Genome Project, and an initial analysis of the human genome was published in February 2001 after twelve years and a cost of $2.7 billion.[21]

Looking for the Genome "Switch"

Unraveling the human genome, however, didn't give us the easy answer scientists anticipated on how to effectively tackle human disease. Claire Fraser, who worked with Venter and Smith at TIGR and collaborated with other scientists on the sequencing of the *Haemophilus* genome to usher in the field of microbial genetics, recalls the high expectations during the early days of sequencing organisms and the Human Genome Project. "I think we believed, maybe not so much for the human genome, but for the first microbial genome, we really believed we would be able to identify all the genes and put them all in pathways. And you'd come up with a pathway map of an organism, and you could flip a switch and see how it all worked."[22]

In the 1990s, scientists estimated the total number of human genes to be around seventy-five thousand to one hundred thousand. Results from the Human Genome Project drastically reduced that number to about twenty-three thousand genes. At that point the paradigm of one gene, one protein, one disease was dispelled. This development left scientists with no clear path to tackle the myriad of human diseases that the Human Genome Project intended to resolve once and for all. Leveraging the massive amount of data and the technological advances that the project engendered, the scientific focus shifted to the human microbiome.

At the time, the TIGR group was working on a microbiome project for the Defense Advanced Research Projects Agency. Although scientists had been collaborating for years on how individual microorganisms interact with the human host to cause infectious diseases, with the shift to microbiome science, things were about to change—drastically.

The Human Microbiome Project

Indeed, it was only with the appreciation of the relatively simple genomic makeup of the human species that scientists hypothesized that the human microbiome could be intricately important, together with the human genome, in defining the balance between health and disease. How little we knew at the time that this line of research would create the foundation for the Human Microbiome Project.

With the completion of the Human Genome Project, Collins assembled the leaders of major genome centers in the United States, along with a number of microbial genomics and ecology experts, to map out what a project on the human microbiome might entail. As Fraser recalls, he was of the opinion that such an effort would be analogous to the Human Genome Project and, with enough microbiome sequencing, the group would be able to reach a defined endpoint. Thinking in terms of ecosystems rather than individual organisms, she wasn't so sure that the findings would be as clear-cut or that it would be obvious when the sequencing would be complete.[23]

"I had begun working with marine microbial ecologists and was taking a more ecosystems approach to the study of microbes. I didn't think that data sequencing alone would be able to explain all the interactions between the organisms and the interactions between the host and the microbiome," said Fraser.[24] Even though her comments were dismissed by some scientists at the time, her "hypothesis" has been borne out with subsequent research.

In the early years of the Human Microbiome Project, two major obstacles to exploiting the technologies developed by the Human Genome Project to study microbiome composition and function were cost and a much higher level of complexity. Before the sequencing of DNA, the world of bacterial pathogens and host-microbiome interaction relied fully on our capability to culture microorganisms that could be seen in a petri dish or under a microscope.

By 2001, the cost of sequencing an entire human genome had dropped from $2.7 billion and twelve years of work to

$100 million and only one year to complete the sequencing. Even though it had become a much cheaper and faster endeavor, it was still not applicable to the much more complex and dynamic target of the human microbiome. At that point in time, clinicians could never have anticipated that genome sequencing would play a role in clinical applicability during their careers.

Fast forward to the present—it costs $700 and a few hours to obtain the entire genome sequence of a human being. To put this in perspective, a laptop computer that cost roughly $4,700 in 2001 dropped to $1,000 in 2019, a fourfold decrease. During that same time period, the cost of human genome sequencing dropped 1.4 million times, a testament to how technology and knowledge in the DNA sequencing domain have evolved during the past few decades, thanks in large part to the Human Genome Project.

Dynamic Nature of the Microbiome

The affordability of genome sequencing made the proposition of sequencing the human microbiome feasible. Funding for the Human Microbiome Project was established in 2008 to provide resources to establish the characterization of the human microbiome and its role in health and disease.

The demonstration that automated DNA sequencing technologies and novel genome assembly algorithms could be used to sequence the entire genome of a single microorganism was accomplished in 1995. D. W. Hood, E. R. Moxon, R. D. Fleischmann, and colleagues sequenced the genome of *Haemophilus influenzae*,[25] and Fraser and colleagues reported on a whole genome sequence of *Mycoplasma genitalium* just a few months later.[26]

Over time, the number of microorganisms for which genome data are available has ballooned to more than one hundred thousand, and each sequencing project now only takes a few hours to complete. The introduction of next-generation sequencing technologies in the early 2000s, together with continued development of assembly algorithms and annotation pipelines, made the sequencing of entire microbial ecosystems living on and within the human host feasible. What we have described so far gives us the basis to analyze the collection of all the microorganisms living in association with the human body, including archaea, eukaryotes, bacteria, and viruses (see figure 4.1).

Through the Human Microbiome Project, we now have a better idea of the microbial communities that colonize several different sites of the human body, including airways, oral cavities,

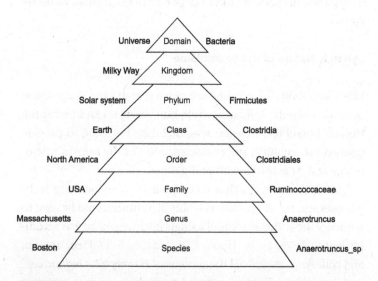

Figure 4.1
Hierarchy of the classification of the bacterial domain.

skin, and the gastrointestinal and urogenital tracts. Initial estimates from the Human Microbiome Project suggested that bacteria in the human microbiome were ten times more abundant than human cells. This figure has been considerably downsized. Rob Knight told the BBC in 2018 that this number is "much closer to one-to-one, so the current estimate is you're about 43 percent human if you're counting up all the cells."[27]

Additional advances in technical and economic feasibility have led to a new level of sophistication in studying the human microbiome. We are shifting from the simple analysis of its composition using 16S rRNA sequencing technologies without the need of cultivation to the metagenomics approach, which allows a comprehensive examination of the whole genetic composition of microbial communities. A final step in the exploitation of the potential of studies on microbiome composition and function is the metatranscriptomic approach.

While metagenomics catalogs what microbes are present and what genomic potential they have, metatranscriptomics tells us about their activity: the genes that are most highly expressed in a specific microbial environment and at a specific time. Thus, metatranscriptomics is the study of the function and activity of the complete set of transcripts (RNA-seq) from specific samples, which allows the identification of possible pathways affected by the host-microbiome interaction leading to clinical outcomes. The metatranscriptomic approach has the potential to allow the identification of genes expressed by the microbiome that are most likely responsible for the activation of specific pathways involved in the pathogenesis of diseases suffered by the host, thereby leading to possible novel therapeutic targets.

What makes the definition of the human microbiome even more complex when compared to the human genome is the dynamic changes in its composition and function over time. The multifaceted human microbiome can generate essential elements such as vitamins, modulate the function of the immune system, help fight off pathogens, and regulate fundamental biochemical pathways that involve bone metabolism, promotion of fat storage, and modification of the nervous system. These are only a few highlights that demonstrate the importance of studying both the composition and the function of the microbiome to better understand its role in health and disease.

Aims of the Human Microbiome Project

When the Human Microbiome Project was conceptualized, five aims were identified as essential to the overall mission for this project. The first aim, namely "the development of a reference set of three thousand isolate microbial genome sequences," has been achieved.[28] But as outlined in this chapter and in many other parts of this book, knowing the composition of the microbiome of an individual (i.e., answering the question of "who" is there) is only one component. It is probably not even the most important element in the study of the potential impact of the microbiome on human health. Based on our current knowledge, simply assessing the composition of the human microbiome will not lead us to link a specific microbiome to the development of disease based on the host's specific genetic background.

The second aim, "to generate an estimate of the complexity of the microbial community at each body site, providing initial answers to the questions of whether there is a 'core' microbiome at each site,"[29] might be even more problematic. To identify the composition and function of a normal or core microbiome does not consider the complexity of the question.

We can perform this core analysis of one hundred people of comparable age, sex, lifestyle, and residence and obtain an apparently chaotic, not homogeneous outcome. We can also take one person and perform this core analysis of their entire microbiome every six months, obtaining similar confusing results. Therefore, it is highly probable that there will be no predictable pattern of core composition in either group.

For example, as discussed in chapter 3, the vaginal microbiome in a single woman harbors a very dynamic microbial ecology, which changes during pregnancy, birth delivery, and menopause, and it is highly affected by environmental factors, including stress and smoking. The bottom line is that focusing on the microbiome's composition rather than its function may represent a misguided approach to achieving the third aim (see below). Very different microbiome compositions may lead to the same functional outcomes, thus creating metabolic conditions that result in similar clinical outcomes.

The third aim, "demonstration projects to determine the relationship between disease and changes in the human microbiome," is the most challenging.[30] The vast majority of current literature on the microbiome is based on cross-sectional studies in which two distinct cohorts, healthy subjects and patients affected by the disease object of the study, are analyzed in terms of microbiome composition and, sometimes, function.

As mentioned later in this chapter, cross-sectional studies, even if perfectly matched for variables—such as age, sex, socioeconomic background, and location—run the risk of assuming that the outcome (i.e., disease) is caused by the detected differences in microbiome composition and function. It is entirely conceivable that the reverse is the case: the disease itself is responsible for changes in the microbiome.

The fourth aim is "development of new tools and technologies," and the fifth aim is "examination of the ethical, legal and social implications to be considered in the study and application of the metagenomic analysis of the human microbiota."[31] Of the five aims, only these two have the potential to directly impact the feasibility of studying the balance of health and human disease as assessed through the human microbiome.

As microbiome hunters continue their global search for human specimens, the ethical, legal, social—and let us add cultural—implications of this race to acquire human samples come more sharply into focus. An article in *Science* by Ann Gibbons addressed some of these concerns for the Hadza as their numbers shrink and their access to game-filled land decreases.[32]

The focus on the Hadza has drawn tourist enterprises, encampments, and road construction as well as alcohol addiction and drug abuse. Gibbons noted that some researchers think scientists have "asked too much" of the Hadza, and some scientists are working to respond to the Hadza people who want more say in research studies and assistance with securing land rights and other issues.[33]

Moving from Genome to "Multi-Biome"

As a key player in both the Human Genome Project and the Human Microbiome Project, Fraser provides insights into the transition between the two projects and how to integrate what has been learned about the human genome into the study of the human microbiome. In the early stages of the Human Microbiome Project, Fraser was learning to think in terms of ecosystems rather than compartmentalized and different organisms sharing the same space.[34] Colwell, who discovered that the causative agent of cholera, *V. cholerae*, inhabited an aquatic environment before moving to its human host, views the bacterium and the human being on a continuum of a larger ecosystem.[35]

But this approach was not initially adopted for the Human Microbiome Project. According to Fraser, the first five years of the Human Microbiome Project were spent setting up the project in much the same way that the Human Genome Project was framed. Initially, the effort focused almost exclusively on 16S rRNA sequencing, with no animal studies included, and looking at the microbiome only in terms of "Who's there?"

"That was important; it was a logical first step," said Fraser. "But what became immediately obvious is that every person is different when you take a 16S rRNA-based snapshot of their microbiome, and no functional data came out of this."[36]

Meanwhile, the European MetaHIT (Metagenomics of the Human Intestinal Tract) project led by S. Dusko Ehrlich was moving ahead, noted Fraser. "They understood it. They were generating metagenome sequences and metatranscriptome sequences and doing animal studies. They included nutritionists on some of their teams, and their approach was holistic right from the beginning."[37] Fraser is a big believer in the idea of "food as medicine" and trying to leverage what we can learn about the microbiome in this regard to prevent and treat human disease.

Launched in 2008 with funding from the European Commission, MetaHIT generated initial results showing that human gut microbes numbered 3.3 million genes compared to the 22,000 in the human genome. MetaHIT generated 540 gigabytes of DNA and catalogued 19,000 gene functions, including 5,000 not previously identified. MetaHIT researchers also were among the first to identify "enterotypes," meaning the classification of humans into groups based on the composition of their gut bacterial communities.[38]

Given the complex and dynamic nature of the gut microbiome, problems with consistency in sampling methods, and the bacterial diversity among different world populations, this approach to gut microbial classification has been challenged. In 2013, Omry Koren and colleagues called for "multiple approaches when calling for enterotypes."[39] In 2018, Fraser collaborated with members of that small research group, including Curtis Huttenhower, Knight, and Ruth Ley, along with Peer Bork from the European Molecular Biology Laboratory in Heidelberg, Germany, and other international colleagues, to revisit the concept of enterotypes. In a paper published in *Nature Microbiology* in 2018, the group cautioned against relying solely on the classification system of enterotypes for the human microbiome; however, they did describe enterotypes as a "useful tool for studying the human microbial community landscape."[40]

Agreeing with Koren's group about using a variety of methodologies, Colwell added that the notion of employing a "gradient" approach provides a more comprehensive view of individual microbiomes.[41] As we advance in human microbiome research, Fraser is calling for a much more diverse sample collection beyond European and Caucasian groups. And, as Fraser predicted, early results showing that every person has a different microbiome should not have been surprising. The study of the interaction between enteric pathogens and human hosts has led to the realization that even if we are dealing with the same pathogen and the same human target, the clinical outcome can be very different from one person to another. Let's look at some examples.

Variations with *Vibrio*

As we discussed earlier, *V. cholerae* causes pandemics with a very
high rate of infection that often leads to death. Colwell has
spent much of her career fighting these outbreaks using innovative and simple techniques, like filtering drinking water through
sari cloth, to reduce the risk of contamination and lower the
mortality rate caused by drinking contaminated water.[42] However, not everyone infected by *V. cholerae* will end up with the
same disease severity and duration. Indeed, some people may
not even have symptoms and may simply become carriers, even
if they have never been exposed to cholera. At the other end of
the spectrum, there are people who become severely dehydrated
and die after being exposed to the microorganism.

While working at the Center for Vaccine Development at the
University of Maryland in Baltimore, coauthor Alessio Fasano was
involved in clinical trials in which volunteers were exposed to
enteric pathogens like *V. cholerae*, *Salmonella*, and *Shigella*. Even
in this very controlled research setting, the clinical outcome was
different from one individual to another. Again, some people
exposed to *Salmonella* never got sick, while others developed the
typical salmonellosis. People exposed to *Shigella* either developed
self-limiting watery diarrhea or suffered the full strength of the
dysentery with bloody diarrhea and severe abdominal pain.

These clear differences in clinical outcomes from exposure
to a single microorganism interacting with a human host could
have suggested that when extrapolated to the entire microbiome community, this much more complex interaction may be
even harder to predict in terms of clinical outcome. Therefore,
it is an unreasonable assumption that the composition of the

microbiome will hold the key to understanding why some people may develop a disease on a given genetic background while others do not.

Looking at the Big Picture

With few exceptions, the human body is not the exclusive ecosystem in which microorganisms need to live and adapt. Rather, we are part of a much more complex global ecosystem, including the soil, the marine environment, and other animal hosts in which microorganisms need to rapidly adapt to survive and reproduce to complete their life cycles. We must frame the knowledge we obtain from metagenomic studies within the concept that the microbiome's composition and function is a dynamic process influenced by many variables, including lifestyle, antibiotic use, and nutrition and diet, to name only a few.

To reiterate, the aim of finding a "normal" microbiome by sequencing the microorganisms at any human site—for example, the gut—is probably not a useful goal. Many microbiome scientists are now convinced that there is no such thing as a normal microbiome. By the time the sequencing is completed, the site's ecosystem could be very different in that individual, subject to the influence of environmental and other factors.

To complicate the matter even further, the human virome can cause lateral transfer of genes from one microbe to another within the same life cycle (see chapter 5). This can create even more variability in the function of a specific site's microbiome over time, even if the composition of the microbiome remains the same.

Fraser thinks the microbiome field has finally broken away from early constraints. But she still sees limitations in the idea that enumerating changes in species is the best way to obtain insights about how the microbiome contributes to health and disease. She asserts that the focus must be on function, which for a complex ecosystem like the gut microbiome must include animal and in vitro studies, and possibly additional integrated studies.

The only acceptable way to link microbiome composition and, most important, function to any given disease is to move from cross-sectional studies comparing cohorts of healthy versus diseased individuals to prospective, longitudinal studies involving subjects who are genetically at risk for a given disease. These types of studies can investigate microbiome composition and function before, during, and after the onset of the disease. By integrating data from the microbiome, metagenomics, metatranscriptomics, and metabolomics with comprehensive clinical and environmental data, we can build a mathematical model of how the environment may affect microbiome composition and function leading to the onset of disease in genetically predisposed individuals. (These concepts will be revisited and expanded in chapters 14 and 17.)

The creation of novel biological computational models and a clear path to move from microbiome association to causation are essential steps in providing a mechanistic approach to gain insights into the development of noninfectious, chronic inflammatory diseases and possible targets for personalized interventions or disease prevention. This can only be done if studies focused on the microbiome are placed in the context of the biological ecosystems in which we live and interface each day. As Colwell noted, "We make mistakes over and over again, because we don't try to understand what the function of this organism is, either in its lower evolved host or in the environment from which it originally evolved."[43]

There is enormous interest in and research effort being put into the many phases and facets of the human microbiome, sparked in large part by the Human Microbiome Project. However, we fully agree with Fraser and Colwell that we are not at the end of the pipeline that will lead us to exploiting the human microbiome for treating a variety of chronic inflammatory diseases. Rather, we believe that we are at the beginning of a very promising path of discovery, assuming we capitalize on the enormous amount of work that the Human Microbiome Project has generated already to bring us closer to using this information to find the balance between health and disease.

5 Beyond Bacteria: Those Other "Omes"

Keeping the Body in Balance

According to the human microbiome hypothesis, the complex communities of microorganisms that make up the gut microbiome help to ensure good health by interacting with the host to maintain physiological homeostasis. However, the model of keeping diverse components or elements of the body in harmony to maintain the unstable and extremely dynamic balance between health and disease is hardly a new one.

Traditional medicine, as still practiced in many parts of the world, is often based on interventions to restore good health to the body by balancing elements that are thought to be in disharmony. In Traditional Thai Medicine (TTM), herbs, massage, and yoga practices are used to correct the balance among the human elements of earth, water, wind, and fire. Unlike many traditional medicine disciplines, TTM was included in the first Thai Medical School founded in 1889, but it was removed from the curriculum in 1916.[1]

The Declaration of Alma-Ata by the World Health Organization (WHO) in 1978 urged member countries to include "traditional practitioners" to respond to the primary health care needs of communities. Following this declaration, TTM once again was included in government policies as Thailand's Ministry of Public Health issued a policy to promote medicinal plants in primary health care programs.[2]

Traditional Chinese Medicine (TCM), practiced for thousands of years by local healers and passed on through oral tradition, was codified by the Chinese government in the 1950s and 1960s. TCM focuses on the concept of holistically restoring and maintaining harmony throughout the body for optimum health, even when disease is present. The earliest written records of this sophisticated medical system exist on tortoise shell and bone dating back approximately three thousand years.[3]

These ancient writings focus on the circular movement of *qi*, which is air or vapor, and *xuè*, which is blood. According to the American College of Traditional Chinese Medicine, ill health results from "stagnation, deficiency or the improper movement of qi and xuè, and may result in an imbalance of *yīn* (陰) and *yáng* (陽)."[4] Chinese migrants who came to work on the transcontinental railroad in the nineteenth century brought TCM with them to the United States, but it was practiced almost exclusively in Chinese communities.

Acupuncture, perhaps the most well-known TCM therapy tool in the West, is only one component of TCM. Other TCM practices include the use of Chinese herbs, Chinese therapeutic massage, qi-gong and tai chi, dietary recommendations, and "cupping," the placement of small glass jars on the skin to create suction.

During the recent COVID-19 pandemic, TCM has been employed, along with nutritional support and prebiotics and probiotics, to regulate the balance of intestinal microbiota and reduce the risk of secondary infection. In 2020, practitioners at the First Affiliated Hospital of Zhejiang University School of

Medicine used TCM to promote disease rehabilitation as part of a treatment strategy called "Four-Anti and Two-Balance." Researchers define it as "antivirus, anti-shock, anti-hypoxemia, anti-secondary infection, and maintaining of water, electrolyte and acid base balance and microecological balance."[5]

TCM has influenced traditional medicine in other countries, including the Democratic People's Republic of Korea, Japan, and the Republic of Korea, with a similar focus on maintaining and restoring harmony. In the Western world, looking back to the Middle Ages, we find similar sophisticated systems of medical treatment and training that emphasize balance and harmony.

Schola Medica Salernitana

Born in Salerno, Italy, coauthor Alessio Fasano knows very well the history of the "Schola Medica Salernitana" (the medical school at Salerno), which has been described as the most important and perhaps most influential source in the Western medical canon on the need to reach balance to maintain health.

The Western world's first medical school, also known as the "Schola Salerni," was founded in the ninth century and rose to prominence in the tenth century. Situated on the Tyrrhenian Sea in the city of Salerno in Southern Italy, it became the most important source of medical knowledge in Western Europe in the Middle Ages. The school, which grew out of the dispensary of St. Nicholas Monastery, achieved its utmost splendor between the tenth and thirteenth centuries, from the last decades of Lombard power to the fall of the Hohenstaufen. The arrival of Constantine Africanus in 1077 marked the beginning of Salerno's classical period. Through the encouragement of Alfano I, archbishop of Salerno, and the translations of medical writings from Arabic into Latin by

Constantine Africanus, Salerno gained the title of "Town of Hippocrates" (*Hippocratica Civitas* or *Hippocratica Urbs*).

People from all over the world flocked to the Schola Salerni, both the sick, in the hope of recovering, and students (both men and women), to learn and to teach the art of medicine. The school was based on the synthesis of the Greek and Latin traditions and was supplemented by ideas from Arab and Jewish cultures. The approach was based on the practice and culture of prevention rather than cure, thus opening the way for the empirical method in medicine.

Following the Four Humors

The therapeutic approach of the Schola Salerni and its botanical studies (at that time the pharmacopeia was mainly focused on the use of herbs) was based on the "doctrine of the four humors," which was itself based on the ancient "theory of the four elements." This theory took shape and gained momentum with Pythagoras of Samos and his followers in the School of Crotone during the sixth century BC. They linked the theory of the four elements with the concept of "harmony," which governs the composition of matter.

With his visionary intuition, Pythagoras conceptualized this harmony as nonstatic and extremely dynamic, based on a continuously unstable equilibrium as the result of the balanced antagonism of opposite forces intrinsically present in any composition of matter. According to Pythagoras, the harmony that governs the universe governs the human being as well, providing a state of health, while perturbation of this equilibrium leads to disease.

But the influence of Pythagoras and his followers in the art of medicine extended far beyond this notion to the conceptualization of life as composed of four elements—air, earth, fire, and water—that correspond to four qualities: dry, cold, warm, and humid. The secretions or "humors" (blood, black bile, yellow bile, and phlegm) correlate to the four elements (air, earth, fire, and water) and possess the same characteristics.

The humors, and therefore the elements, are in a direct relationship with the so-called "primary qualities" that characterize them (dry, cold, warm, and humid) and are linked to the four bodily humors: the blood is humid and warm; the phlegm is cold and humid; the yellow bile is warm and dry; and the black bile is dry and cold. According to this theory, the combination of these four humors determines the "temperament" of the person, their mental qualities, and their state of health.

Based on this theory, the human body is governed by the presence of these four humors, and their disequilibrium causes a pathological status to arise in a person. The "disease," conceptualized as the abundance of one humor compared to another, therefore needs to be treated using a "product" (either "simple" or "complex") of the nature opposite to the humor in excess.

In this approach, it is important to classify the simple herbs with the same criteria used to study the human humors. There are simple herbs that are warm and humid, warm and dry, cold and humid, and cold and dry. Along with this classification, there is an equally important classification for the potency of the herbs based on their concentration. This concentration was the classification criteria described in *De Simplici Medicamine* by Constantine Africanus.[6] In this textbook, he outlined four degrees of strength; the strongest herb had side effects that included mortality.

Modern-Day Medicine in Salerno

All these concepts and the classification of the concentration of the medical herbs have been faithfully recreated in the Botanical Garden of the Giardini della Minerva of the modern Schola Medica Salernitana, located close to the former St. Nicholas Monastery in Salerno. Professional medical training returned with the opening of the School of Medicine at the University of Salerno in 2011, exactly two hundred years after the school was closed in 1811.

This historic site is also home to the European Biomedical Research Institute of Salerno (EBRIS), founded in 2013 as an international research enterprise supported by local municipalities and MassGeneral Hospital for Children (MGHfC) in Boston, Massachusetts. The modern research facilities at EBRIS are housed in the former monastery and boast architectural details such as glass floors that showcase the archaeological remnants of the thermal baths from before the ninth century. These magnificent monuments on the cliffs of Salerno overlooking the Tyrrhenian Sea are an important stop for anyone interested in the history of medicine.

Contributions of Hippocrates

The association of Hippocrates of Kos with the concept of four bodily humors as related to four natural elements is perhaps more familiar to many Western-trained clinicians. As the "father of modern medicine," Hippocrates relied on clinical observations and rational conclusions rather than religion or superstition.

In Hippocratic medicine, the physician had to reinstate the healthy balance of black bile, yellow bile, phlegm, and blood by "facilitating the healing work of 'benevolent nature.'" Greek neurologist Christos Yapijakis reminds us that modern ethical practices in Western medicine, as well as most clinical terms used today, originated with Hippocrates.[7] It is hard to believe now, but the theory of four bodily humors stood unchallenged from 500 BC until the Virchow revolution in 1858 when the "Father of Pathology," Rudolf Virchow, enumerated cell theory with his publication of *Cellular Pathology: As Based upon Physiological and Pathological Histology*.[8]

In a parallel development to the four humors, Ayurvedic medicine has been practiced on the Indian subcontinent for more than five thousand years. It focuses on maintaining or restoring a person's unique balance of three "humors," *vata*, *pritta*, and *kapha*, which are a combination of ether (space), air, fire, water, and earth. In Ayurveda, the universal interconnectedness of all things, the body's condition, and its life force, are all considered in the delicate balance of mind, body, and spirit that maintains health and wellness. As the ancients valued balance as an arbiter of health, so does the balance between various components of the gut microbiome form a critical part of our discussion about human health.

An Interwoven Microbial Tapestry

The concept of regaining balance to maintain health applies very closely to the human microbiome. By more closely examining the totality of the human microbiome, which includes diverse components other than bacteria, it has become clear that studying the human microbiome is much more complex than just sequencing and identifying communities. As Rita Colwell said, "There is an integrated fabric and an interwoven connectivity. We need to understand that we humans are simply another species in this massive evolved environment on planet Earth."[9]

As we delve more deeply into the composition and function of microorganisms and the interactions among them, we are finally moving beyond bacteria to identify the other communities that make up our human microbiome. In an indication of the bewildering complexity of the task, there is currently no consensus on what exactly comprises the gut microbiome, by far the most-studied human community of microorganisms.

Depending on the source, the gut microbiome is described as including bacteria, viruses, fungi, archaea, and parasites; yeast and protozoa are sometimes included. One of the next great challenges for microbiome researchers is to use "omics" technologies—which include metagenomics, transcriptomics, proteomics, and metabolomics—to unravel this multifaceted ecosystem.

As the microbiome field advances through metagenomic studies, scientists have expanded their focus from bacteria to other components in the microbiome, including the "virome." Viruses are the most abundant living entity on Earth, and after bacteria, they are perhaps the component of the human microbiome that contemporary researchers study most. Scientists identified viruses for the first time more than a century ago, and well-known and deadly viruses such as smallpox, human immunodeficiency virus (HIV), swine flu, Ebola, and now SARS-CoV-2 have been closely intertwined with human history for more than one hundred years.

According to Antonio Gasbarrini and colleagues, we can now reconstitute viral particles from single genetic sequences from almost every environment.[10] Considered the most abundant and diverse organisms on Earth, virus-like particles on our planet have been calculated to number 10^{31}, which would stretch for one hundred million light years when laid end to end.[11] This figure will likely change as large sequencing surveys of marine environments are completed and our analytical tools become more precise. And the depiction of viruses as opportunistic killers in the microbial wars inherent in human physiology is shifting to a more complex portrait that includes bacteriophages (a type of virus) as builders as well as destroyers.

Redefining the Role of Viruses

Bacteriophages are viruses that do not infect mammalian cells, including human cells, but exclusively infect bacteria and sometimes archaea, hence the name "bacteriophage." There are two main types of phages: lytic, which hijack bacterial cell machinery to replicate and eventually destroy the cell as they spread to other bacteria; and temperate, which integrate their genome into the bacterial chromosome, replicate with the host bacteria, and then repackage themselves as phages and escape their host to infect another bacterium.

Temperate phages play a very important role because they are responsible for lateral gene transfer, which occurs within the same generation of bacteria; vertical transfer occurs from one generation to the next. Lateral gene transfer is one of the ways in which bacteria acquire antibiotic resistance.

As with the bacterial component of the gut microbiome, metagenomic DNA sequencing has revealed a rich viral community composed of animal-cell viruses that cause transitory infections, bacteriophages that infect both bacteria and archaea, endogenous retroviruses, and viruses that can cause both lasting and latent infection.[12] Viral metagenomics analysis has also been used to detect novel human-associated viruses in the gut.

In a recent outbreak of unexplained acute gastroenteritis in the Netherlands, metagenomic analysis of fecal samples showed novel members of the families Anelloviridae, Picobirnaviridae, Herpesviridae, and Picornaviridae.[13] But in the race to identify unknown pathogens, epidemiological detectives have not always had such sophisticated tools at their disposal.

Viral Detectives

In the spring of 1993, according to *Washington Post* reporter Steve Sternberg, Merrill Bahe, a twenty-year-old Navajo cross-country runner in excellent health, was looking forward to his upcoming wedding in Little Water, New Mexico. But on May 9, 1993, his fiancée, Florena Woody, twenty-one, died from an unknown respiratory ailment. Traveling to her funeral service several days later, Bahe fell gravely ill with respiratory distress.[14]

His relatives drove him to the Gallup Indian Medical Center, where he died of respiratory failure several hours later. As chief of internal medicine at the center, Bruce Tempest saw the young man soon after he was admitted. When Tempest heard about the earlier death of Bahe's fiancée, he contacted colleagues and learned of three other young, healthy Navajo adults who had died from acute respiratory distress within the previous six months. He contacted a state medical investigator who attended the young woman's funeral to ask about conducting an autopsy for the couple, a highly unusual request made of members of the Navajo Nation, which was readily granted.

Tempest suspected bubonic plague, not uncommon in New Mexico at the time, but test results from the state health department were negative. He ruled out common pneumonia and Strep-A pneumonia and puzzled over why the mysterious respiratory illness did not strike the couple's infant son. Consequently, the Centers for Disease Control (CDC, now called the Centers for Disease Control and Prevention); its Special Pathogens Branch; the state health departments of New Mexico, Colorado, and Utah; the Indian Health Service; the Navajo Nation; and the University of New Mexico came together to work on identifying the cause of the unknown illness.

More fatalities occurred, and the illness was dubbed "the Navajo flu" by a local newspaper, fueling anxiety and prejudice. By late May, more than thirty collaborators had narrowed the cause down to an unrecognized hemorrhagic fever, an atypical influenza, or an unrecognized environmental toxin. A non–Native American living close to the Navajo tribal area succumbed to the illness, and cases emerged in Arizona, Colorado, Nevada, and California. According to the CDC, although the condition was unknown in epidemiological and medical communities, evidence showed that the Navajo people had recognized it decades before, and Navajo medical beliefs align with public health recommendations to prevent the disease.[15]

In early June, the president of the Navajo Nation, Peterson Zah, convened a meeting of Navajo traditional healers or "medicine men" to consider the situation. Ben Muneta, a Navajo and a Stanford-trained physician, learned some tribal history at that meeting that helped to uncover the cause of the mysterious illness. The only other physician at the meeting was Ron Voorhees, New Mexico's deputy commissioner of health.

According to Muneta's account in the *Washington Post*, the medicine men said there was great disharmony in the world because people were straying from traditional practices. And when disharmony occurs, death follows, they said. With some prodding by Muneta, the elders recalled similar disease outbreaks in 1918 and 1933, when heavy spring rains resulted in a bumper crop of pine nuts from piñon trees. There were many mice in those years when "many young Navajo died."[16]

Revered in Navajo legend as the sower of the seeds of life, the mouse is also dreaded as a disease carrier. According to the CDC, the deer mouse population was ten times its normal size in 1993, as several years of drought were followed by heavy snowfall and spring rains.

A team of CDC epidemiologists gathered rodent tissue samples from the Navajo tribal area and started an antibody search for a matching virus from the patients. The antibodies reacted only with a family known as hantaviruses, which cause hemorrhagic fever. Researchers suspected that a rodent-borne hantavirus, inhaled through virus-like particles in mice feces, might have caused the mystery illness.

While CDC field workers began collecting mice and mice stool from Navajo and other households, C. J. Peters and his team started work on extracting hantavirus genes from victims' tissue. At this stage, a newly developed technology, polymerase chain reaction to amplify a specific region of a DNA strand, proved vital to identifying a new hantavirus that had 139 base pairs.

Mice specimens from the four states whose boundaries intersect at Four Corners—New Mexico, Utah, Arizona, and Colorado—arrived at the CDC in Atlanta on June 12. Four days later the researchers had identified the genetic sequence and counted out 139 base pairs, confirming that it was the same hantavirus that caused the "Four Corners" viral infection found in Merrill Bahe's lungs.[17] In a familiar pattern in the development of medicine, technological advances drove the discovery of a new illness, just as the leap in DNA sequencing and technologies is driving the microbiome revolution.

Viral Architects

One of the most tantalizing remaining mysteries of the Four Corners story is why the illness was fatal to seemingly healthy young people, but it bypassed children and the elderly. This runs counter to the long-held, popular concept of viruses as indiscriminate killers; perhaps part of the answer might be found in the composition of the gut microbiome.

Most human infectious diseases originate in animals, with a fairly recent recognition that the majority of emerging enteric pathogens originate in wildlife, according to C. K. Johnson, Jonna Mazet, and colleagues.[18] Like Ebola and Zika, hantaviruses are zoonotic "spillover" viruses, which are first transmitted from animals to humans and then by human-to-human transmission.

In the Four Corners hantavirus, the healthy young people mounted a very strong immune response when the unknown virus entered their systems, which was spread through airborne contact with mice feces. A similar response, called a "cytokine storm," emerged in many cases of COVID-19, resulting in increased epithelial and endothelial hyperpermeability, vasculitis, and thrombosis leading to multiorgan failure and death if the concentrations remained high.[19]

This hyperimmune response to a "weird" virus is pretty common, said Forest Rohwer, a pioneer in the field of metagenomics and its applications to microbial ecology. "Spillover viruses are not really adapted to humans, and that's why they are so weird—for the most part, they're not attenuated. Most of our viruses, not all of them, but most of the ones that we interact with all the time, don't cause us that much trouble."[20]

But new viruses acquired by zoonosis, like the novel coronavirus SARS-CoV-2, find a new host in humans that is naive of previous encounters. This makes humans more susceptible to serious disease from the infection that can result from the transmission.

The emergence of COVID-19 in late 2019 also outlined another aspect of our disease susceptibility to new viruses, namely the role that the resident microbiome (in this case the lung microbiome) plays in the immune reaction in different stages of this disease. The University of Pennsylvania's Ronald Collman is among researchers looking into the microbiome of the human respiratory tract. According to Collman, some scientists have shown that the amount of bacteria in the lung "may actually set the level of natural immune response." He conjectured that regulation of the immune system in the lung by the microbiome might affect the way people respond to COVID-19.[21]

Rohwer investigates the role of viruses, mainly bacteriophages, in both macroenvironments and microenvironments. His work includes challenging current beliefs on microbial diversity through decades of genomic mapping of coral reefs in the central Pacific and studying the role of the virome in cystic fibrosis. Rohwer's lab at San Diego State University (SDSU) has been instrumental in discoveries leading to the recognition that viromes are unique to each individual, with more diversity within an individual virome than between individuals. His group has also identified phages used in successful phage therapy treatment, a topic that will be discussed further.

He views humans not as isolated individuals but as "walking ecosystems," with viruses as the resident predators—and creators—that help balance the ecosystem's energy flow. As the dominant microorganism on the planet, viruses are more successful than cells, noted Rohwer, and we ignore the role they play in maintaining health in the human microbiome at our peril.[22] As noted earlier, a recent estimate of the total number of viruses worldwide is 10^{31}, with 30 billion bacteriophages or "phages" traveling around the human host daily to seek out bacterial prey and protect us from pathogens.[23]

The current debate over whether viruses are "living" organisms does not interest Rohwer as much as their ability to move information and create new blueprints through manipulating genetic material. Phages infiltrate their bacterial host, hijacking the cell machinery to replicate themselves. Rohwer said viruses excel at creating diversity through genetic manipulation of the genetic blueprint of their prey's bacteria—sometimes inadvertently, sometimes directly. They propagate more genetic material than all other so-called living organisms combined, yet Rohwer lamented that viruses seem to have no place in the three branches of Bacteria, Archaea, and Eukarya on our traditional phylogenetic tree.[24]

As described earlier, phages exploit their host either through lytic action, which results in many virus particles released prior to cell death, or through temperate action, which results in the integration of the viral genetic material into the host's chromosome. According to Rohwer, incremental progress in the study of the virome during the last two or three years shows that every bacterium in the gut microbiome is a lysogen carrying at least one prophage, and usually more. Rohwer and Cynthia Silveira proposed a new model of bacteria-phage interactions they called "Piggyback-the-Winner" (PtW) in which "lysogeny predominates at high microbial abundance and growth rates."[25]

Prophages are involved in bacterial wars and bringing in genes that are important to an individual for adaptation, said Rohwer. One dramatic example is that when someone has a chronic illness and is continually exposed to antibiotics, the viruses move around the antibiotic-resistant genes, and as a result, the individual does not lose their microbiome. A subtler role, Rohwer noted, is when the virus brings in components involved in carbohydrate metabolism. And discoveries led by former SDSU postdoctoral student Jeremy Barr have uncovered the existence of what Rohwer called "basically a new, microbiome-derived immune system" from viruses protecting the host tissue by forming a barrier in the mucus.[26]

Bacteriophage Adherence to Mucus

In 2013, Barr, Rohwer, and other SDSU researchers discovered that bacteriophages adhere to the mucus layer found in samples from marine life to humans. Bacteriophages formed bonds with sugars within the mucus when placed on top of a layer of mucus-producing tissue. When *E. coli* was introduced to the mucus cells, the researchers observed that the bacteriophages attacked and killed the *E. coli* in the mucus by forming an antimicrobial barrier for the host.[27]

When the researchers introduced bacteriophages and *E. coli* to non-mucus-producing cells, the rate of cell death was three times as high in the nonmucus samples. They called the new model "bacteriophage adherence to mucus (BAM)," describing it as one in which "phages provide a non-host-derived antimicrobial defense on the mucosal surfaces of diverse metazoan hosts."[28]

Barr said the finding could change the way doctors treat a number of diseases. "The research could be applied to any mucosal surface. We envision BAM influencing the prevention and treatment of mucosal infections seen in the gut and the lungs, having applications for phage therapy and even directly interacting with the human immune system."[29] With the discovery of this phage-metazoan symbiosis, there is a recognition of "the key role of the world's most abundant biological entities in the metazoan immune system," according to Barr and colleagues.[30]

Furthering the BAM model, Rohwer and Silveira propose that the PtW and BAM models are closely interrelated and play an important role in microbiome development. "In BAM, phages produced by the microbiome attach to mucins and protect underlying epithelial cells from invading bacteria. Spatial structuring of the mucus creates a gradient of phage replication strategies consistent with PtW." Rohwer and Silveira speculate that lysogenic phages are more prevalent in the top mucosal layer and lytic phages are favored in the intermediary layers with few bacteria. With this hypothesis, lysogeny "confers competitive advantage to commensals against niche invasion and the lytic infection eliminates potential pathogens from deeper mucus layers."[31]

How Phages Get Around

Rohwer and Barr agree that the BAM discovery raises many questions, with much more research needed into the concept of a new immune system and the role it plays in the mucosal microbiome, specifically the gut microbiome. "It's a very broad picture, and we don't really know how it's working. We know the mechanism better than we know the actual specifics per person; that's going to take a while," said Rohwer.[32]

The gut microbiome is the largest reservoir of phages in humans, and there is evidence that phages use different mechanisms to move from one district to another. For example, increased intestinal permeability or the "leaky gut" route is a sort of shortcut in the mucosal and vasculature layers that allows phages to bypass the epithelial and endothelial layers. (See the discussion of intestinal permeability in chapter 6.) Other proposed mechanisms are the Trojan Horse, in which a bacterium infected by phage enters epithelial cells; a phage display in which homing ligands are engineered onto viral capsids; and the free uptake of phage particles via endocystosis, which is the process of the cell engulfing material within its plasma membrane.

Following up on Barr's theory, Rohwer and his collaborators uncovered another mechanism that bacteriophages use to cross epithelial cell layers. In experiments with apical and basal incubation chambers separated by a human cell layer, researchers observed epithelial cells taking up and transporting phages across the cell and releasing active phages on the opposite cell surface.[33]

Through transcytosis, phages gain access to so-called "sterile" regions, including the blood, the lymph system, organs, and the brain. The researchers determined that phages showed a preference for apical-to-basal transport, and chemical inhibitors "suggest that phages transit through the Golgi apparatus before being exocytosed." From their results, Barr and colleagues estimate that approximately thirty-one billion phages move from the gut into the body every day through the mechanism of transcytosis.[34]

Rohwer noted that the discovery of phages in the blood can provide a tool for phage therapy in the blood and treatment for sepsis. But it is still "early days," he cautions. "It's one of those things where we're going to think it's this way, and we do a bunch of experiments and say, ah, we're completely wrong—it's actually doing it this way. It's the frontier where all the cool discoveries are being made, but we don't really see how they all go together yet."[35]

Phage Therapy

A marine ecologist with a broad viewpoint, Rohwer sees the human microbiome as a rich sediment similar to a lake filled with rich nutrients. When a baby is born, viruses arrive soon after the meconium is passed, said Rohwer, and breast milk has mucus-covered globules from the cream fraction that help the viruses stick to the mucus. Things get more interesting after weaning, according to Rohwer, with the development of complex microbial systems, when phages interact with microbes and set up a "sediment sort of community" with viruses unique to each person.[36]

Through horizontal or lateral gene transfer, viruses "give the cellular world a way to explore a whole bunch of sequence bases that they would not normally be able to explore," said Rohwer.[37] Phages also play a role in decreasing inflammation through cytokine reduction and other processes. According to Rohwer, this could occur through the phage interacting directly with the cell, but also by killing off bacteria—and there is evidence for both roles. Directly harnessing lytic phages for phage therapy in drug-resistant infections is an emerging field that holds many challenges as well as promise in personalized treatment.

In 2017, a group of researchers from academia, industry, and the U.S. Navy and other government agencies collaborated to develop a method to treat a sixty-eight-year-old diabetic patient. The patient had necrotizing pancreatitis that was complicated by a multidrug-resistant (MDR) *Acinetobacter baumannii* infection. With no effective antibiotics, researchers from the U.S. Navy and Ry Young's laboratory at Texas A&M identified a phage with lytic activity effective against an *A. baumannii* isolate from the patient.

Rohwer's group removed endotoxins like LPS from the phage, making it safe to inject into the patient. The patient's downward trajectory was reversed when bacteriophages were introduced intravenously and percutaneously into the abscess cavities. Not only did the patient (Thomas Patterson) return to good health with the clearance of the *A. baumannii* infection,[38] he coauthored a book about the experience with his wife (Steffanie Strathdee), an epidemiologist at the University of California, San Diego.[39]

With the recent COVID-19 pandemic added to antibiotic-resistant and MDR infections rising at an alarming rate, phage therapy is under increasing scrutiny as a possible therapeutic intervention. The challenge with phage therapy, said Rohwer, is that it works some of the time, and it doesn't work the next time. "We do know many of the reasons that there is such variability, and there would be a technical workaround that would make it a little better. It's a big challenge, though, and it's going to take a lot of money. To really develop phage, we're going to need to spend billions of dollars—but the potential payoffs are gigantic."[40]

Successful Phage Targeting of a *Shigella* Strain

Those payoffs could include drastically reducing child morbidity and mortality in the developing world through targeting specific strains of enteric pathogens including *E. coli*, *Shigella*, and *Salmonella*. WHO has recently included these pathogens, which predominantly affect children under five years of age in the developing world, on a list of antibiotic-resistant priority pathogens that are in urgent need of new therapies.

Our researchers at the MIBRC, led by Christina Faherty, in collaboration with Tim Lu's lab at MIT, recently isolated a bacteriophage to target a specific strain of *Shigella* in a proof-of-concept analysis. Using a novel human intestinal organoid as a monolayer infection model, the researchers determined that "bacteriophages specific to *S. flexneri* can effectively kill the bacteria in numerous growth and infection conditions, potentially without harming commensal bacteria."[41]

The intestinal organoids, or "mini-guts," are cultivated from intestinal-crypt stem cells, which are derived from human tissue taken during routine colonoscopies or other screenings. After "infecting" the intestinal organoid, the *Shigella*-targeting bacteriophage killed several strains of *S. flexneri*, which included a "strain harboring an antibiotic resistance cassette." Researchers also demonstrated the bacteriophage's ability to kill *Shigella* in traditional assays, including various "broth kill curve" assays, which are dose-response experiments with increasing amounts of antibiotics, and infection assays with HT-29 cells as well as in the intestinal organoid model.[42]

This nascent research from Faherty's group presents a potential alternative to antibiotics in the fight against enteric pathogens that cause childhood diarrhea. This condition can lead to environmental enteropathy and severe damage to the gastrointestinal tract, stunted growth, delayed development and cognition, and an impaired gut microbiome (see chapter 11 for a detailed discussion). By employing the novel human intestinal organoid monolayer model, the group has developed an efficient and safe model to evaluate bacteriophage treatment of enteric pathogens.

Phage and Autoimmunity

Viruses have been associated with autoimmune diseases. One example is the Epstein-Barr virus leading to an increased risk of multiple sclerosis or systemic lupus erythematosus.[43] Prospective studies of children at risk of autoimmunity have shown phage-dependent lateral gene transfer, which suggests that phages play a role in changing the host immune system and dictating the switch from tolerance to immune response.

If the mucosal membrane is intact, the mucosa will tolerize the phage, said Rohwer. If the phage is coming in with no LPS, or no "danger signal," there is nothing to activate the immune system. But if there is some kind of insult—an ulcer, increased intestinal permeability, or another type of breach—the body will mount an aggressive response to the phage because of the increased amount of LPS.

Using fecal samples from eleven children at risk for T1D, scientists defined the intestinal virome from birth to the development of autoimmunity. They found the intestinal virome to be less diverse in children at risk of developing T1D than in controls and noted changes in disease-associated DNA segments linked to specific components of the bacterial microbiome. Five of the subjects developed T1D, and the researchers concluded that specific components of the individuals' viromes were "both directly and inversely associated with the development of human autoimmune disease."[44] The eukaryotic viruses and bacteriophage contigs identified in this study could provide targets in the virome for protection against T1D diabetes.

Another innovative approach employed by scientists at Rohwer's lab and other collaborators relied on a practice that is as old as humans: eating food. Using 117 common foods, chemical additives, and plant extracts, Lance Boling, Rohwer, and other researchers found that certain foods increase the production of bacteriophages, viruses that mitigate the levels of harmful bacteria in the gut. "This shows we could sculpt the human gut microbiome with common dietary compounds," said Rohwer. "The ability to kill specific bacteria without affecting others makes these compounds very interesting."[45]

Small but Mighty Mycobiome

As we unravel the many threads that make up the human microbiome, the focus has been overwhelmingly on bacteria, with viral particles taking a distant second. But there are other actors—archaea, fungi, and parasites that make up the human microbiome, including the gut microbiome. We are only starting to tease out the roles that these other organisms play in health and disease.

Databases and computational tools for analysis of these other human inhabitants are not yet robust. According to the National Center for Biotechnology Information at the United States National Library of Medicine, 3,520 fungal genome assemblies have been collected, compared to 162,834 bacterial genomes as of September 2018.[46] Studies of the "mycobiome," or fungal communities in the human microbiome, have uncovered a marked diversity among individuals and body sites, including skin, lungs, oral, and gut, similar to bacterial studies.

Andrea Nash and colleagues expanded our quantitative snapshot of the gut mycobiome by sequencing 317 samples from subjects from the Human Microbiome Project cohort. The researchers described the human gut mycobiome as low in diversity and "dominated by yeast including *Saccharomyces*, *Malassezia*, and *Candida*."[47] Characterizing the mycobiome faces technological challenges, and the ability of fungi to exist in two states, hyphal or yeast form, has led to misclassification of some genetically identical fungi as distinct microorganisms, thus corrupting the public database.[48]

Along with bacteria *E. coli* and *Serratia marcescens*, researchers led by Mahmoud Ghannoum found *Candida* at significantly higher levels in patients with Crohn's disease when compared to the gut microbiome of their healthy family members. Together the three microorganisms created a biofilm capable of exacerbating intestinal inflammation, which can result in symptoms of Crohn's disease. The researchers also found strong similarities in the gut microbial profiles of the twenty people with Crohn's disease, which were markedly different from the individuals without Crohn's disease.[49]

Ghannoum is hopeful that the findings will provide direction for new therapeutic approaches to IBD through the use of antifungals and probiotics to maintain a healthy gut microbial balance that includes fungi as well as bacteria and other components. *Saccharomyces boulardii* has been effectively employed in the treatment of postinfectious and postantibiotic diarrhea.[50] Other applications of yeast include the use of *Saccharomyces cerevisiae* as a genetic and chemical screening platform in anti-aging research. Yeast-based studies have also led to the discovery of the polyphenol resveratrol, found in red wine and other components of the Mediterranean diet, and other potential anti-aging substances.

Finally, the Parasites

When it comes to the role of protozoans and worms, it is a mixed research bag with enormous potential for future treatment interventions. Some scientists are debating whether the parasite *Blastocystis* is associated with gut dysbiosis, while other researchers have demonstrated interactions between the human host immune system and helminths and protozoans.

The reintroduction of intestinal helminths, parasites that have coevolved with our immune system, into the gut microbiome as a treatment for decreasing the risk of allergic and inflammatory disease has been proposed. Human-parasitic nematodes are being investigated for treatment of autoimmune disorders, including T1D, multiple sclerosis, ulcerative colitis, and insulitis.[51]

We agree with Valeria Marzano and colleagues that systems biology–based profiles of the gut "parasitome" can help contribute to the control of infectious diseases. Recent studies show that an increased abundance of *Lactobacillaceae* and *Clostridiaceae* family members has been linked to increased activity of regulatory T cells (Tregs).[52]

An Ancient Scourge

In 2017, according to WHO, nearly half of the world was at risk of contracting malaria, a parasitic infection transmitted through the bite of an infected female anopheles mosquito. That same year saw 219 million cases and 435,000 deaths, with more than 90 percent of both cases and deaths occurring in the WHO African Region. Those most at risk of serious illness and death from malaria are children under the age of five, pregnant women, people with HIV/AIDS, and nonimmune migrants, mainly in sub-Saharan Africa.[53]

As with many pathogenic viruses, the epidemiology and history of combatting the malaria parasite is intimately intertwined with complex cultural, economic, social, and political factors. The U.S. Centers for Disease Control and Prevention (first called the Communicable Disease Center) grew out of the Malaria Control in War Areas program in 1946. Eradicating malaria from the Southeastern United States was a primary goal of the new agency, and according to its historical account of malaria, the disease was considered "eliminated from the U.S." by 1951.[54] But in tropical and subtropical areas of the world, according to WHO, one child under the age of five dies every two minutes from malaria.[55]

Although there are five parasite species that cause malaria, *Plasmodium falciparum* and *P. vivax* are the predominant killers. According to WHO, *P. falciparum* accounted for 99.7 percent of cases in the African Region; 63 percent in South-East Asia; 69 percent in the Eastern Mediterranean; and 72 percent in the Western Pacific. *P. vivax* was responsible for 74 percent of malaria cases in the Americas.[56]

There has been some good news on the global front, with increased prevention and control measures leading to a 29 percent reduction in global mortality rates since 2010. The mortality rate of children under the age of five fell by an estimated 35 percent between 2010 and 2016. Vector control, which includes the use of insecticide-treated mosquito nets and residual indoor spraying, is the main way to prevent and reduce transmission. Antimalarial drugs have been in widespread use for decades, but drug resistance has been "one of the greatest obstacles in fighting malaria." Resistance to earlier generations of drugs arose in the 1950s and 1960s, reversing gains in child survival.[57]

Chinese scientists have subsequently gone on the hunt for new antimalarial compounds extracted from traditional Chinese herbs. Looking back to a traditional remedy used for more than two thousand years, scientists isolated the antimalarial principle of qinghao (*Artemisia annua*).[58] Current treatment recommended by WHO is an artemisinin-based combination therapy. But the parasite is becoming resistant to the second generation of drug therapy, making a search for other alternatives an urgent initiative for the elimination of malaria worldwide.

Midgut Microbiota of *Anopheles*

With the current emphasis on intestinal microbial communities, it is not so surprising that some scientists are exploring the gut microbiome for possible avenues to drug-resistant malaria treatments. What is unexpected is that they are not only looking at the human gut but also at the midgut microbial colonies of mosquitoes. Malarial parasites have complex developmental transitions within their insect vector, and intestinal microbial flora are thought to contribute to parasite losses during the development of oocysts on the basal lamina during the invasion of the midgut epithelia.[59]

As in human guts, the microbiota found in insects' guts can have a substantial impact on the development of pathogens. Certain Gram-negative bacteria in the midguts of wild and laboratory mosquitoes have been associated with an inhibitory action on the development of *Plasmodium* parasites. Through a comprehensive, functional genomic approach from the Bloomberg School of Public Health at Johns Hopkins University, researchers have shown that the midgut microbiota of the *Anopheles gambiae* mosquito can "modulate the anti-*Plasmodium* effects of some immune genes."[60]

The researchers treated mosquitoes with antibiotics to create a group of "aseptic" insects, which were significantly more susceptible to *P. falciparum* infection compared to the "septic" or untreated mosquitoes. Not surprisingly, they found great natural microbial variability in mosquitoes from the same wild colony, which plays an important role in regulating mosquito "permissiveness" to *Plasmodium*. Using RNAi gene silencing, the researchers report that the natural microflora and artificially induced microflora negatively affected development of the malarial parasites through a mechanism that seems to implicate the innate immune system; *Plasmodium* is not killed directly by bacteria.[61]

From their findings, Yuemei Dong, Fabio Manfredini, and George Dimopoulos suggest that future studies on gene-specific, anti-*Plasmodium* action "should also consider the complex interplay between the microbiota and the mosquito's immune defenses against the *Plasmodium* parasite."[62] It seems that the complex developmental role that *Plasmodium* undertakes before it can become successful in infecting a human host is paralleled by the complex role played by the *Anopheles* midgut microbiome in microbial and genomic immunology.

Traveling Different Paths to the Same Destination

Viewing the totality of the microorganisms living within human districts, including the human gut, leads us back to the concept of the ecosystem and expanding our minds to consider much more complex and multidimensional explanations of the drivers of human health and disease. The final destination might be the same clinical manifestation, but the journey to arrive there can be made along dissimilar paths.

Examining the complex interactions of bacteria, viruses (both lytic and temperate), fungi, yeast, and parasites is a necessary step toward a deeper understanding of the myriad interactions—environmental, genomic, dietary, economic, social, and cultural aspects of human health—and the interconnected health of the other species on our planet. In part III of this book, we will explore the implications of this deeper understanding in the discussion of personalized and preventive medicine. In the next chapter, we delve more deeply into the role of the gut microbiome and its epigenetic effect on the immune system.

6   The Microbiome Hypothesis: The Epigenetic Role of the Microbiome

Getting Your Hands Dirty

Our early childhood years were spent on farms: Alessio Fasano on his family farm along the Amalfi Coast in Italy, and Susie Flaherty on a dairy farm in the low rolling hills of Carroll County, Maryland. As farm kids, we were exposed to a diverse microbial world that included microorganisms from the fur and feces of farm animals and pets, microbes from local water sources, including marshes, streams, and the Tyrrhenian Sea, and microbes from the soil and air.

We ate seasonal vegetables from family and farm gardens after picking the crops under the watchful eyes of our elders. We spent hours every day outside and gathered every night for large family dinners. Generally healthy, we had limited use of antibiotics and enjoyed close social circles of family and friends. Unbeknownst to us, in the early years of our rural upbringings, we were building a diverse microbiome.

A variety of studies comparing both children and adults in urban and rural settings show distinct differences in microbiome composition, with a less diverse microbiome found in urban settings. A cross-sectional survey by Josef Riedler and colleagues determined that a lack of early-life and long-term exposure to farm animals and farm milk can promote chronic inflammatory disorders including allergies and asthma.[1]

Michelle Stein and colleagues showed a four to six times lower rate of asthma and allergic sensitization prevalence for Amish children on single-family dairy farms when compared with Hutterite farm children living on highly industrialized farms with little contact with animals.[2] Using a standardized laboratory psychosocial stressor, Till Böbel and colleagues found that young healthy participants in urban settings without pets demonstrate a greater inflammatory response than young healthy participants in a rural setting with farm animals.[3]

In a reprise of Blaser's missing microbes hypothesis (see below), Susan Lynch and Homer Boushey argue that increased inflammation in urban environments due to impaired immunoregulation is related to reduced exposure during early life to the ancestral microorganisms that we have coexisted with for millennia.[4] Böbel and colleagues describe microbial diversity, developed through contact with commensal and environmental microorganisms present throughout mammalian evolution, as "progressively diminishing in high-income areas, particularly in urban areas."[5] No matter how this occurs, as our collective microbial diversity decreases, the implications for chronic inflammatory diseases, and especially autoimmunity, are troublesome.

Alternatives to the Hygiene Hypothesis

As described in chapter 2, improved hygiene leading to a reduced exposure to microorganisms has been implicated as one possible cause for the epidemic of chronic inflammatory diseases, particularly autoimmune and allergic diseases, in industrialized countries during the past three to four decades. This phenomenon has coincided with the drastic reduction of human infectious disease due to our advanced knowledge of infectious diseases and bacterial pathogenesis, and most important, the advance of antibiotics.

To revisit David Strachan's theory, he hypothesized that the lower incidence of infection in early childhood could be the reason why we witnessed such a rapid increase in chronic inflammatory diseases in the latter half of the twentieth century. Strachan also blamed changes in social lifestyle, including family size and moving from rural to urban settings, as well as improved household sanitation, as additional factors supporting his "hygiene hypothesis."[6]

A different reading of the same epidemiological phenomenon has been provided by other investigators like Graham Rook, who in 2003 proposed a complementary, if not alternative, hypothesis. Rook argues that lack of exposure to "old friend" microbes, which have been an integral part of our coevolution with microorganisms, rather than decreased exposure to pathogens, is the real culprit in the increased prevalence of chronic inflammatory diseases.[7]

This theme will recur frequently throughout this book, as there are some scientists who support the notion that dramatic lifestyle changes in the Western world have led to the extinction of the old friend microbes. Scientists contend that these missing microbes can still be isolated from hunter-gatherer populations, even though they are no longer present in Westernized and modern populations.

Along with the extinction of these ancient microorganisms, the increased density of modern populations has given rise to several new species of pathogens that, according to the old friends hypothesis, do not interact with their host in line with our coevolution plan. Rather, these pathogens, which are responsible for common childhood infections, either kill or immunize their host. Conversely, the ancient microbes evolved with the mammalian immune system and developed symbiotic interactions with mutual benefit to both microbes and host. These include hospitality and nutrients for the microorganisms in exchange for training the immune system to protect the host against deadly infections.

Besides bacteria, ancient viruses and parasites could create those chronic infections or a carrier status that, rather than leading to disease, would create a state of tolerance as the consequence of a path of nonbelligerence, or an immune regulatory relationship. When humans evolved from living in small, isolated groups in primitive, often muddy environments with an open exchange from soil to human mammalian host, to a more domesticated and sanitized lifestyle with clean water, proper sewage disposal, and increased population density, the microbial ecosystem shifted radically from its original symbiotic composition with the human host.

Whether it is the hygiene hypothesis or the theory of the old friend microbes in play (and they are not mutually exclusive), it is indisputable that increased hygiene and an improvement in sanitation led to a tangible increase in life expectancy, which has nearly doubled since 1900.[8] What humans did not anticipate was that we would live longer, but not necessarily healthier, lives. In other words, in the Western world, we don't die quickly from infectious diseases, but instead we die slowly from chronic inflammatory diseases that negatively impact our quality of life.

Adapting to Pathogens

Several studies have shown that a variety of chronic inflammatory diseases are much less frequent in developing countries
than in the Western world, although this seems to be shifting
rapidly to an increased rate of certain conditions in developing
countries.[9] However, immigrants moving from developing countries to industrialized countries are subject to the same risk as
Western-born subjects, and their risk increases with the length
of time since migration.[10] How can we mechanistically interpret
this epidemiological phenomenon?

If we project ourselves back in time to when our human
ancestors first appeared five to seven million years ago, there
were most likely few interactions among these early hominid
groups, who were typically organized into extended family
units. The isolation of our early species meant that pathogens
had a reduced opportunity for infecting large groups. Also, the
nomadic lifestyle and change of seasons provided different interactions with the surrounding environment, including soil and
water, and therefore exposure to a multitude of organisms that
would thrive under different conditions.

The result of this hunter-gatherer lifestyle was conducive
to an enrichment of symbiotic microorganisms and rare exposure to pathogens, with limited spread of disease agents from
one group to another. With the advent of agriculture approximately ten thousand years ago, the domestication of crops and
animals drastically changed this microbial exchange from the
environment to the mammalian host and vice versa. Changes in
nutrition, close interaction with animals, and increased population density revolutionized the biological plan of evolution

concerning human hosts and microbial interactions. Now pathogens had many more opportunities to spread.

Zoonosis, meaning infections acquired from animals, like the hantavirus discussed in chapter 5 or the SARS-CoV-2 virus responsible for the COVID-19 pandemic, started to appear. The genetic adaptations needed to face the reality of this new ecosystem also began to appear. One classic example of beneficial genetic adaptability of the human host to more frequent exposure to pathogens is the selection of the gene for thalassemia, a condition that changes the shape and biology of red blood cells. With thalassemia, red blood cells can be less easily infected by flies carrying falciparum malaria, and the risk of contracting malaria is decreased.

The human immune system also had to adapt to this new reality. Recent comparative genomic studies have shown that genes that control the immune response are typically more influenced by a positive selection from microorganisms that coevolved with the human species. Probably the best example of this phenomenon is the helminths, introduced in chapter 5. Compared to bacterial and viral pathogens, these parasitic worms can exert a strong pressure on specific human genes, encoding interleukins and interleukin (IL) receptors. The rationale for this worm-specific focus of our immune system may be explained by the fact that helminths are as old as our adaptive immune system, which was originally exposed mainly to parasites.

Viruses, Helminths, and Th1 and Th2 Responses

T lymphocytes are small white blood cells that produce cytokines, which are the chemical messengers of the immune system that direct effects like cell-mediated immunity and allergic response.

The subgroup of lymphocytes called T helper cells (Th) was broken down into two further subgroups of Th1 and Th2 by Tim Mosman and colleagues in 1986, based on which cytokines are secreted. In general, these components of the adaptive immune system also produce different immune responses: Th1 cells respond to intracellular threats like bacteria and viruses and prolong autoimmune response; Th2 cells respond to helminths and extracellular dangers and are associated with allergic diseases.[11]

Strategies based on mounting a Th2 immune response, in contrast to exposure to bacterial or viral pathogens that elicit a Th1 immune response, have shown that a Th2-dominated phenotype is more often associated with a modern lifestyle, including a high use of antibiotics, a Western-style diet and lifestyle, an urban environment, and reaction to dust mites and cockroaches. Conversely, the Th1 phenotype seems to be associated with rural living, contact with animals, large family size, early day-care attendance, and the presence of older siblings.

Also intriguing is the fact that Th1 and Th2 responses can negatively influence each other; when one is active, the other is suppressed. Based on these observations, the original hygiene hypothesis anticipated that insufficient Th1 stimulation of the immune system led to the overreacting Th2 immune arm, which, in turn, led to the development of allergic diseases. While this observation can explain the early focus of the hygiene hypothesis on the epidemics of asthma and hay fever, it surely cannot explain the much more complex epidemic of other chronic inflammatory diseases such as autoimmune and neuroinflammatory diseases, which are both fueled by the Th1 immune response.

What seems to be the explanation for this apparent dichotomy is the tight control that a specific subgroup of T cells, called T regulatory cells or "Tregs," exert on this Th1-Th2 "yin-yang" balance. The maturation and function of Tregs seem to be highly dependent on the microbiome's composition and function. This observation, along with the fact that it is not entirely true that our modern practices of cleanliness and hygiene may have any impact on rising rates of chronic inflammatory diseases, poses some challenges to the hygiene hypothesis and the old friends hypothesis as the driving force to explain the epidemic of chronic inflammatory diseases.

Even if cleanliness has reduced our exposure to microorganisms, it seems to have had very little impact in reducing the overall load of microbes in a specific environment. No matter how obsessed we are with sanitizing our environment, microbes rapidly repopulate any given space through air, dust, or any other environmental exposure that is typically involved in spreading microbes. Therefore, reducing hygiene will not decrease the risk of chronic inflammatory diseases but will significantly increase the risk of infectious disease.

What Triggers Chronic Inflammation?

Until about thirty years ago, when the Human Genome Project was still in the planning stages, the general hypothesis was that to develop a chronic inflammatory disease, genetic predisposition and exposure to an environmental trigger were both necessary and sufficient. No matter which disease we consider—infectious diseases, allergic diseases, neuroinflammatory/neurodegenerative diseases, cancer, or autoimmune diseases—these two elements are always in play.

While the genes and the environmental triggers may vary from one group of diseases to another, what seems to remain constant is the mismanagement by the immune system of the exposure to environmental triggers. This leads to the onset of chronic inflammation, the common denominator of all chronic inflammatory diseases affecting humankind.

Parallel to the early hypothesis of "genes plus environment equals chronic inflammatory disease" was the epidemiological observation that major epidemics of these conditions coincided with the declining rate of infectious disease in the Western world. This generated the hypothesis that we had become too clean for our own good, and people in the developed world would die slowly of chronic inflammatory diseases instead of succumbing rapidly to infectious disease.

If these two elements, genetics and environment, are the only ones in play (and that was the premise), this interpretation of these epidemics of chronic inflammatory diseases can only lead to two possible conclusions. If we take a pessimistic approach and see the glass as half empty, these epidemics should be interpreted as the consequence of the drastic changes in our environment caused by human intervention. These changes materialized too quickly (in a matter of four to five decades) for us as a human species to adapt through genetic mutations, which would take much longer than one or two generations. This is surely part of the equation.

Conversely, if we take a more optimistic approach and view the glass as half full, these epidemics of chronic inflammatory disease teach us a very different lesson. The supposed pathogenesis of these diseases, which was based on the classic theory that the expression of one gene equals the encoding of one protein which equals the development of one disease, was the premise that justified the launch of the Human Genome Project. We thought that if we unlocked the mystery of the human genome, we would be able to solve all the diseases of humankind. In other words, if you have the genes for the disease, you have the destiny to develop that disease, no matter how you behave or conduct your lifestyle.

What we learned when the Human Genome Project was completed is that we are genetically much more rudimental than we had previously thought. The premise of one gene, one protein, and one disease cannot explain the complexity of the balance between health and disease. Twenty-three thousand genes are insufficient to explain all the permutations of human pathophysiology, including if, when, and why we develop diseases.

Rather, it is the interplay between us as individuals and the environment in which we live that makes us play our genetic cards in such a way that we can stay healthy and "win the game" despite unfavorable genetic cards. Or we can play the game poorly and develop disease, including autoimmunity and other chronic inflammatory conditions. Otherwise, how do we explain these escalating epidemics?

If the genes don't mutate, we're doing something else wrong here. Let's view the glass as half full, meaning that we have some control over our health outcomes. In that case, if you have the genes for Parkinson's disease or breast cancer, whether you develop the disease depends on your lifestyle; it is not just genetic destiny.

Over time, coauthor Alessio Fasano came to realize that there are five elements or "pillars" leading to chronic inflammation. Besides the two we have identified thus far (genetic predisposition and exposure to environmental factors), the loss of barrier function, a hyperbelligerent immune system, and an unbalanced microbiome all need to be in play for the individual to develop chronic inflammation. Genes and environmental triggers are still necessary, but not sufficient, players in this game of chronic inflammation.

Interactions in Chronic Inflammation

The third of the five pillars needed to develop the problem of chronic inflammation is the loss of barrier function, mainly in the intestine. At approximately eight to nine meters in length, the human intestine (both small and large) provides the largest physiological interface between our body and the outside world. Tightly packed single layers of cells called epithelial cells cover the external surfaces and line the body cavities of the human body.

The skin, which allows little or no exchange with the outside environment, is the most visible interface between us and our external environment. Contrast this with the intestine, which represents a much larger and more permissive interface with the outside world. If it is stretched out on the floor, the gut surface of an adult intestine can cover a doubles tennis court! Although it is not visible from the outside, it plays a pivotal role through its dynamic interactions with a variety of factors coming from our surrounding environment, including microorganisms, nutrients, pollutants, and other materials.

The intestinal epithelium maintains a barrier against the external environment through a single layer of epithelial cells. In a healthy intestine, the first two pillars of genome and environment are physically segregated by the gut barrier. This barrier allows for the absorption of nutrients and defends against toxins, antigens, and other invaders.

"Tight junctions" (see below) are part of a complex protein network that connects the very top of adjacent epithelial cells, sealing the space between them. Under normal circumstances, foodstuffs move through the small intestine and are digested, with nutrients moving into the bloodstream and waste products into the large intestine. If these tight junctions become permeable, and there is a loss of this segregation, the unregulated movement of promiscuous materials from the outside world within our body can instigate inflammation in a person genetically predisposed to develop a chronic inflammatory disease. This loss of barrier integrity is sometimes called "leaky gut."

Inflammation is not always detrimental. Indeed, it is an essential part of the protective and healing processes when the body is exposed to infection or injury. In an appropriate inflammatory response, factors produced by white blood cells also protect our cells against bacteria and viruses until the enemy is defeated, when it is time to turn the inflammation off.

Like inflammation, modulation of intestinal permeability is an essential physiological function that has been unnecessarily vilified. For reasons that are still unclear, intestinal permeability became associated with the term leaky gut and has been since connected with systemic negative effects on the human body. "Leaky gut syndrome" has been blamed for all kinds of human diseases and conditions, ranging from chronic fatigue syndrome to cancer. This presents a very narrow view of the story, which is much more complex and will be revisited in more depth later in this chapter.

A hyperbelligerent immune system is the fourth pillar in this recipe for chronic inflammation. In this case, the immune system produces inflammation constantly, not just when it is appropriately unleashed—for example, in response to exposure to a pathogen. Once this happens, the immune system is no longer able to turn the inflammation off even if the initial trigger has been eliminated.

The fifth pillar is the human microbiome, which is perhaps the most important element. Since *H. sapiens* appeared on the evolutionary stage approximately three hundred thousand years ago, we have coevolved with the collective microorganisms that inhabit our bodies. Therefore, it should be no surprise that this coevolution requires constant adjustments to establish a pacific, symbiotic relationship, which can be achieved only through communications mediated by epigenetic influence from the microbiome.

Returning to the parallel of how our destiny is influenced by how we play our genetic cards, the microbiome dictates the game strategy by deciding if and when our genes should be expressed or repressed, and so it determines if and when we shift from genetic predisposition to clinical outcome. It should also be emphasized that these three additional pillars—gut permeability through the loss of barrier function, a hyperactive immune system, and the microbiome's composition and function—are constantly interacting and influencing each other. A change in the status quo in one domain (gut dysbiosis) may affect another domain (increased gut permeability).

Prenatal, perinatal, and postnatal environmental factors are instrumental influences on the composition and function of the human microbiome. Most likely, these changes help us to quickly adapt to our surroundings through the microbiome's epigenetic modulation of our genes. However, with the drastic modifications to lifestyle and environment that have occurred in the past two to three decades, this plasticity has shifted from a superb asset to a major liability. This premise, which has been examined by the experts interviewed for this book, gives rise to the "microbiome hypothesis," the central theme of this book.

Innate and Adaptive Immunity

The microbiome hypothesis, which was put forward approximately ten years ago, is now gaining momentum. Results from the Human Genome Project and a growing number of recent publications focused on human genetics show that the genes that control intercellular barrier function are integral genetic components of a variety of diseases. This growing evidence suggests that loss of mucosal barrier function, particularly in the gastrointestinal tract, may substantially affect antigen trafficking, ultimately causing chronic inflammation in genetically predisposed individuals.

Up to this point, we have been mainly focused on the adaptive component of the immune system in the development of chronic inflammation. An adaptive immune response is created when a foreign substance, or what is viewed as a foreign substance, enters our body. The adaptive immune system mounts a slow but antigen-specific and potent response to this invader, which will be triggered the next time this particular invader enters the body, thanks to immune memory generated by the first encounter.

Adaptive immunity, which scientists think has developed relatively recently along the timeline of biological evolution, is found only in vertebrates and is not passed on from parents to offspring. When you get a tetanus shot or other vaccines, adaptive immunity is activated along with innate immunity, unleashing a prompt immune response that leads to mobilization against the infectious pathogen at the next encounter.[12]

We have only recently come to recognize that the innate immune response is as important as, and perhaps more important than, the adaptive immune response and its sophisticated mechanisms of defense regarding treatment and prevention. Already present in the body, the innate immune response is activated at the first sign of injury or invasion. This defense machinery is the "first responder" with weapons that include chemical and physical barriers, plasma proteins, phagocytic leukocytes, and dendritic and natural killer cells. If you cut yourself shaving, the innate immune system sends white blood cells to fight bacterial infection, causing the classical redness and swelling characteristic of an innate immune response.

As research into the gut microbiome advances, it has become clear that examining the earliest stages of the immune response—and the cross-talk between microbiota and immune components—is essential to developing effective treatment and prevention for chronic inflammatory diseases. The early stages of this work in the field of microbiome studies, in other words, learning how important the microbiome is and what microbial communities make up the human microbiome, are shifting to more sophisticated research on what function, through the interaction with the human genome, these organisms perform. We are now looking at how epigenetic pressure on the genes of a host can shift genetic predisposition to a disease to a clinical outcome.

Developing Therapeutic Targets

The five pillars contributing to chronic inflammation that are listed in this chapter provide five different possible targets to mitigate that inflammation. The first approach is the editing of the human genome. As we have seen from recent events in China, after gene editing was practiced on human embryos before two live births, there are enormous ethical, cultural, political, and social concerns with this approach. Even if these issues were satisfactorily resolved, the multifactorial nature of chronic inflammatory diseases makes genetic editing unfeasible. Hundreds, if not thousands, of genes are involved, and they can be different in each individual despite exerting the same function even when affected by the same disease.

The second pillar of environmental influences also is not a pursuable target. We can't even agree as a society on how to tackle global warming, so imagine what it would take to try to change our environment to mitigate all potential triggers of inflammation. The way we farm, the way we raise our animals, the way we prepare and process food, the way we travel, the way we treat our air, soil, and water, and the way we generate energy are only a few considerations. This would require a dedicated, coordinated, and international effort that is not currently feasible.

This leaves us with three targets: modulation of the immune response; changes in microbiome composition and function; and mitigating loss of intestinal barrier function. Let us examine these three possible targets separately, starting with the breach of barrier function. To eliminate gut permeability defects, we have a steep learning curve to determine how all these pathways interact. At this point, there are not many options for treatment with the exception of a few pathways, which include the zonulin pathway.

The Structural Components of Mucosal Barriers

As mentioned earlier, the human body interfaces with the environment in several areas. The gut mucosa encompasses the largest interface between the human body and the external environment, with numerous implications for health and disease. The gut mucosa acts as a semipermeable fence. It permits nutrient transport and immune detection, while strongly restricting the passage of potentially harmful microbes and environmental antigens from the lumen to the systemic circulation.

This physical barrier is made of several components. The two most important are the epithelial monolayer with its intercellular tight junctions and the mucus layer composed mainly of

proteins. These include mucin (MUC) proteins, the most important and abundant of which are MUC2, produced by goblet cells.
The main function of these two components is to segregate the
luminal gut microbiome from the host immune cells present in
the lamina propria.

This separation promotes bacterial clearance and prevents
inappropriate instigation of the immune cells from exposure to
microorganisms that may lead to inflammation. In the colon,
where the microbiome is much richer than in other tracts of the
gastrointestinal system, there are many more goblet cells, creating a thicker mucus layer, which is organized in an inner, firm
mucus layer and an outer, loose mucus layer.

The outer mucus layer is inhabited by a large number of
intestinal microbes, while the inner mucus layer is free of germs,
thanks to the action of defensins and secretory IgA present in
the intestinal lumen. The gut microbiome utilizes polysaccharides of MUC2 as an energy source; therefore, the absence of
dietary fiber, which is the main energy source of many intestinal bacteria, leads to the expansion of mucin-degrading species,
resulting in the increase of inner mucus degradation, potentially
jeopardizing the segregation from the host immune cells.

If the mucus layers are compromised, the glycocalyx covering the surface of epithelial cells would provide the next line of
defense before microorganisms are able to reach the surface of
enterocytes. At this level, the final line of defense is represented
by the junctions in between epithelial cells. Each intestinal epithelial cell is linked to its neighbor cell by junctional proteins,
thus preventing uncontrolled paracellular transport of bacteria
and large molecules.

Three molecular complexes constitute this junctional system
(starting at the upper part of the cell facing the intestinal lumen

and going down toward the lamina propria): tight junctions, adherens junctions, and desmosomes. Tight junctions are the most apical junctional complex, and they create a polarized system separating the apical and basolateral cellular poles.

When coauthor Alessio Fasano was in medical school, he learned the dogma that the space between intestinal epithelial cells was closed and no substances could move between them. He thought it was like an impenetrable brick wall. In the 1960s, when researchers used "tight junction" terminology for the first time, they were convinced that, as the consequence of the fusion of the membrane of the two adjacent cells, the space in between neighboring cells was completely sealed.

In 1981, this paradigm was upended by the finding that cyclic adenosine monophosphates altered the permeability of epithelial tight junctions in the gallbladder.[13] Following this discovery, there was still much debate regarding the mechanisms by which paracellular permeability was regulated, and the structure of tight junctions was still unknown.

Finally, in 1993 the first tight junction protein, occludin, was discovered by Mikio Furuse and coworkers in Japan.[14] This discovery literally opened the door to a new paradigm in both science and medicine. Since then, almost 150 proteins creating tight junctions have been identified. These proteins exhibit extensive redundancy in function, indicating that these structures oversee an essential physiological task in the human body.

The Zonulin Family of Tight Junction Modulators

In 2000, Fasano and his research team at the University of Maryland School of Medicine discovered zonulin, the only currently known physiological modulator of gut permeability. Through proteomic analysis of blood sera from patients affected by autoimmune diseases (celiac disease and T1D), the first member of the zonulin family was identified by Fasano's group as pre-haptoglobin (HP) 2 and described as the inactive precursor of HP2, one of the two genetic variants (together with HP1) of human HPs.[15]

Phylogenetically speaking, mature human HPs are very ancient plasma glycoproteins composed of two subunits: α (alpha) and β (beta). While the β chain (36 kDa) is constant, the α chain exists in two isoforms, that is, α1 (~9 kDa) and α2 (~18 kDa). The presence of one or both α chains produces the three human HP phenotypes: HP1–1 homozygote, HP2–1 heterozygote, and HP2–2 homozygote.

Despite this multidomain structure, the only function assigned to HPs to date is to bind hemoglobin (Hb) to form stable HP-Hb complexes, thereby preventing Hb-induced oxidative tissue damage. In contrast, no function has ever been described for their precursor forms. The primary translation product of mammalian HP mRNA is a polypeptide that dimerizes co-translationally and is proteolytically cleaved while still in the endoplasmic reticulum.

Conversely, zonulin is detectable in an uncleaved form in human serum, adding another intriguing aspect to the multifunctional characteristics of HPs. HPs are unusual secretory proteins in that they are proteolytically processed in the endoplasmic reticulum, the subcellular fraction that contains the highest concentration of zonulin.

Since the discovery of zonulin, its upregulation, which coincides with the loss of gut barrier function, has been shown to be associated with a host of chronic inflammatory disorders. These include autoimmune diseases, neurodegenerative diseases, infections, metabolic disorders, and even cancer (see table 6.1).

Table 6.1
Chronic inflammatory diseases to which zonulin has been linked as a biomarker of gut permeability

Disease	Model
Aging	Human
Ankylosing spondylitis	Human
Attention deficit hyperactivity disorder	Human
Autism spectrum disorder	Human
Celiac disease	Human
Chronic fatigue syndrome/Myalgic encephalomyelitis	Human
Colitis—Inflammatory bowel diseases	Human
Colitis	Mouse
Environmental enteric dysfunction	Human
Gestational diabetes	Human
Giloma	Human and Cell
Human immunodeficiency virus	Human
Hyperlipidemia	Human
Insulin resistance	Human
Irritable bowel syndrome	Human
Major depressive disorders	Human
Multiple sclerosis	Human and Mouse
Necrotizing enterocolitis	Human
Non-alcoholic fatty liver disease	Human
Non-celiac gluten sensitivity	Human
Obesity	Human
Schizophrenia	Human
Sepsis	Human
Type 1 diabetes	Human and Rat
Type 2 diabetes	Human

Adapted from A. Fasano, "All Disease Begins in the (Leaky) Gut: Role of Zonulin-Mediated Gut Permeability in the Pathogenesis of Some Chronic Inflammatory Diseases," *F1000Research* 9 (January 31, 2020): 69, https://doi.org/10.12688/f1000research.20510.1.

Structural Characterization of Zonulin and Subunits

Phylogenetic analyses suggest that HPs evolved from a complement-associated protein (mannose-binding lectin-associated serine protease, MASP), with their α chain containing a complement control protein (CCP: the domain that activates the complement), whereas the β chain is related to chymotrypsin-like serine proteases (SP). Nevertheless, the SP domain of HP lacks the essential catalytic amino acid residues required for protease function; structure-function analyses have implicated this domain in receptor recognition and binding.

Although not a serine protease, zonulin shares about 19 percent amino acid sequence homology with chymotrypsin; their genes both map to chromosome 16. The active site residues typical of the serine proteases, histidine-57 and serine-195, are replaced in zonulin by lysine and alanine, respectively. Due to these mutations, zonulin most likely lost its protease activity during evolution, even though zonulin and serine proteases evolved from a common ancestor. Therefore, zonulin and the serine proteases represent a striking example of homologous proteins with different biological functions.

Other members of the MASP family include a series of plasminogen-related growth factors such as epidermal growth factor (EGF), hepatocyte growth factor (HGF), and other factors involved in cell growth, proliferation, differentiation, migration, and disruption of intercellular junctions. Another element supporting the conclusion that zonulin belongs to the MASP family is that it shares the same receptor (EGF receptor, EGFR; see below) as other members of this family related to the plasminogen-related growth factors.

Evolutionary and Structural Biology of the Zonulin Family: Properdin as the Second Member

The α2 chain gene (and therefore zonulin), which is found only in humans, originated approximately two million years ago. This was five hundred thousand years before the human and chimpanzee lineages split through a chromosomal aberration in a humanoid in India.[16] Since Fasano's laboratory has detected zonulin in other mammals, it is likely that frequent zonulin polymorphisms secondary to a high mutation rate during evolution led to a family of structurally and functionally related zonulins. A new member of the zonulin family has been recently identified as properdin, another MASP protein.[17] Once released from neutrophils, T cells, and macrophages in response to acute microbial exposure, properdin produces the chemotactic anaphylatoxin complement (C) 3a and C5a with subsequent formation of immune complexes that increase endothelial permeability.

Intriguingly, zonulin as pre-HP2 also generates C3a and C5a, with subsequent increased vascular permeability in several organs, including the lungs, and with subsequent onset of ALI as observed in the most severe cases of COVID-19 infection. Indeed, another striking similarity between zonulin and properdin is that both are associated with viral respiratory tract infections. Severe cases of COVID-19 infection are characterized by the development of interstitial pneumonia causing ALI, which is mainly responsible for the infected patients' mortality. ALI is characterized by a leakage of plasma components into the lungs, compromising their ability to expand and optimally engage in gas exchange with blood, resulting in respiratory failure.

Fasano and his collaborators previously reported that zonulin is involved in ALI pathogenesis in mouse models, and that its inhibitor AT1001 ameliorated ALI and subsequent mortality by decreasing mucosal permeability to fluid and extravasation of neutrophils into the lungs.[18] Thanks to the crystallographic resolution of the SARS-CoV-2 main protease (M^{pro}), an enzyme fundamental in the viral lifecycle, Fasano and colleagues have generated preliminary data demonstrating that AT1001 can be a strong M^{pro} inhibitor, since it docks extremely well in the enzyme catalytic domain. Preliminary in vitro studies have confirmed AT1001 antiviral activity. Combined, these studies suggest that AT1001, besides its well-demonstrated effect in ameliorating mucosal permeability in ALI/ARDS, may also exert a direct anti-SARS-CoV-2 effect by blocking the M^{pro}.

Stimuli That Release Zonulin in the Gut

Among the several potential intestinal-luminal stimuli that can trigger zonulin release, Karen Lammers, Fasano, and coworkers identified an unbalanced microbiome (dysbiosis and/or small intestinal bacterial overgrowth) and gluten exposure as two more powerful triggers.[19] Enteric infections have been implicated in the pathogenesis of several pathological conditions, including allergic, autoimmune, and inflammatory diseases, by increasing the paracellular permeability of the intestinal barrier.

The Fasano team has generated evidence showing that the small intestine secreted zonulin when it was exposed to enteric bacteria. This secretion was independent of either the animal species from which the small intestines were isolated or the virulence of the microorganisms tested. It occurred only on the luminal aspect of the bacteria-exposed small intestinal mucosa and was followed by an increase in intestinal paracellular permeability, which was coincident with the disengagement of the scaffold protein ZO-1 from the tight junctional complex. This zonulin-driven opening of the paracellular pathway may represent a defensive mechanism that "flushes out" microorganisms, contributing to the innate immune response of the host against bacterial colonization of the small intestine.

Besides bacterial exposure, the Fasano lab has shown that gliadin also affects intestinal barrier function by releasing zonulin. Gliadins, key components of gluten, are complex proteins unusually rich in prolines and glutamines that are not completely digestible by mammalian intestinal enzymes. The final product of this partial digestion is a mix of peptides that can trigger host responses (increased gut permeability and innate and adaptive immune response) that closely resemble those instigated by exposure to potentially harmful microorganisms. Moreover, gliadin immediately and transiently enhances zonulin-dependent, increased gut paracellular permeability, irrespective of disease status.

This observation led to the identification of the chemokine receptor CXCR3 as the target intestinal receptor for gliadin. Data from Fasano's lab demonstrate that in the intestinal epithelium, CXCR3 is expressed at the luminal level, is overexpressed in celiac patients, and co-localizes with gliadin. This interaction coincides with the recruitment of the adapter protein MyD88 to the receptor (zonulin release is MyD88-dependent).

Lammers, Fasano, and colleagues also demonstrated that the binding of gliadin to CXCR3 is crucial for the release of zonulin and the subsequent increase in intestinal paracellular permeability, since CXCR3-deficient mice failed to release zonulin and disassemble tight junctions in response to a gliadin challenge. Using an α-gliadin synthetic peptide library, they identified two α-gliadin 20mers that bind to CXCR3 and release zonulin.[20]

Using the same machinery, but in a much more complex way, an unbalanced microbiome follows the same trail. If our microbiome is in balance, we have a tightly controlled antigen trafficking system.

Zonulin Transgenic Mice: Intimate Interplay among Gut Permeability, Immune System, and Microbiome

As mentioned earlier in this chapter and elsewhere in this book, the epithelial barrier, immune system, and microbiota "interactome" maintains gut homeostasis, while disruption of this interplay may lead to inflammation. The zonulin transgenic mouse (Ztm), genetically engineered to express murine pHP2, is a model of zonulin-induced increased intestinal permeability. The Ztm model represents a unique and extremely valuable tool to understand how primary loss of control of antigen trafficking may influence gut microbiota composition and immune system development. This Ztm model has been shown to express zonulin and exhibit constitutive increased gut permeability in vivo.[21]

Intriguingly, gut permeability defects in the Ztm do not have any phenotype consequence under baseline conditions; the animals grow well, reproduce without difficulty, and do not develop any spontaneous disease. However, the zonulin-dependent intrinsic increase in gut permeability seems to have a critical effect on both the gut microbiota composition and the immune system.

Akkermansia versus Rikenella

The analysis of stool microbiota done in Fasano's lab showed dysbiosis in the Ztm gut. The markedly reduced presence of *Akkermansia* (sp. *muciniphila*) versus a significantly more abundant *Rikenella* genus in the Ztm gut was striking.[22] In healthy human adults, *A. muciniphila* comprise 1 to 3 percent of the gut microbiota. By feeding on mucin, a primary component of the mucus layer, *A. muciniphila* contribute to the integrity of the enterocyte mono-layer and the overall strengthening of the gut epithelial barrier.

Low levels of *A. muciniphila* have been associated with many human diseases and have been reported as an intestinal microbi-ota feature of both genetic and diet-induced obese and diabetic mice. Drug and surgical approaches aimed at reducing diabetes have both been associated with an evident increased abundance of *A. muciniphila*, suggesting a protective effect with respect to diseases associated with low-grade inflammation.

A. muciniphila are indicators of a healthy gut. This contrasts with findings that show a high abundance of *Rikenella* in obe-sity and diabetes, both in humans and in murine models, which associates *Rikenella* with a state of low, noninfectious, and chronic inflammation. Overall, the low abundance of *A. muciniphila* and the enriched *Rikenella* presence in the gut of Ztm suggest that the Ztm gut microbiota is skewed toward a more maladaptive and pathogenic profile.

It is well known that commensal bacteria are involved in the development of the immune system and the establishment of oral tolerance. Germ-free animals exhibit an underdeveloped immune system whose abnormalities are reversed by coloniza-tion with commensals supporting the role of the microbiota in shaping the immune system. Both dysbiosis and increased gut permeability have been observed in patients and experimental models of chronic and acute intestinal diseases, like IBD (Crohn's and ulcerative colitis) and celiac disease. This evidence supports the involvement of gut permeability and the increased trafficking of microbial or dietary antigens in the dysregulated immune response and break of tolerance.

The balance between proinflammatory IL-17$^+$ helper Th17 cells and anti-inflammatory FOXp3$^+$ Tregs in the mucosa plays a key role in the development of chronic inflammatory diseases. While IL-17-producing CD4$^+$RORγt$^+$Th17 cells are involved in host defense by eliminating pathogens and inducing tissue inflammation, intestinal CD4$^+$CD25$^+$FOXp3$^+$Treg cells suppress Th17 cell–induced inflammation, thereby preventing dysregulated mucosal immune responses and establishing oral tolerance. It has been shown that specific components of the commensal microbiota induce Th17 cells in the lamina propria of the small intestine, and that antibiotic treatment prevented such differentiation.

Reversing Negative Outcomes in Ztm Mice

In the Ztm model, despite altered baseline antigen trafficking, major immune cell subsets involved in the adaptive immune response appear unaffected. Conversely, these animals show an increase in proinflammatory, IL-17-producing innate and innate-like cells in the intestine, as well as in secondary lymphoid sites located in the descending colon, suggesting that the threshold of immune reactivity in these mice might be altered. This renders them more susceptible to losing tolerance to nonself-antigens and incurring the subsequent onset of inflammation.

In keeping with this hypothesis, Ztm are more susceptible to environmental stimuli, like dextran sulphate sodium (DSS), which induce colitis in mice. When exposed to DSS, these animals showed increased morbidity and mortality (up to 70 percent compared to 0 percent in wild-type mice) associated with elevated zonulin gene expression and more severe impairment of the gut barrier function.[23]

This adverse clinical outcome was completely rescued by the oral administration of the zonulin synthetic peptide inhibitor AT1001, Larazotide acetate, which is now in clinical trials as a supplemental treatment to the gluten-free diet for celiac disease. This outcome supports the role of the zonulin pathway in controlling gut permeability and causing loss of tolerance and the onset of inflammation.

Taken together, these data indicate that zonulin-dependent increased gut permeability along with antigen trafficking in the gut orchestrates the development of the immune system and the microbiota composition toward a proinflammatory status in Ztm. This is reflected in increased IL-17-producing cells and the lack of gut-protective microbial species (i.e., *Akkermansia*) versus an overrepresentation of proinflammatory strains (i.e., *Rikenella*).[24]

This low-grade, proinflammatory-skewed status renders Ztm critically susceptible to losing tolerance and developing overt inflammation leading to disease in the presence of exogenous stimuli (i.e., DSS). The same chain of events can be ascribed to chronic inflammatory diseases in which exogenous factors that trigger epithelial barrier dysfunction and activation of the immune system in genetically predisposed individuals could selectively promote the growth of microbial species in the gut. These developments could exacerbate changes in the intestinal tissue that result in the breaking of mucosal tolerance and the onset of chronic inflammation and clinical symptoms.

Slowing Immune Response

When we look at attempts to mitigate a hyperimmune response, the fourth pillar of chronic inflammatory disease, this has been in practice for the last few decades, but with fairly gloomy results. Think of cortisone, immunosuppressants, and biologics—these are all drugs employed to try to put the brakes on the immune system. These are not effective for everyone, and in the subgroups for which they are effective, the drugs do not always work in the long run. Patients pay a dear price in terms of side effects, some of which can be more detrimental than the condition for which the treatment was given.

This should not come as a surprise. Putting the brakes on inflammation for someone with Parkinson's disease or multiple sclerosis is like giving acetaminophen to someone with bacterial pneumonia. Of course, the fever will go away, but only to return after a few hours. With acetaminophen we treat the symptom (fever), but we do not eradicate the cause (infection). To do this, we use a targeted antibiotic treatment to resolve the problem.

Some people respond to immunosuppressants early in treatment, but the drugs lose their effectiveness over time. This shows that targeting the immune response may be too far downstream from the chain of events that lead to clinical outcomes. For example, if we put a roadblock on the path of inflammation by using a biologic like a TNF-α receptor inhibitor such as infliximab, our biology will find a way to go around this obstacle. Our sophisticated immune system is trained to deploy the program it uses when we are exposed to an enemy, which is to get rid of the enemy at almost any cost.

Manipulating the Microbiome

This leaves us with the fifth and final pillar of chronic inflammatory disease: an unbalanced microbiome. Manipulating the microbiome—its composition and, most important, its function—is probably a more promising but also a more challenging target to mitigate these epidemics of chronic inflammatory disease. We have incorrectly, if unwittingly, increased the rate of chronic inflammatory disease by having an unbalanced microbiome, which epigenetically changes the genes controlling the immune response that causes chronic inflammation.

We can equally target the microbiome to make sure that those genes are not changing their expression so that chronic inflammation can be mitigated. We can change the microbiome and leave the genes alone. Discovering how to do that and which tools will allow us to reach this target is the main objective of this book. Strategies to manipulate the microbiome, including the use of prebiotics, probiotics, synbiotics, fecal microbial transfer (FMT), and other therapies, will be discussed in part III.

Turning Off Inflammation

The logical interpretation of what has been discussed so far implies that the historical supposition that chronic inflammatory disease is a one-way path, and that once tolerance is broken it cannot be reestablished, is probably an erroneous assumption. What would be the advantage, in terms of evolution, to have genes whose exclusive function is to generate chronic inflammation?

We can appreciate that a gene mutation like thalassemia (see below) provides an advantage to the affected individual by decreasing susceptibility to becoming infected by the malaria bug *P. falciparum*. A chronic inflammation like the development of an autoimmune disease, such as diabetes or multiple sclerosis, delivers no advantage to the host. So, if susceptibility to chronic inflammatory diseases like autoimmune diseases is merely the result of genetic mutations, these individuals would not survive through the natural selection of evolution, because this is a great disadvantage. Conversely, the "five pillars proposition" described earlier suggests instead that this chronic inflammation is the consequence of a continuous exposure to stimuli that instigates the immune system to fight. If we remove the stimuli, that chronic inflammation can be turned off.

Celiac disease is a typical example of a chronic inflammatory condition. It is the only autoimmune disease for which we know the trigger, namely, gluten. In most cases, once gluten is eliminated from the diet, the autoimmune response is shut down, autoantibodies disappear, symptoms resolve, and the autoimmune intestinal damage heals. Some classical immunologists will argue that if celiac disease can be treated, then by definition it cannot be listed among autoimmune diseases.

Until recently, autoimmunity has been defined as a one-way street in which tolerance cannot be reestablished once a break of tolerance occurs. Based on this traditional, self-serving prophecy of autoimmunity as irreversible, we have been trying to resolve it using the old gene-environment interaction paradigm. This approach has not led to any effective cure, despite decades of intense research.

Taking into consideration all of the aforementioned factors and based on our current knowledge of the biology and immunology of the human species as an integral part of a much more complex ecosystem, we can formulate the microbiome hypothesis as a more logical explanation for the current epidemic of chronic inflammatory disease. In part II, we delve more deeply into the microbiome hypothesis as it relates to a variety of human diseases and conditions.

II   The Microbiome's Role in Disease

7 The Microbiome and Gut Inflammatory Disorders

Intestinal Inflammatory Conditions

The largest body of literature on the human microbiome is focused on the gut microbiome, so it is not surprising that there has been a growing interest in the relationship between dysbiosis and intestinal inflammatory diseases. In this chapter we will review those intestinal inflammatory conditions, both functional and organic disorders, in which the link between the microbiome and disease pathogenesis has shown stronger evidence.

Gut Functional Disorders

Irritable Bowel Syndrome: Then and Now

Functional gastrointestinal disorders are extremely frequent, multifactorial conditions affecting a large proportion of the general population worldwide. Irritable bowel syndrome (IBS) is the most common functional gastrointestinal disorder, affecting more than 10 percent of the general population. IBS is clinically defined as a heterogenous condition characterized by recurrent abdominal pain associated with irregular bowel habits.

Due to the lack of specific and sensitive biomarkers, the diagnosis of IBS is still based on clinical criteria that have been outlined as Rome criteria (Rome IV criteria are the most recent). These criteria classify IBS into three main groups depending on bowel habit: IBS diarrhea (IBS-D); IBS constipation (IBS-C); and IBS mixed (IBS-M), alternating diarrhea with constipation.[1] People affected by IBS report a high burden of the disease with a negative impact on their quality of life and work productivity, along with the frequent utilization of health care resources, which in the United States has been described as more than $30 billion annually in direct and indirect costs.[2] It also has been reported that IBS patients miss an average of two days of work per month with an average cost to the health system of $7,000 to $10,000 annually for each patient.[3]

Classically, IBS was regarded as a prototype form of gut-brain axis dysfunction. Miscommunication between neuroendocrine signaling and effector nerves within the gastrointestinal tract, caused by psychosocial stressors and leading to visceral hypersensitivity and altered motility, was thought to have a pathogenetic role. It is now appreciated that this model is too reductive, and new evidence suggests the involvement of additional factors, including exposure to environmental stimuli such as food, infections, antibiotic treatment, and psychosocial events.

These factors, combined with genetic predisposition and epigenetic changes, may lead to modifications in the function of the epithelial gut barrier. This can cause increased permeability, excessive passage of nonself-antigens and endotoxins, and the activation of both intestinal and brain immune and neuroendocrine responses. Combined, this chain of events triggers a low-grade, noninfectious inflammation and changes in gut microbiota composition and function that ultimately lead to inappropriate secretory and sensorimotor output in the gut, causing classic IBS symptoms.

Even in the heterogenous condition of IBS, we can appreciate the theme that recurs throughout this book, namely that there are highly interconnected and mutually influenced changes in gut permeability, mucosal immune response, and microbiome composition and function. Given the limited efficacy of current treatments for IBS and increasing evidence of a pathogenetic role of the gut microbiome in IBS, there is growing interest in exploiting this new knowledge to find additional therapeutic targets involving manipulation of the intestinal microbiome.

Connecting the Gut Microbiome with IBS

Clinical evidence showing that the onset of IBS is often preceded by gastrointestinal infections, irrespective of whether it is caused by bacterial, viral, or parasitic infections, already suggests the possible involvement of dysbiosis in its pathogenesis. Additional evidence shows that the presence of microorganisms in the gut lumen is instrumental for proper motility function in the gastrointestinal tract.

When compared to animals raised in environments where they are regularly exposed to microorganisms, germ-free animals show inappropriate maturation of motility patterns within the gastrointestinal tract. This is characterized by delayed gastric emptying, slow intestinal transit time, and a reduced migratory motor complex due to changes in the expression of genes that control the maturation and function of the gut's enteric nervous system. As suggested in other chronic inflammatory diseases, the overall hypothesis linking IBS to the gut microbiome postulates that gut dysbiosis leads to increased gut permeability and the passage of macromolecules including endotoxins, with the subsequent activation of the gut-associated lymphoid tissue (GALT) that leads to low-grade inflammation.

There is a fast-growing body of literature, which sometimes shows the opposite result, linking changes in the abundance of specific components of the gut microbiome with IBS. The general trend suggests an increased abundance of "proinflammatory" bacterial phyla, genera, and species, including *Enterobacteriaceae*, which is paralleled by a decrease in beneficial microorganisms such as *Bifidobacterium* and *Lactobacillus* genera. However, other reports show an increase in the *Lactobacillus* genus in IBS. This reiterates the concept that studying the microbiome at a high hierarchical level (family or genus) may lead to conflicting results, since different species or even strains within the same genus may exert opposite functions.[4]

These results also reiterate the importance of the microbiome's function over composition in linking the microbiome with clinical outcomes. In line with these considerations, it is interesting to outline the link between a methane-producing microbiome and IBS. In the typical yin-and-yang way of biological systems, it has been reported that methane production is reduced in IBS-D and increased in IBS-C.

Methane and Gut Motility

Methane is a metabolite produced by methanogen microorganisms belonging to the Archaea domain, and Methanobacteriales is the order most capable of converting hydrogen into methane. What makes these associations intriguing is that methane has been mechanistically linked to a direct effect on gut motility; it is capable of slowing intestinal transit time. Also suggestive of this link is the observation of an unbalanced microbiome characterized not only by the enrichment of Methanobacteriales-related, methane-producing *Clostridiales* and *Prevotella* species, but also by increased methane exhalation correlating with the severity of IBS symptoms.

Intestinal dysbiosis is not exclusively associated with changes in the microbiome's composition and function; it can also be secondary to a different microbial distribution within the gastrointestinal tract. Under physiological circumstances, the vast majority of the gut microbiome (more than 70 percent) is confined within the large intestine, while in the proximal small intestine the microorganisms' load is rather limited, with an increasing gradient from proximal ($\sim10^1$–10^3 colony-forming units (CFU)/ml in the stomach) to distal ($\sim10^{11}$–10^{13} CFU/ml in the colon). This gradient is maintained by a variety of anatomical and functional factors that prevent colonization of the proximal intestine.

These include gastric acidity that eliminates the vast majority of microorganisms orally ingested, gastric and biliary juices that exert an antibacterial effect, and peristalsis that prevents the colonization of microorganisms in the small intestine. Other factors are glycocalyx and mucin that maintain the microorganisms' distance from the surface of mucosal epithelial cells and prevent direct contact; mucosal humoral and cellular mechanisms of defense; specific antimicrobial peptides, such as defensins; and anatomical "checkpoints," such as the ileo-cecal valve, which prevents the backflow of distal, intraluminal colon content, which is much richer in microorganisms, into the small intestine.

Additionally, external factors may affect the abundance of microflora in the small intestine, including diet, prescription or nonprescription drugs (including opioids) that affect gut motility, acid suppression (H_2 blockers and proton pump inhibitors), a high consumption of alcohol, and stress that affects peristalsis. The impairment of intrinsic protective factors and/or exposure to extrinsic factors favoring colonization of the proximal intestine may lead to small intestinal bacterial overgrowth (SIBO).

Small Intestinal Bacterial Overgrowth

SIBO is characterized by quantitative ($>10^3$–10^5 CFU/ml) and qualitative (the presence of both symbiotic and pathogenic species) microbial alterations in the small intestine. SIBO diagnosis is not an easy task. The direct culture of duodenal aspirate is invasive and technically challenging. And the use of glucose or lactulose sugar probes to measure either hydrogen production through anaerobic fermentation or bacteria or methane from methanogenic bacteria by breath hydrogen test is poorly standardized and not highly accurate. For this reason, an accurate number for the prevalence of SIBO in the general population remains elusive, ranging from zero to 20 percent. Similarly, the prevalence of SIBO reported in IBS patients, particularly among IBS-D patients, is generally much higher and ranges between 5 and 80 percent, reflecting the limitations of current diagnostic tools.

The possible pathogenetic role of SIBO as a form of dysbiosis in IBS is supported by the efficacy of treating this clinical condition with certain probiotics or poorly absorbed antibiotics such as rifaximin, or both. This treatment resets microbial diversity and abundance in the small intestine and decreases fermenting microorganisms, with a subsequent reduction of symptoms (bloating, cramps, and irregular bowel habits) directly linked to the reduction of gas generated by bacterial fermentation. Nevertheless, it should be pointed out that by altering gut peristalsis, IBS may cause a secondary SIBO, posing the question of whether

SIBO is the cause or consequence of IBS. Or are these conditions so tightly interconnected in a vicious cycle that it is difficult to establish which one precedes the other?

Gut Organic Disorders

Infant-Specific Conditions

Necrotizing Enterocolitis   Based on the already described importance of the gut microbiota development at birth and the influence of several prenatal, perinatal, and postnatal factors on its development (see chapter 3), it should not be surprising that proper microbial engraftment is especially relevant for very preterm infants. The chance that "something can go wrong" as the consequence of the first engraftment of these early colonizers of a structurally, functionally, and immunologically immature gut mucosa is much higher compared to a physiologically full-term birth. Of all possible complications of this initial encounter between an immature gut and the colonizing intestinal microbiome, necrotizing enterocolitis (NEC) is a life-threatening disease characterized by severe morbidity and mortality.

Despite major improvements in neonatal intensive care, the incidence of NEC and its morbidity and mortality rates have remained unchanged during the last two decades. Given the undisputed role of the gut microbiome in the pathogenesis of NEC, there has been a major effort to identify the "microbiome signature" associated with the onset of NEC in very low birth weight infants. Barbara Warner and coworkers have performed a well-controlled, cross-sectional study in a prospective birth cohort that described such a "signature" microbiome of very preterm infants who eventually went on to develop NEC.[5]

The microbiome analysis of these infants showed a positive association with Gammaproteobacteria (a class of Gram-negative bacilli that range from aerobic to facultative or obligate anaerobic metabolism) and a negative association with the Clostridia and Negativicutes classes of strictly anaerobic bacteria. The explanation of the presence of these "hostile" settlers within the gastrointestinal tract of infants who develop NEC remains elusive. Most likely, many factors contribute to their presence and engraftment, including atypical oral intake (delayed enteral feeding, tube feeding, and formula supplementation) and the ubiquitous use of broad-spectrum antibiotics typical in neonatal intensive care units.

Feeding very premature infants with predominantly human breast milk has been shown to decrease the risk of NEC, possibly due to the maternal microbiome transferred in human milk, the role of HMOs, immune mediators, or more likely a complex interaction of all of these factors. The unusually clean environment of the neonatal intensive care unit, common C-section deliveries, limited breastfeeding, and the lack of continuous close physical contact with parents may further influence the establishment of an unbalanced, proinflammatory microbiome.

Nevertheless, the presence of an atypical gut microbiome may be necessary but not sufficient to develop NEC. Most likely, the immature defense mechanisms typical of preterm babies, including inefficient peristalsis and a decreased expression of intercellular epithelial tight-junction proteins, increase the likelihood that both bacteria and their endotoxins, which are normally confined to the intestinal lumen, may cross the gut barrier and reach systemic organs and tissues. This uncontrolled passage of microorganisms (and their byproducts) triggers the activation of an exaggerated inflammatory response that further compromises the gut barrier. This hypothesis is based on a multitude of studies using animal models, fetal intestinal xenograft transplants, fetal intestinal organ cultures, and a fetal primary intestinal cell line, all showing that an abnormal response to gut-colonizing bacteria seems to contribute to NEC susceptibility.

Specifically, there is growing evidence suggesting that the immature human enterocyte reacts to colonizing intestinal bacteria with an enhanced inflammatory response. Toll-like receptors (TLRs) aimed at recognizing microorganisms on mucosal surfaces have been implicated as key molecules in promoting inflammation. Specifically, TLR4 and other signaling factors connecting TLR4 to nuclear factor-κB (NF-κB) and activator protein transcription factor–mediated inflammation have been found to be upregulated on the surface of fetal enterocytes, whereas genes that inhibited these signaling pathways were downregulated. Together this evidence suggests that an exaggerated innate immune response to colonizing commensal bacteria mediated by TLR activation is mounted by immature intestinal epithelial cells, which could contribute to the pathogenesis of NEC.

This hypothesis also has been supported by gene expression studies performed on human fetal gut organoids showing that they clustered according to their developmental age in two distinct groups: early (less than fifteen weeks gestational age) and late fetal organoids (from sixteen to twenty-two weeks gestational age), with the latter more closely resembling adult gut organoids.[6] In line with published research in nonhuman models and in cell lines, genes involved in maturation, gut barrier function, and innate immunity were responsible for these differences. Organoid-derived monolayers exposed to either LPS or commensal *E. coli* showed that late fetal organoids activated gene expression of key inflammatory cytokines, whereas early fetal organoids did not, owing to decreased expression of NF-κB–associated machinery. Combined, these data suggest that during fetal development, there is an early phase in which the gut mucosal machinery needed to mount an innate immune response is not operative.

This is most likely one of the reasons why, below a certain gestational age, fetuses are not compatible with extrauterine life. At a later phase, the innate immune system is operative and hyperactive, mounting a more robust immune response compared to full-term and adult gut mucosa. This response, combined with an atypical microbiome composition, can add to the pathogenesis of NEC.

Environmental Enteric Dysfunction   A high proportion of children ages four to twenty-four months living in impoverished areas around the world often develop stunted growth, a problem always considered secondary to malnutrition. Active but ineffective nutritional campaigns have cast doubt on the lack of nutrients as the only driving force in the suboptimal growth of these children. A more in-depth analysis of the possible causes of stunted growth, coupled with increasingly sensitive detection methods, has highlighted that multiple and recurrent intestinal infections (with or without diarrhea) are even more common in young children living in impoverished areas in which adequate sanitation or clean water is lacking.

These heavy pathogenic burdens may be associated with growth faltering or cognitive impairment, concomitant with serious social, political, and economic consequences in developing countries. Undernutrition is also common and may further worsen the impact of enteric infections on growth and development, leading to a potential "vicious cycle" of malnutrition and enteric infections.

Animal models have confirmed that the combination of undernutrition and exposure to enteropathogens, such as *Cryptosporidium* or enteroaggregative *E. coli*, or even certain organisms from mixed Bacteroidetes and Proteobacteria phyla, can further worsen growth impairment and intestinal damage in undernourished states. In animal models, these conditions have led to growth failure and microbiome alterations, but without intestinal damage unless certain mixed microbiota were added. Environmental enteric dysfunction is discussed in more detail in chapter 11.

Food Allergies   Allergic diseases are chronic inflammatory conditions characterized by a Th2 immune response. They are activated by exposure to environmental triggers that are harmless for the vast majority of people not genetically predisposed to allergies. These conditions range from modest reactions, including skin rash or allergic colitis, to life-threatening conditions like anaphylaxis.

Like many other chronic inflammatory diseases, allergic diseases are rapidly rising in frequency in the Western world. These diseases are becoming more frequent, and their clinical course is radically changing. They are more prolonged and severe, and they rarely resolve spontaneously within the first one to two years of life, as happened formerly, particularly with food allergies.

Recent statistics from Australia suggest that one out of three Australians will develop allergies at some time in their life; one in twenty will develop a food allergy; and one in one hundred will have a life-threatening allergic reaction. Based on Australian hospital admission records, anaphylaxis cases doubled during the decade from 1994 to 2004, with rates five times higher in children under five years of age compared to older children and adults.[7] This suggests that the development of allergies is increasing at a faster rate in young children compared to adults. While the cause (or causes) for this epidemic remains unclear, recent studies have shown that the composition of the gut microbiome of infants and young children who develop a variety of allergic diseases is different compared to their healthy peers.

Among all allergies, food allergies have been the object of a large body of studies, particularly concerning the possible role of the intestinal microbiome in their pathogenesis. Food allergies affect approximately 5 percent of America's children and are on the rise, resulting in significant economic, psychological, and medical burdens. Therefore, the mechanisms involved in the balance between tolerance and allergic immune response to foods are under increasing investigation.

Oral tolerance is the result of a complex interaction among food proteins, the microbiome inhabiting our gut, and the intestinal mucosa with its multitude of immune and nonimmune cells. The mechanism behind the breakdown of these interactions, leading to clinical food allergy, remains poorly understood. Children with adverse reactions to various foods have increased intestinal permeability measured at baseline, which is worsened by allergic reaction. Tolerance to food is mainly acquired by the highly integrated interplay among dendritic cells, intestinal epithelial cells, and the gut microbiome. A subset of dendritic cells is capable of inducing Tregs that express anti-inflammatory cytokines. Tregs play an important role in the acquisition of oral tolerance, and children with dysfunctional Tregs have an increased risk of food allergies.

The importance of the timing and nature of antigen exposure was demonstrated in a landmark clinical trial with findings that show, contrary to our previous dogma concerning appropriate

timing for exposure to food allergens, that early peanut ingestion decreased rates of clinical food allergy in a high-risk group.[8]
However, we have limited comprehension of the mechanisms
involved in inducing tolerance rather than an allergic immune
response. Our knowledge also remains limited about when antigens are presented early in life and their applicability to the general population.

Research has shown that children raised on farms with an
increased diversity of the microbiome in their homes have a
decreased risk of asthma and atopic diseases, and that antibiotic
exposure early in life increases the risk of developing allergies.
This epidemiological observation supports the potential role of
the microbiome's composition and function in inducing oral
tolerance, a possibility that is only beginning to be appreciated.

Limited data show that specific commensal microbial species
may promote oral tolerance to food proteins, while dysbiosis is
often detected in food allergies. For example, data in antibiotic-
treated mice suggest that *Clostridium* species promote oral tolerance via mechanisms of both intestinal permeability and innate
lymphoid cell function.[9] Certain families of commensal bacteria,
such as *Lachnospiraceae* and *Ruminococcaceae*, are known to produce SCFAs, of which butyrate has been the most studied. Butyrate has multiple immune cell-mediated effects in the context of
food allergy and has been shown to increase the number of Tregs
in the colon in mice.

Translating these data to human biology remains a question-
able proposition since, as mentioned in other parts of this book,
extrapolating microbiome-related results from mice to humans
could be deceptive. In a study performed in children with food
allergies, no major changes were detected in the overall diversity of the microbiome, while changes in the abundance of

several bacterial phylotypes resulted in association with food allergies.[10]

Specifically, *Clostridiaceae* I organisms were prevalent at the family level in infants with food allergies compared to healthy controls. The apparent dichotomy between data in mice showing a protective effect of *Clostridium* species against food allergy and data in children showing just the opposite can be easily explained by the fact that mechanisms leading to a break of tolerance in humans and mice can be different. However, a more logical explanation is related to the fact that the two microbiome analyses differ deeply in terms of their hierarchical level. The human study showed data only at the family level, while the mouse study included data from the species level.

It is entirely possible that a family such as *Clostridiaceae* may occur in greater abundance in children with food allergies, while one of its genus or species involved in oral tolerance may be deficient. This observation stresses once again the importance of critically reading the microbiome literature for proper comparison of different studies. The same study showing an increase in *Clostridiaceae* reported that infants with IgE-mediated food allergy had increased levels of *Clostridium sensu stricto* but decreased levels of *Clostridium* XVIII genus.

Caution should be exercised in interpreting these data in terms of the possible pathogenetic role of specific microbiota components, and therefore their potential therapeutic exploitation. Many conflicting studies have shown alterations in the microbiome of children with food sensitization or allergies, but a unifying pattern has been difficult to elucidate. This is likely attributable to the limitations of small sample size, cross-sectional analyses, differing approaches, and the apparently chaotic infant microbiome shifts detected early in life.

For this reason, more mechanistic studies are needed to demonstrate the link between microbiome composition and food allergies. One example of this kind of research is the work from Taylor Feehley and colleagues on germ-free mice colonized with feces from healthy infants or cow's milk–allergic infants. This study showed that germ-free mice colonized with bacteria from healthy infants were protected against anaphylactic responses to a cow's milk allergen, while those colonized with bacteria from cow's milk–allergic infants were not.[11]

These disparities were correlated to differences in bacterial composition between healthy and cow's milk–allergic infants, and these differences persisted in the engrafted mice. Interestingly, healthy mice and cow's milk–allergic colonized mice also showed unique transcriptome signatures in the ileal epithelium. Correlation of ileal bacteria with genes upregulated in the ileum of healthy mice or cow's milk–allergic colonized mice led to the identification of a species of bacteria in the Clostridia order, *Anaerostipes caccae*, which protected against an allergic response to food.

Studies like this one teach us a few lessons concerning the proper way to exploit the microbiome as a possible therapeutic target, which include the need for translational studies. We also need to deepen microbiome analysis to the species and, ideally, to the strain level if we want to mechanistically link microbiome studies to the pathogenesis of diseases. A final lesson is that animal models remain useful for microbiome studies, as we have seen in this case, in which murine humanized models are used to link a specific microorganism to the instigation of or protection against a chronic inflammatory process, including an allergic response, as previously outlined.

All-Ages Conditions

Celiac Disease   Celiac disease is unique among autoimmune diseases. There is a strong association with HLA DQ2 or DQ8 or both, the environmental trigger (gluten) is known, and disease-specific autoantibodies have been identified as robust biomarkers of the break of tolerance to gluten with the subsequent onset of the autoimmune enteropathy that characterizes this condition. Therefore, exposure to the environmental trigger can be carefully traced based on the time of introduction of gluten into the diet, and frequent prospective screening against the autoantibody tissue transglutaminase can determine precisely when the loss of tolerance to gluten occurs.

Among all of the chronic inflammatory diseases affecting humankind, we did not put celiac disease at the top of the list of diseases in which the microbiome could play a possible pathogenic role. Until the recent past, genetic predisposition and the ingestion of gluten-containing grains in the diet were considered necessary and sufficient to develop the disease. However, epidemiological observations at odds with this paradigm (already discussed in other chapters) led to the hypothesis that, even in celiac disease, changes in microbiome composition and function may play a role.

In vitro studies suggest that microbes can influence the digestion of gliadin, the production of cytokines in response to gliadin, and the increased intestinal epithelial permeability induced by gliadin. The vast majority of research describes differences in the composition, structure, and diversity of the fecal and small intestinal microbiota in patients with celiac disease on the basis of age, disease status, and associated signs and symptoms. Associated metabolic activity, as measured by patterns of SCFAs in the stool, is altered in patients with active celiac disease and is linked to the described dysbiosis. Differences in specimen collection, analysis techniques, age of the study population, and disease status, however, make it difficult to compare studies.

Underlying differences in the colonization of the gastrointestinal tract, related to genetic makeup, also may contribute to an infant's future risk of developing celiac disease. Microbial communities in infants carrying the DQ2 haplotype were found to have a greater abundance of Firmicutes and Proteobacteria compared to control infants. Infants with a first-degree family member with celiac disease and a compatible genotype had a decreased representation of Bacteriodetes, a greater abundance of Firmicutes, and overall delayed maturation of the microbiota in the first two years after birth.[12]

Gluten ingestion is necessary for the development of celiac disease. Introducing gluten into an infant's diet at twelve months of age compared with six months transiently delays the onset of celiac disease but does not prevent its development. Exposing infants to small amounts of gluten between sixteen and twenty-four weeks has not been shown to influence the prevalence of celiac disease. Although clinical studies evaluating the timing of gluten introduction have yet to unveil a meaningful pathological or preventive time point, we know that the introduction of gluten alters the host microbiome in infants at risk for celiac disease. The introduction of gluten leads to notable changes in the abundance of the phyla Firmicutes and Proteobacteria. Infants who developed autoimmunity had high lactate signals in their stools preceding the first detection of celiac disease antibodies, which corresponded to a greater relative abundance of *Lactobacillus* species.[13]

Looking at the trajectory of gut microbiota early in life can provide some clues about why some genetically predisposed individuals will develop the disease and others will not. A nested case-control study conducted by the Yolanda Sanz research group showed that the microbiota of infants who remained healthy was characterized by an increase in bacterial diversity over time, with increased Firmicutes families. These changes were not detected in the infants who developed celiac disease. Furthermore, an increased relative abundance of *Bifidobacterium longum* was associated with the group of control children, while increased proportions of *Bifidobacterium breve* and *Enterococcus* species were associated with celiac disease development. These findings suggest that alterations in the early trajectory of gut microbiota in infants at risk for celiac disease may lead to the onset of the disease.[14]

Combined, these findings open up the possibility of identifying a specific shift in microbiome composition and function as predictors of autoimmunity onset. This will be discussed in chapter 14, which is focused on the value of birth cohort studies in shifting the field of microbiome studies from association to causation.

Inflammatory Bowel Disease   The involvement of microorganisms in the pathogenesis of IBD has been postulated for many years. Despite the great effort spent in search of the pathogens that trigger the chronic inflammatory process that characterizes IBD, an identification of the microorganisms causing IBD has remained elusive. The parallel effort of searching for the genetic traits linked to IBD initially led to promising results with the identification of specific mutations in key pattern recognition receptors, including the *NOD* gene.

However, this initial excitement was partially tempered by the appreciation that these genetic mutations involved only a subgroup of patients with IBD, and that genetic makeup can only explain part of the risk of developing these conditions. This realization suggests that, in addition to genetic predisposition, the immune dysregulation that characterizes IBD must be driven by additional factors.

In the past few years there has been a growing body of literature suggesting that the pathogenesis of IBD could be the consequence of an inappropriate immune response to commensal microbiota. This exaggerated response seems secondary to the combination of genetic mutations and dysbiosis. Disparities in methodological approaches, including different techniques used to analyze the gut microbiome, disease activity, site of inflammation, and different sites of microbiota sampling (stools versus mucosa), however, make comparison across reported studies difficult.

Nevertheless, a common theme has emerged suggesting that this dysbiosis is characterized by a reduction in biodiversity (α-diversity) and an altered representation of several taxa. Some studies have reported an increase in Bacteroidetes and Firmicutes in Crohn's disease; however, the number of ribotypes of Firmicutes was decreased compared with healthy controls. When disease activity was considered, opposite results were reported in patients with Crohn's disease, showing a decreased diversity of Bacteroidetes during the acute phase of the disease compared with patients analyzed during remission. Even the site of the disease inflammation may influence the microbiota composition, with a reported increased diversity of Firmicutes in patients with colonic Crohn's disease and decreased diversity in patients with ileal Crohn's disease.[15]

The site of sampling can also highly influence microbiome analyses. Increased abundance of *Enterobacteriaceae* (mainly *E. coli*) and a decreased diversity of *Faecalbacterium* were detected in patients with Crohn's disease after their mucosal biopsies were compared with fecal samples.[16] Gut dysbiosis is often associated with specific dysfunctions of microbial metabolism and bacterial protein signaling, including involvement of oxidative stress pathways, decreased carbohydrate metabolism, and amino acid biosynthesis counterbalanced by an increase in nutrient transport and uptake.

Even though these changes suggest a possible mechanistic link between modifications in microbiota composition and IBD pathogenesis, these studies are mainly associative and descriptive in nature. They are also limited because they are primarily focused on fecal microbiome samples, while for IBD, the mucosal microbiome can be even more relevant in linking specific species to disease pathogenesis. This factor of site sampling is relevant to many microbiome studies, and it may affect our progress toward an accurate and complete characterization of the microbial ecosystem that would lead to effective therapeutic interventions.

Under physiological circumstances the colonic micromilieu is relatively lacking in oxygen, which is mostly produced by the epithelium, so the growth of aerobic bacteria is favored only if the bacteria live in close proximity to the mucosa. During inflammation, there is an increase in oxygen that is secondary to an increased blood flow. These changes cause a shift in microbiome composition characterized by a proliferation of aerobic microorganisms and a depletion of the anaerobic species. However, it needs to be pointed out that this shift is the consequence and not the cause of IBD, even if it can perpetuate the inflammatory process in a kind of "vicious circle."

Data on gut dysbiosis in pediatric patients with IBD are more limited. Microbiota analysis at the species level in the tissues of children affected by IBD led to the identification of two previously unreported strains of adhesive-invasive *E. coli*, which were mechanistically linked to the upregulation of carcinoembryonic, antigen-related cell adhesion molecule 6; tumor necrosis factor (TNF)-α; and interleukin-8 gene/protein expression. The decreased diversity of the fecal microbiome, including of Firmicutes, Verrucomicrobiae, and Lentisphaerae, was more pronounced among corticosteroid-nonresponsive children with ulcerative colitis.[17]

The recent use of metagenomics analysis combined with the host-transcriptome-based approach has provided a more mechanistic understanding of how changes in the microbiome's function can affect the clinical outcome of the host. Using this approach, Dirk Gevers and colleagues have analyzed a cohort of children recently diagnosed with IBD who had received no therapeutic intervention. They identified a variety of mucosal bacteria differentially present in children affected by IBD compared to those who were not affected by IBD, thus demonstrating that this set of microorganisms can predict clinical outcomes in Crohn's disease.[18]

Studying a subset of these children, they were also able to demonstrate by host RNA-seq analysis that there was a positive correlation between antimicrobial dual oxidase (DUOX2) expression and the expansion of Proteobacteria that was increased in Crohn's disease, while the expression of the lipoprotein gene APOA1 was positively correlated with the abundance of the Firmicutes that was decreased in these patients.[19] The combination of a signature of increased DUOX2 and decreased APOA1 expression, which favors oxidative stress and a Th1 phenotype, has been shown to be more frequent among patients with severe mucosal injury.

Of course, these associations need to be validated through large-cohort studies to confirm their pathogenic role. Nevertheless, these kinds of studies provide solid foundations for future work aimed at integrating host genomics and metagenomics datasets. This knowledge will allow us to incorporate a full appreciation of the interactions between host immune responses and the microbiome that are linked to disease pathogenesis.

8 The Microbiome and Obesity

Rising Rates of a Worldwide Epidemic

WHO defines overweight and obesity as "abnormal or excessive fat accumulation that presents a risk to health."[1] For decades, clinicians have debated whether obesity is a true disease or simply a risk factor for disease, or the result of lifestyle choices, including limited exercise and excess caloric intake. In the United States, the American Medical Association made the controversial decision to define obesity (body mass index [BMI] of more than 30) as a chronic medical disease in 2014,[2] and clinicians employ codes to diagnose various forms of overweight and obese conditions.

But the popular conception remains of obesity as a moral or personal failing and, all too often, a source of shame—as evidenced when overweight and obese children are bullied much more frequently than lean children. Just who and what is responsible for prolonging this epidemic continues to spark considerable debate in academic and general venues, with recent advances in genetic risk assessment and microbiome science shaping part of that debate.

One aspect of obesity that is not open to interpretation is the alarming rate of its increase during the past several decades. Obesity has nearly tripled worldwide since 1975, according to WHO, and most of the world's population lives in countries where obesity and overweight kill more people than underweight or malnutrition. In 2016, more than 1.9 billion adults were overweight and 650 million were obese. An especially alarming aspect of the obesity epidemic is its pediatric component, with 41 million children under the age of five listed as overweight or obese in 2016, according to WHO.[3]

In the classical view of obesity, the quantity of food consumed is the driving factor. In other words, if you eat more than you can metabolize, you become obese. Now we know it is not that simple. Other biological and environmental aspects of obesity are equally important, including our genetic makeup. We all know those fortunate folks who seem to be able to eat anything they want and remain trim, while some of us seem to gain weight by simply looking at food!

Until the recent past, we have interpreted these differences as being related to our genetically determined metabolic profiles that dictate our ability to burn calories, which keeps us in shape, or accumulate them, which translates into a propensity to obesity. The gut microbiome has recently been added to the mix of factors that can influence our metabolism, and therefore our risk of developing obesity. There are many lifestyle factors that can ultimately affect the composition and function of the gut microbiome, which is highly influenced by the quantity and, more important, by the quality of the food we ingest.

For insights into the impact of the gut microbiome on obesity, we turned to one of our colleagues at MGH. Lee Kaplan is a clinically trained hepatologist who treats patients with overweight and obesity and conducts research into the causes of this condition at the Obesity, Metabolism and Nutrition Institute at MGH. One of his favorite quotes comes from a presenter at an early talk he attended on obesity; it defines his views on the topic: "The study of obesity isn't rocket science; it's much more complicated." Kaplan shared his thoughts with us on the role that the human microbiome plays in this chronic inflammatory disease and other obesity-related topics.[4]

A Complex and Challenging Disease

Before moving to the role of the microbiome, Kaplan disputed some of the suppositions that have underpinned the classical model of obesity. He challenges the presumption that for the first time in world history, we have an excess of food, and now we eat more calories than ever before, which results in obesity. He questions the idea that the current obesity epidemic is simply the result of the shift from chronic food shortages throughout most of human history to the more recent excess of food in industrialized countries.[5] But as Kaplan argues, if humans have not generally suffered from limited alimentary resources, then what is causing the obesity epidemic, now that we have a surplus of both food and choice?

The notion that "it's just calories" doesn't fit in well with the concept of an exquisitely accurate regulatory system aimed at achieving an optimal metabolic equilibrium to maintain our ideal weight, said Kaplan. As an example, when children become sick, they might lose a few pounds, which are quickly regained once the disease is resolved. To Kaplan, this phenomenon suggests that there is a protective mechanism to ensure that our body mass is safeguarded.[6]

Genome-wide association studies (GWAS) have identified more than 250 genes associated with both obesity and body mass index.[7] Kaplan noted that there is a lot of redundancy, suggesting we have an evolutionarily determined plan for weight homeostasis.[8] The fact that genetics can play a role in obesity is supported by evidence that shows the recurrence of obesity in the same family. It can be argued that family members also have similar unhealthy lifestyle habits that potentially contribute to obesity. However, animal studies and GWAS analyses looking at obesity through modeling on large cohorts point the finger toward genetic predisposition as well.

Kaplan and colleagues from MGH, the Broad Institute, Harvard Medical School, the University of Bristol in the United Kingdom, and several other institutions examined a study performed by a research group in 2015 on approximately three hundred thousand individuals, ranging in age from newborn to middle age.[9] Using computational algorithms and large datasets, Kaplan and colleagues derived, validated, and tested a new polygenic predictor using more than two million common gene variants for polygenic prediction of weight and obesity trajectories from birth to adulthood.[10]

The bottom line from this complex study is that the authors developed a kind of "genetic blueprint" based on the presence of multiple genes that can predict the risk of becoming obese. This finding contradicts the idea that obesity is nothing but a lack of human willpower to "get our act together" to eat less. Nevertheless, regarding the epidemics of noninfectious, chronic inflammatory diseases mentioned in other chapters, it is difficult to blame our genes for the soaring rates of obesity registered during the past thirty years, since genetic mutations would take a much longer time to translate into clinical outcomes.

Aside from genetics, many researchers agree with Kaplan's supposition that there is something in the modern environment driving the worldwide obesity epidemic. He calls obesity a problem related to modernization (see below) that is affected by several complex factors, and we can add an unbalanced gut microbiome to that list. During the last two decades, pioneering microbiome researchers have focused on the extremes of obesity and malnutrition to study the role of the microbiome in weight and calorie homeostasis, and we turn to that discussion for insights into the complex nature of obesity.

Evolutionary Biology and Obesity

At various times in their life span, many humans have not been able to eat for varying reasons: they become sick or incapacitated, or they are subject to famine through natural calamities or displacement caused by war or other conflicts. When that happens, the body uses up its reservoir of fat. Evolutionarily speaking, when humans become sick or have to face a food shortage, they metabolize any stored fat for energy. We have to have an adequate amount of fat to get us through a food emergency; this is how the body is programmed.

Let's look at conditions that impede proper food intake, such as infections or other illnesses, from which we can recover. One idea is that the amount of fat that a person's body has, or any animal has, is enough to survive during the amount of time it takes to recover from a long bone fracture, said Kaplan.[11] Whether you're a human, a mouse, or a lion, if you're injured, you can no longer be a predator. You go and hide so you don't become prey, and you can't eat much, if at all. You have to have a certain amount of fat stored to survive such emergencies. Humans and other animals are ingeniously designed to survive under most circumstances, including a long period of not being able to procure food.

Therefore, if you have too little fat, you are evolutionarily disfavored. You need a certain amount to survive emergencies. But if you have too much fat as a consequence of the activation of genes aimed at protecting against losing fat, you also are evolutionarily disfavored. Like zebras, chipmunks, and giraffes, humans have evolved with an exquisite regulatory mechanism to maintain the precise amount of fat to protect the survival of the species.

But during the past two hundred years, particularly with the advent of the Industrial Revolution, we have drastically altered that delicate balance. Kaplan argues that our genes have not evolved sufficiently, using this new equation of food availability, to recalibrate the balance between enough fat storage and too much fat. "The way we are protected from obesity is to bring ourselves out of the range of the body storing too much fat. We have to have enough to protect the body in the case of injury or prolonged illness, but not too much to the point that fat storage can become a liability."[12]

Kaplan used the example of pregnancy to illustrate the pathophysiology of fat regulation. In pregnancy, the body stores more fat, exploiting its normal coordinated program of fat storage. After pregnancy and lactation have ended, the body drives that accumulation back down, getting rid of excess fat. Kaplan noted that this is the opposite of what happens after acute illness, when the body accumulates fat that has been lost.[13] Puberty is another example of the natural loss of body fat.

If obesity can be defined as an abnormal accumulation of fat, then an effective treatment should mimic what happens at the end of lactation or the beginning of puberty, that is, getting rid of the excessive accumulation of fat. Obesity and pregnancy are almost identical as "circuitry models," said Kaplan; however, there are fundamental differences in terms of pathophysiology, particularly concerning certain environmental factors, including stress and disruption of circadian rhythms, or other factors affecting the composition and function of the microbiome.[14]

Compelling evidence suggests that shifts in the function of the gut microbiome in the proximal small intestine can cause increased gut permeability.[15] This allows passage of endotoxins and other macromolecules, leading to a noninfectious, low-grade chronic inflammation. This inflammation is secondary to an activation of the innate immune response, which can spread via portal circulation to the liver, ultimately causing inflammation of that organ and metabolic derangement typical of fatty liver and T2D.[16]

Nevertheless, these metabolic changes are within the framework of a much more complex metabolic circuitry that is characterized by redundancy and compensatory mechanisms. Not only does this make the study of the relationship between obesity and the composition and function of the gut microbiome more challenging, but it also complicates the identification of possible targets for intervention. The robust and collaborative research portfolio of one of the pioneering giants of the microbiome research movement provides us with some insights into obesity and function in the gut microbiome.

Obesity Opens Door into Gut Microbiome Research

Jeff Gordon is considered by many to be the "father" of the research movement into the human microbiome. He pioneered studies in the early 2000s related to the role of the gut microbiome in the pathogenesis of obesity by using germ-free mouse models developed at Washington University. Gordon's team showed that mice with a gene mutation resulting in obesity had a different mix of gut bacteria than their lean littermates, even when they shared the same diet. Working with Gordon, Ruth Ley and her group reported that obese mice had more Firmicutes and less Bacteroidetes than lean mice.[17]

Peter Turnbaugh, another member of Gordon's team, transplanted the gut microbiota from genetically obese mice into germ-free normal mice, and the normal mice became fatter compared to the germ-free mice that received the gut microbiota from lean animals. In a paper published in *Nature* in December 2006, Gordon's group showed for the first time that gut microbiota composition was a cause and not a consequence of obesity.[18] In human studies, Gordon, Ley, Turnbaugh, and Samuel Klein showed that the gut microbiota of obese humans changed when they consumed a low-fat diet. When these results were published in the same 2006 issue of *Nature*, Gordon and his team set forth the idea that microbial communities can cause disease.[19]

In further studies, Gordon's group used mice humanized with stools obtained from human twin pairs to determine the role of the gut microbiome in establishing an obese phenotype. Using stool samples from identical twin girls, one of whom was lean and the other obese, his group transplanted the gut microbiota of these subjects into germ-free mice. The researchers note that "microbiota from the obese co-twin transmitted features of obese humans—increased body fat and metabolic abnormalities—to these mice, while recipients of the lean co-twin's microbiota were leaner and metabolically healthy."[20]

The researchers also noticed that when the mice shared the same cages and bacterial strains (mice typically eat their fecal matter), the microbes associated with the lean phenotype became established in the gut microbiome of the mice that had received the obese gut microbiome. Since the mice were consuming the same amount of food, the researchers concluded that features related to obesity were associated with differences in the microbiome, not the number of calories consumed.[21] This natural microbiome transfer experiment demonstrated that the lean donor's microbiome performed functions missing from the obese donor's microbiome, which seemed to prevent obesity.

With the findings from Gordon's lab, the paradigm of "calories in" equals "energy out" experienced a major shift. Kaplan said the ELEM model, or "Eat Less, Exercise More," is clearly not effective for many people who are overweight or obese.[22] As we learn more about the functions of the gut microbiome, as well as its composition, the exquisite nature of the energy balance influenced by the gut microbiome will become clearer, with the potential of successful therapeutic interventions to treat and prevent obesity.

Obesity from the Inside Out

Kaplan noted that there are changes in the environment that lead to internal signaling changes inside the body that, in turn, cause modifications in the regulatory system that controls body fat.[23] How this circuitry to maintain adequate energy source and balance works and where the "master board" of this circuitry is located have been the focus of intense research.

Eighty percent of the approximately two hundred genetic loci associated with obesity are primarily expressed in the brain. This strongly suggests that the brain may represent the master regulator of the circuitry coordinating the interaction between our metabolic pathway and the external environment, with the ultimate goal of controlling body mass index. This observation adds another layer of complexity to the possible cross-talk between the gut and the brain (gut-brain axis), if the microbiome-dependent obesity outlined earlier is linked to changes in the brain's master regulation of BMI. We already have examples illustrating how other peripheral signals can influence body metabolism through brain circuitry.

While we do not have the complete mapping and function of the hundreds of genetic loci in the brain that control body mass, we do know that one of the most important interfaces controlling this function is represented by the skeletal muscles that elaborate several cytokines communicating with the brain. One of these cytokines (or myokines) is irisin, named for Iris, the Greek goddess of the rainbow and messenger to the gods on Mount Olympus. Irisin, which has been shown to increase the expression of brain-derived neurotrophic factor, favors the expression of a variety of genes controlling muscle regeneration and hypertrophy, leading to the mediation of the beneficial effects following physical exercise.[24]

Irisin is found naturally in muscle cells, and when we exercise, our levels of irisin rise, boosting the amount of energy expended and controlling blood sugar levels. This myokine also has the benefit of helping to convert "bad," calorie-storing white fat, which typically sits around a person's waist, into "good," calorie-burning "brown fat," which is found in babies and young children but largely disappears in adults. Brown fat is particularly beneficial to health because it burns off more excess calories than exercise alone.

This signaling pathway can be an example of how the brain can process environmental changes, such as shifting from an active to a more sedentary lifestyle, into metabolic rearrangements that can lead to obesity through the decreased production of irisin with the subsequent increase in calorie-storing white fat. The bottom line is that obesity is not simply a matter of the accumulation of fat secondary to decreased exercise or increased calorie intake, or a combination of both. As Kaplan tells us, it is a much more complex process that involves brain-periphery communication, which affects overall caloric balance and body mass homeostasis.[25]

This novel concept is at odds with our desire to seek a simple, straightforward solution to obesity. For example, we all know that one way to control BMI is through the food we eat. Our body has various methods to discharge calories. If we change the composition of the calories, we have sensors on the gut mucosa that can adapt to the new caloric composition to maintain homeostasis. For example, those sensors on the gut mucosa can move away from absorbing saturated fat or favoring the metabolism of certain sugars, or SCFAs, or starches, or whatever else is the source of calories.

Nevertheless, another major variable that seems to dictate our metabolic functions is the composition and, most important, the function of the gut microbiota, a factor that is highly influenced by our diet. In other words, our gut microbes eat what we eat. Changes in diet can change microbiome functions even if the microbiome composition remains the same; the microbiome gene expression profile, known as the metatranscriptome, changes if the diet changes.

How these changes can epigenetically influence our metabolism is still a dynamic field of research with many unknown factors. We know that metabolic homeostasis can change in different ways, depending on the host biology. There are changes in the environment, changes in the composition of the diet, and changes in the structure and function of the microbiome, all of which make the interpretation of how we can control BMI much more complex than we previously imagined.

Moving to the Mechanistic Part of the Story

So, what is needed to move microbiome studies related to obesity to a more mechanistic level? Let's revisit Gordon's studies with mouse models that transfer the metabolic trait for obesity. His work suggests that microbiome function has a great deal to do with scavenging calories from the diet and influencing host metabolic pathways that favor obesity.

However, as Kaplan reminds us, most studies of the mouse microbiome are descriptive, and researchers cannot conduct the functional studies in humans that they can perform in germ-free mice. Germ-free studies are not feasible in humans, as any microbiota transfer in human studies occurs in an environment that cannot be germ-free.

Results from mouse studies offer solid science, but the findings do not necessarily translate into parallel discoveries in human biology that result in clinical applications. Let's look at the comparative biology of humans and mice in regard to the homeostasis of BMI. Physiologically, mice and humans are exceedingly similar, and the wiring diagrams in mice that control weight, including fat storage, are indeed very similar to those of humans.

But when we take a closer look, we see significant differences in how wiring diagrams are used. For example, humans have very tightly controlled wiring to maintain blood glucose levels within certain limits. Humans regulate blood glucose levels by transporting sugar in muscle cells, while mice use the liver for gluconeogenesis and glycolysis. Both humans and mice have these circuits, but the difference is related more to which of the two pathways is preferentially used.

Why all this detail? Kaplan noted that if you want to understand wiring diagrams, the mouse is a good model for studying its general functioning. But if you want to develop clinical interventions to affect that wiring, and then look at structural changes, there is very little similarity between mice and humans. "The primary discovery work in the microbiome, working with probiotics, prebiotics, etc., has got to be in the human," he said.[26]

Socioeconomic Considerations Regarding Obesity

Another major difference between animal models and humans is that research in mice does not provide any insight into the socioeconomic factors in obesity. If we want to seriously tackle the challenge of obesity with effective preventive or therapeutic approaches, these factors must be included in the equation. Kaplan reminds us that although lower socioeconomic groups in the United States have higher rates of obesity than higher socioeconomic groups, once again, obesity confounds us with conflicting scenarios around the globe.[27]

Historically and geographically speaking, the relationship between socioeconomic status and obesity fluctuates, with overweight as a sign of wealth in many societies and cultures. In certain Middle Eastern and African societies, obesity or overweight, particularly for women, is associated with wealth and a higher socioeconomic status. *Mbodi*, or "bride fattening," is a rite of passage still practiced in southern Nigeria, where prospective brides are confined in "fattening rooms" for six weeks to gain weight by loading up on carbohydrates and fats and moving as little as possible.

In the northwestern African nation of Mauritania, one of the poorest and least developed countries in the world, the practice of gavage or force-feeding young girls is known as *leblouh*. In 2012, researchers from Tulane University estimated that 20 percent of Mauritanian women reported experiencing the phenomenon, with almost a quarter of women reporting being force-fed as children, often accompanied by beatings and breaking of fingers.[28]

In an article by Nacerdine Ouldzeidoune and colleagues, respondents' attitudes toward the practice are split, with 40 percent of women and 30 percent of men reporting that gavage increases a girl's beauty and 25 percent stating that it increases the family's social standing in the community. Conversely, 40 percent of women and 55 percent of men said it has no advantage, and the most common reason not to practice gavage was improved health. The authors call for relevant intervention and enforcement strategies "to challenge these cultural norms and protect the rights and welfare of children in Mauritania."[29]

If we view obesity across a geographic spectrum, what do obese populations around the world have in common? Some are urban, some are rural, and this distribution can change over time and with location. But one thing that obese populations in different global regions around the world have in common is the modernization of society, said Kaplan. "It's not Westernization—but it's modernization."

He lists seven factors related to our modern lifestyle that are associated with obesity: sleep deprivation, disruption to circadian rhythm, chronic stress, changes in food composition, labor-saving devices, drugs (including antibiotics), and endocrine disruptors.[30] Once we depart from our evolutionary plan as a species, we pay a price—in this case, obesity—that can severely affect both our well-being and our longevity.

Thomas Edison and the Microbiome

It seems to be no coincidence that each and every one of these seven factors can alter the composition and function of the gut microbiome. Some factors, like the overuse of antibiotics, strongly demonstrate the concept of a powerful interconnection among evolutionary biology, lifestyle, microbiome composition and function, epigenetics, and clinical outcomes.

One of the lifestyle factors that Kaplan has examined in relationship to weight gain is sleep deprivation. According to Kaplan, since the invention of the light bulb, humans have experienced a two-hour-per-night decrease in sleep. "In all of humanity, that's huge—20 percent less sleep across humanity." He adds jokingly, "If you really want to blame someone for obesity, blame Thomas Edison."

Each person responds differently to the various lifestyle factors identified by Kaplan, who said we have the "perfect storm" for obesity throughout our modern society. One setting that can exacerbate conditions for obesity is the medical internship. "During residency training, everyone seems to gain or lose weight, and five of those seven lifestyle factors occur."[31]

Kaplan noted that in the U.S. population, stress might be a bigger factor than changes in the food supply. It could be different in another society, and this is all driven by genetics. Obesity can result from a genetic response to those environmental factors, and signals to the brain become disrupted. How are those environmental factors transmitted to the brain?

The gut microbiome is perfectly positioned to transduce these environmental signals to alter the regulatory function of the brain around obesity and metabolic disease. Hence, the microbiome is the mediator or the transducer; it is no longer simply the cause. When we look at the gut microbiome, it is uniquely positioned, physiologically speaking, between the environment and host with the capacity to mediate these signals to the brain. It is pretty clear that it plays a critical role in the interactions of the gut-brain axis, a topic we address more fully in chapter 10. Brain-gut researcher Emeran Mayer and colleagues at the University of California, Los Angeles, have examined the rapidly increasing research in this area. Mayer and his fellow researchers state:

> The past decade has seen a paradigm shift in our understanding of the brain-gut axis. The exponential growth of evidence detailing the bidirectional interactions between the gut microbiome and the brain supports a comprehensive model that integrates the central nervous, gastrointestinal, and immune systems with this newly discovered organ. Data from preclinical and clinical studies have shown remarkable potential for novel treatment targets not only in functional gastrointestinal disorders but in a wide range of psychiatric and neurologic disorders, including Parkinson's disease, autism spectrum disorders, anxiety, and depression, among many others.[32]

Like Mayer, Jeff Gordon equates the significance of the discovery of the human microbiome to the discovery of a new human organ.[33] Kaplan helps us to put this emerging field of research into perspective: "Imagine if we had just discovered the liver last week, and suddenly we had to catch up and learn everything about the liver and its functions. That's how it is with the microbiome. Our understanding of the microbiome is very preliminary."[34]

Instead of treating the microbiome as the cause of obesity, it can be considered the signal transmitter, the amplifier, or "the communicator," said Kaplan. "Everything that comes forward would be to explicate what is the breadth and nature of the regulatory role of those signals. Every environmental factor may influence the microbiome, which is perfectly capable of being the transducer."[35]

In the case of obesity, the data from both mice and humans outlined earlier seem to support this hypothesis. These considerations provide us with potential opportunities to slow down the obesity epidemic but also point out liabilities in the search for the "holy grail" of an obesity cure through the manipulation of the microbiome's composition and function. We know that certain changes in our modern lifestyle, such as adding two more hours of sleep and eliminating chronic stress, are not always feasible. Can mitigating these factors by adding probiotics or prebiotics really fix the problem?

Kaplan takes issue with such a simplistic model. "First and foremost, no one has shown that it works." With bariatric surgery and microbiome transfer, Kaplan has been able to transfer a lean microbiome to a postbariatric mouse and make it lose weight. "We haven't been able to replicate this in human studies, and there are such studies under way at MGH," said Kaplan.[36]

Treating Different Types of Obesity

There are more than one hundred different types of obesity, all of which have different regulatory systems, noted Kaplan.[37] Obesity has all the levels of complexity that cancer has in cell growth and differentiation in cancer pathology, and numerous genes and tumor suppressants come into play to create dozens of different pathways. Different therapies are needed because the mechanism to create cancer as the end result is different in each individual.

A similar personalized approach is needed to treat obesity. When Kaplan treats patients, he seldom sees results by targeting just one factor. "There could be just one mechanism at play in any given patient. For example, some people respond just to exercise or by decreasing stress, but these cases are rare, since typically more than one mechanism is at play." A combination of different therapies is needed for the vast majority of patients with obesity, said Kaplan.[38]

A shift in paradigm is needed from thinking of obesity as a simple metabolic status linked to excessive eating and limited exercise to thinking of it as a multifactorial disease like cancer or autoimmunity. Until then, said Kaplan, we cannot properly develop strategies to effectively treat this condition, whether it is through microbiome manipulation or other strategies already in use. Kaplan noted that "wholesale change is needed, and we need to do it early."[39]

Intervening early means we need to find a mediator, which could be the microbiome, but more data also are needed. "We're years away from a probiotic fix to treat obesity," said Kaplan. "And, furthermore, these probiotics or combination of probiotics would be different for people who need more sleep, or a change in diet, more exercise, etc."[40]

The lesson that the microbiome seems to be teaching us again and again, that there is no "one size fits all" in treatment, is also reflected in obesity. Because of the heterogeneity of this disease, there is no probiotic, or prebiotic, or drug that will treat all types of obesity, just as there is no one drug that will treat all types of cancer.

Big Thinking on Obesity

A theme that recurs throughout this book is that humans are not made equal when it comes to genetics and physiological mechanisms. Identifying and validating biomarkers to stratify the heterogeneity of patients affected by any of the chronic inflammatory diseases linked to the microbiome, including obesity, is the first step we need to move from descriptive to mechanistic studies leading to novel therapeutic interventions, said Kaplan. Developing a strategy based on multiomics—including gene expression, proteomics, metabolomics, and transcriptomics, along with physiological strategies—is essential to try to understand how to stratify the population.

"Without that, we don't have a clue," said Kaplan. If he had unlimited resources, this is where Kaplan would put his funds in the next several decades. "This is what we've been doing in cancer for seventy-five years, and there are lots of different biomarkers to stratify the different kinds of cancers."[41]

His second step would be determining the comprehensive program at work in the body and establishing the model for modulating it. According to Kaplan, there is good evidence to support the model that what the body "cares most about" is how much fat it has. If there is too much fat, disarrangements occur. If the regulatory systems and physiological systems fail, and there is the wrong kind of fat or the wrong placement or too little, the body must have a program to correct these mistakes. For example, if the body senses that there is too little fat, it must have a program to increase it; if it senses that there is too much fat, it needs a program to reduce it.[42]

So, how is that program driven and coordinated? It has to be coordinated: consider factors like the emotions of eating, satiety, thermogenesis, and physical activity. All of them are coordinated in some way. And what is the coordinating program that is needed to store additional fat after a person becomes sick and then recovers from an illness? There must be a coordinated program to get rid of excess calories and another coordinated program to store them in the form of fat deposits.

As Kaplan noted, fluid regulation is coordinated throughout the body; think of the kidneys, the liver, the sweat glands, and the brain. If someone is dehydrated, this program makes them drink more and urinate less. If someone is overhydrated, there's a coordinated program that makes that person sweat more and urinate more. Thinking along parallel lines, what is the program that moves the fat mass up or down? If we understood the molecular physiology of those coordinated genes, proteins, and signals, we would have the tools to test every condition in every individual.

Future Treatment of Obesity

Ten years from now, said Kaplan, we may have twenty drugs to test as therapeutics for obesity. "We can run a trial for six months or a year and take into consideration the placebo effect. We want to see what makes the drug work more effectively, whether it is improved sleep or a different diet. But we cannot spend two years testing different drugs on different diets, i.e., the low-fat, the all-plant, the paleo, etc., along with other environmental factors," said Kaplan.[43] If we have the biomarkers that drive the biology, we can use them to tell us which specific pathway is activated, so that it can be targeted for intervention.

"Then we could run through, in as little as twenty-four hours, every therapy and find the right therapy for the right person. And we want to find out what mitigates obesity along with these drugs: is it better sleep, or diet, or low-fat diet, plant diet, etc.? This is the promise of personalized medicine," said Kaplan, "and it's a very complementary approach. If we understand the biology and the heterogeneity, then we can get to the correct treatment."[44]

Unlocking the role of the microbiome in chronic inflammatory diseases is a primary goal of personalized and, ultimately, preventive medicine. Teasing out both the gut microbiome composition and functional pathways found in autoimmunity is an important piece of the puzzle that we examine next in chapter 9.

9 The Microbiome and Autoimmunity

The Rising Global Epidemic of Autoimmunity

When discussing the potential role of the microbiome in the pathogenesis of noninfectious, chronic inflammatory diseases, autoimmune disorders are definitely the most intriguing and, at the same time, controversial conditions to be linked to this paradigm. Autoimmune diseases are a family of more than eighty chronic, often extremely debilitating conditions that stem from an autoattack by the host immune system against its own organs, tissues, and cells. While many of these diseases are rare, collectively they affect more than twenty million people in the United States and, like many other noninfectious, chronic inflammatory diseases, their prevalence in industrialized countries is on the rise.

Due to an incomplete understanding of their pathogenesis, we currently have no effective cures for autoimmune diseases. Patients face a lifetime of illness and palliative treatments aimed mainly at putting a brake on the immune system in order to ameliorate the inflammatory process. Therefore, besides suffering debilitating symptoms related to the loss of organ function, patients experience side effects related to the therapeutic immune suppression. These side effects can result in a decline in their quality of life, reduced productivity at work, and high medical expenses. Most autoimmune diseases disproportionately afflict women; these disorders are among the leading causes of death for young and middle-aged women.

Autoimmunity is defined as a process in which an adaptive immune response involving both T and B lymphocytes is mounted against an antigen normally present in the body of the host. In the classic view, this process is triggered by either external nonself-antigens (typically from microorganisms) that are similar but not identical to self-antigens (antigen mimicry) or by an insult of the host cells with the subsequent exposure of sequestered antigens (the bystander effect). In both cases, a dysfunctional immune system mounts a permanent, nonreversible immune attack responsible for the autoimmune insult, which no longer depends on exposure to the initial trigger.

As already mentioned in chapter 6, this classic view is at odds with evolutionary biology, which tends to eliminate detrimental genetic traits, and with the fact that genetics cannot explain the recent "epidemics" of autoimmune diseases. While genes related to autoimmunity, particularly those associated with the human leukocyte antigen (HLA) system, are widely present in the general population, only a minor subgroup of the population (approximately 10 percent) develops autoimmune disorders. This suggests that the environment plays a crucial role in the pathogenesis of these conditions, which currently remains largely unknown.

What is indisputable is that the immune system is involved. In many parts of this book, we have already presented growing evidence of the role of the microbiome in shaping the maturation and function of the immune system. Therefore, it is logical to direct our attention to the potential role of the microbiome, with the main focus on the impact of the gut microbiome and its epigenetic influence on genes that control the function of the immune system as it shifts the host's genetic predisposition to clinical outcomes.

It is important to again acknowledge, in the perennial search for the proper balance between tolerance and immune response, that the microbiome also influences epithelial biology. This includes barrier function and sensing capability to perceive the intraluminal environment in the continuous balance between recognition and sensing, transepithelial trafficking, and immune response of the GALT. Anything that perturbs this dynamic triangulation among the gut microbiome, the epithelial barrier, and the immune system may cause an imbalance in the host's immune homeostasis, leading to systemic immune hyperactivation and disease onset.

In autoimmunity, like any other chronic inflammatory disease, this perturbation can be initiated by any environmental trigger that affects microbial balance and ecology, since the microbiome can produce a wide variety of biochemically active compounds. These compounds include neurotransmitters, polyamines, SCFAs, and tryptophan-derived metabolites, which may affect the maturation and activity of the immune system, particularly during the first one thousand days of life (a concept expanded on in other chapters), thereby increasing the risk of autoimmunity in genetically susceptible individuals. Furthermore, there is increasing evidence that changes in the microbiome's composition and function later in life may lead to the alteration of the metabolic network that influences specific pathways associated with the onset of a variety of local or systemic autoimmune diseases.

The link between the microbiome and certain autoimmune diseases has already been discussed (see chapter 7 on IBD and celiac disease). In this chapter, we review the current evidence in the literature linking composition of the microbiome to other autoimmune diseases, keeping in mind that the vast majority of information currently available is based on circumstantial or indirect evidence.

Type 1 Diabetes

T1D is an autoimmune disorder caused by the destruction of the insulin-producing β (beta) cells of the pancreas. Although it is widely accepted that adults can also develop T1D, the highest incidence rate is found in adolescents. Epidemiological studies have estimated the worldwide prevalence of T1D to be less than 1 percent. Recent evidence, however, suggests that its incidence has been increasing at a rate of 3 percent annually. Some large-scale studies, such as The Environmental Determinants of Diabetes in the Young (TEDDY), the Diabetes AutoImmunity Study in the Young (DAISY), and TrialNet, have been designed to identify potential environmental triggers and biomarkers for T1D so that intervention may be possible to delay or even prevent the development of this condition.

The disease's first manifestations develop when a lack of insulin prevents cells from adequate glucose uptake, which is necessary and vital to cell function. Classic symptoms include polyuria, polydipsia, weight loss, fatigue, and hyperglycemia, which, if left untreated, can lead to a coma and ultimately to death. Diagnosis of diabetes includes fasting blood glucose higher than 126 mg/dL (milligrams per deciliter), any blood glucose of 200 mg/dL, or an abnormal oral glucose tolerance test. Since 2009, the American Diabetes Association has modified the guidelines for diabetes diagnosis to include the measurement of glycated hemoglobin levels. This reflects the amount of blood glucose attached to hemoglobin, and it is considered positive if it is higher than 6.5 percent on two occasions.

An important serological component that characterizes T1D and distinguishes it from T2D is the presence of autoantibodies against β-cell autoantigens. Islet cell antibodies (ICA) were the first autoantibodies found to be associated with the development of T1D. In addition to ICA, more than 90 percent of T1D patients have insulin autoantibodies, glutamic acid decarboxylase antibodies, or protein tyrosine phosphatase-like protein. These autoantibodies are also used to identify subjects at high risk of developing the disease, since they are present for months to years before symptom onset and can be detected in serum at as early as six months of age in genetically susceptible individuals.

The exact pathogenesis of T1D is not completely understood, but it is well accepted that both genetic and environmental factors play a role. The genetic locus with the highest association with T1D is, like for celiac disease, the HLA locus, which accounts for about 50 percent of the genetic load. HLA DQ2 and DQ8 loci are the strongest determinants of susceptibility to diabetes, and HLA DR4 and DR3 have been shown to be associated with T1D. The heterozygous DR3/DR4 genotype is associated with the highest risk of disease onset, followed by DR3/DR3 and DR4/DR4 homozygosity.

Other genes that also have been associated with T1D are IL-2 receptor α, cytotoxic T lymphocyte antigen, protein tyrosine phosphatase nonreceptor 22, intercellular adhesion molecule 1, and the insulin gene. In addition to genetic components, it is suggested that environmental factors may play a key role in T1D onset. The incidence of diabetes is increasing faster than can be explained by genetics alone, which is likely due to environmental changes.

Although many possible contributing environmental factors have been identified, to date none has been confirmed as a clear causative agent of T1D. The most frequently proposed candidates are viruses, such as enterovirus, rotavirus, and rubella. In the last few decades, it has been proposed that changes in the microbiome's composition also have been involved in the development of T1D by altering intestinal permeability and subsequently modifying immune system regulation.

The autoimmune attack directed against β cells occurs several years (five years or more) before the clinical presentation of diabetes. Even after the diagnosis of diabetes, there is still significant β-cell function. This progressively declines, so that many patients require full replacement doses of insulin within a few months. Immunological interventions have been directed after diagnosis (new-onset T1D) to prevent decline in residual β-cell function. Antigen-specific therapies (insulin and glutamic acid decarboxylase) have not been successful to date.

Non-antigen-specific therapies, including broad-spectrum, immunosuppressive agents such as cyclosporine, azathioprine, prednisone, and antithymocyte globulin, which deplete or inactivate pathogenic T cells, also have been tried in new-onset T1D. The goal was to prolong clinical remission, but the effect was modest, and these drugs have both short- and long-term toxicity. Consequently, researchers continue to search for treatment modalities that not only preserve residual β-cell function but also halt disease progression or even reverse the disease.

Increased Permeability in BioBreeding Diabetes-Prone Rats

An improved understanding of the complex immunological pathogenesis of T1D has recently led to a paradigm shift suggesting that continuous nonself-antigen exposure fuels the inflammatory process, at least during the initial phase of the autoimmune insult. This finding points to the importance of increased gut permeability and antigen trafficking in the pathogenesis of T1D. We and other investigators have shown that zonulin upregulation is involved in T1D pathogenesis.

The studies we performed in a rodent model that develops T1D spontaneously, BioBreeding diabetes-prone (BBDP) rats, showed an increased permeability of the small intestine (but not of the colon) that preceded the onset of diabetes by at least one month.[1] Furthermore, histological evidence of pancreatic islet destruction was absent at the time of increased permeability, but it was clearly present at a later time. Also, the development of clinical diabetes in patients with T1D is often preceded by intestinal dysfunctions such as increased permeability, immune activation, and ultrastructural abnormalities of the gut epithelium.

Similar to the animal model, we have shown increased intestinal permeability in at-risk individuals with ongoing islet autoimmunity, which precedes the onset of clinical diabetes.[2] We also demonstrated that the altered intestinal permeability that occurs two to three weeks before clinical signs of diabetes in the BBDP rats is zonulin-dependent.[3] Together these studies provided evidence that increased permeability occurs at early stages in T1D pathogenesis, before either histological or overt clinical manifestations of disease take place.

We also have generated data suggesting that continuous antigen trafficking rather than an irreversible break of tolerance fuels the immune process that characterizes T1D, and that inhibiting the zonulin pathway ameliorates the inflammatory process.[4] Gut dysbiosis is the strongest stimulus for upregulation of the zonulin pathway; figure 9.1 postulates an overall hypothesis

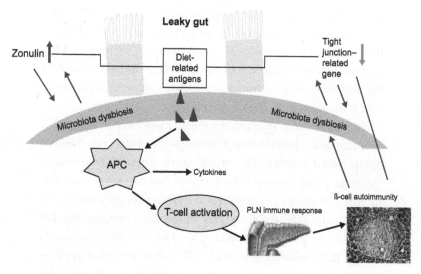

Figure 9.1

Gut dysbiosis causes switching from controlled antigen trafficking to zonulin-dependent increased antigen trafficking that is secondary to changes in tight junction–related gene expression. These changes lead to T-cell activation, production of proinflammatory cytokines, and onset of β-cell autoimmunity. Blocking the zonulin pathway will ameliorate inflammation by blocking antigen trafficking, with subsequent preservation of residual β cells.

linking gut dysbiosis, gut permeability, and immune activation in the onset of T1D, while figure 9.2 shows how gluten may cause activation of the zonulin pathway in T1D.

The causative role of gut dysbiosis through an increase in gut permeability in T1D pathogenesis has been corroborated by a multitude of studies conducted both in animal models and in humans. There is now evidence suggesting that there are shifts in the microbiome before T1D clinical onset; however, exactly how these changes mechanistically lead to T1D has not yet been

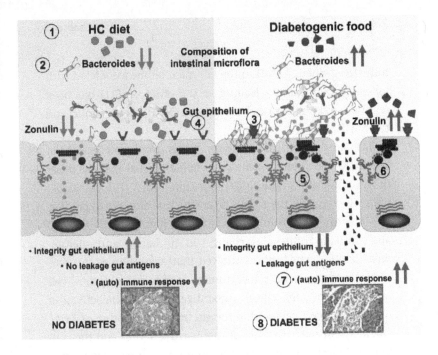

Figure 9.2

Diet affects the composition of the intestinal microflora. A hydrolyzed casein (gluten-free) diet reduces the number of *Bacteroides* species within the microflora, while the diabetogenic (gluten-containing) diet favors a high titer of *Bacteroides* [1]. Colonization of the gut with an imbalanced microflora in which the *Bacteroides* species is favored over other species, such as *Bifidobacterium* and *Lactobacillus*, activates the zonulin pathway [2]. In parallel, gliadin, a component of the grain protein gluten, binds to the chemokine receptor CXCR3 (expressed on intestinal epithelial cells) and induces MyD88-dependent activation of the zonulin pathway [3]. A hydrolyzed casein diet prevents activation of the CXCR3-zonulin pathway [4]. Activation of the zonulin pathway leads to increased zonulin release [5]. The released zonulin binds to zonulin receptors on the surface of the intestinal epithelium and causes disassembly of tight junctions by causing changes in tight-junction dynamics, including phosphorylation of occludin and ZO-1, changes in occludin-ZO-1 and ZO-1-myosin IB protein-protein interaction, and actin polymerization [6]. The disassembly of tight junctions leads to the impairment of the barrier function, leading to increased passage of luminal antigens into the lamina propria, where they are taken up and processed by mucosal antigen-presenting cells and presented to T cells [7]. The cascade of immune events eventually leads to autoimmune disease [8].

fully determined. A shift in the gut microbiome associated with increased permeability leading to the onset of T1D has been reported in both the nonobese diabetic (NOD) mouse model and the BBDP rats.[5]

The most convincing evidence of this paradigm comes from Marika Falcone's group, which has demonstrated that the onset of autoimmunity in the NOD mouse model of T1D is associated with alterations in the mucus layer structure and loss of gut barrier integrity. Confirming once again the highly interconnected and mutually influencing changes in gut barrier and microbiome function, the authors showed that disruption of gut barrier integrity in BDC2.5XNOD mice carrying a transgenic T-cell receptor specific for a β-cell autoantigen leads to the activation of islet-reactive T cells within the gut mucosa and onset of T1D.[6]

The intestinal activation of islet-reactive T cells and the subsequent development of autoimmunity was ameliorated when mice were depleted of endogenous commensal microbiota by antibiotic treatment. In an elegant and mechanistic way, Falcone's group showed that loss of gut barrier continuity can lead to the passage of microbiome-derived molecules, which can then activate islet-specific T cells within the intestinal mucosa, with the subsequent onset of autoimmune diabetes.[7]

In both NOD and BBDP murine models, changes in diet have been shown to facilitate a shift in microbiome composition that affects the onset of T1D.[8] Specifically, our group has shown that BBDP rats fed a gluten-free, hydrolyzed casein (HC) diet showed a reduced incidence of autoimmune diabetes compared to those fed a regular gluten-and-casein-containing chow. In BBDP rats, prediabetic gut permeability was mediated by zonulin and was negatively correlated with the moment of autoimmune diabetes onset. The improved intestinal barrier function that was induced by the HC diet in BBDP rats was associated with decreased serum zonulin levels and barrier function in the small intestine. The HC diet modified ileal mRNA expression of Myosin 9 beta, claudin (Cldn) 1, and Cldn2, key tight-junction proteins involved in barrier function.[9]

These results were further corroborated by another study performed in the same animal model, which showed that the peripheral inflammatory state correlated with autoimmune T1D susceptibility was kept under control by the same HC diet. More specifically, this study showed that BBDP rats fed a standard cereal diet presented a proinflammatory transcriptional expression profile characterized by the presence of the proinflammatory islet transcriptome and cell chemokine expression consistent with microbial antigen exposure. This inflammatory profile was ameliorated by weaning these animals onto an HC diet; it was reignited by introduction of gluten to the HC diet. The microbiome analysis of these animals showed a recurrent motif of previously reported changes that consisted in a shift in the Firmicutes/Bacteroidetes ratio, which was reduced in BBDP rats fed a regular diet and enhanced in those fed an HC diet, secondary to the relative abundance of lactobacilli and butyrate-producing taxa.[10]

Reduced Bacterial Diversity in Human Trials

Similar evidence of changes in the Firmicutes/Bacteroidetes ratio has been reported in humans, with a decreased Firmicutes/Bacteroidetes ratio and shifts in microbiome functions preceding the onset of T1D, which are linked to lifestyle. In a study by Marcus de Goffau and colleagues, at-risk children in their pre-T1D stage (positive islet autoantibodies but still glucose tolerant) showed different bacterial diversity compared to low-risk autoantibody-negative children.[11] Studies performed in the United States and Europe have reported microbiome perturbation mainly characterized by a reduced bacterial diversity prior to the onset of T1D.[12]

Similar results have been reported in China,[13] where the prevalence of T1D has been rising rapidly during the last decade. This supports the recurring idea that radical environmental changes, mainly adopting a Westernized lifestyle, are responsible for these changes. Many of these studies have the limitation of analyzing the gut microbiome at the phylum level, describing changes in Bacteroidetes/Firmicutes mainly driven by an increase in *Bacteroidetes* species and sometimes associated with an increase in *Clostridiaceae*.

The mechanism by which microorganisms belonging to the Bacteroidetes phylum favor the onset of T1D has been the object of several research hypotheses, including their effect on altering the gut mucosal barrier function and affecting the differentiation of immune cells into Tregs, the subgroups responsible for "putting the brakes" on autoimmunity. Less speculative but much more mechanistic are the studies focused on metagenomic or, even better, metaproteomic analysis to link microbiome composition and function to possible pathways involved in T1D pathogenesis. One of these studies has been performed on new-onset T1D, comparing the results for host-derived and microbial-derived proteins with autoantibody-positive subjects (preclinical T1D), low-risk autoantibody-negative at-risk subjects, and healthy controls.[14]

The results showed significant changes in the prevalence of host proteins associated with exocrine pancreas output, inflammation, and gut mucosal barrier function. Intriguingly, when the researchers performed a host-microbiome integrative analysis, they found the depletion of specific microbiome taxa associated with host proteins involved in maintaining the function of the mucous barrier and exocrine pancreas in patients with new-onset T1D. Taken together these results suggest that specific dysfunctions linked to intestinal barrier competency, inflammation, and pancreatic exocrine activity are the result of the interplay between a genetically predisposed host and an unbalanced microbiome, and that these dysfunctions are detectable in the preclinical phase of those individuals who will progress to T1D.

Multiple Sclerosis

Multiple sclerosis (MS) is an autoimmune, neuroinflammatory demyelinating disease of the central nervous system (CNS). The disease, which affects more than 2.5 million young adults worldwide, is the consequence of the inappropriate activation of immune cells targeting the CNS. Clinically, the disease most frequently presents with acute inflammatory flare-ups followed by periods of remission, the so-called relapse-remitting form of MS (RRMS). Over time, due to the progressive accumulation of neurodegeneration, patients go on to develop the secondary progressive form of the disease, in which remission phases become less frequent and then disappear. More rarely, MS presents with a steady increase in disability secondary to a steady progression of the neuroinflammatory process without periods of remission, the so-called primary progressive form (PPMS).

These different clinical presentations, together with novel neuroimaging and neuroimmunological and neuropathological technologies, suggest that MS is not a single clinical entity but rather a spectrum of neurodegenerative disorders. Given its clinical heterogenicity and the fact that, among all autoimmune diseases, MS is the one in which the genetic component seems most modest, environmental factors have emerged as the driving force in its pathogenesis. In the absence of a firm understanding of the mechanisms underlying MS, researchers have suggested a combination of environmental factors are involved in this disease, including age, sex, family history, infections, race, climate, environment, and smoking. Nevertheless, the exact mechanisms responsible for this neurodegenerative disease still remain poorly defined.

The best clues to the key steps involved in MS pathogenesis come from animal models, mainly the experimental autoimmune encephalomyelitis (EAE) animal model, the most widely used and best model for clinical MS. EAE is a well-established mouse model of MS that recapitulates the main immune response at play in MS, namely a CD4+ T cell–mediated autoimmune attack on the myelin protective layer covering the neurons of the CNS.

As noted frequently, the intestinal microbiome plays a pivotal role in shaping host mucosal immune maturation and function. Indeed, mice raised in germ-free conditions show an immuno-compromised phenotype due to lack of the microbial stimulation needed to promote maturation of the GALT, including CD4+ T helper cell differentiation and balance. As a consequence, mice raised in a germ-free environment or treated with antibiotics showed a less aggressive disease.[15]

The mechanistic role of specific microbiome components in causing EAE is supported by the consequences of exposing these animals to specific bacterial species or their bioproducts or both. For example, it has been shown that germ-free mice engrafted with segmental filamentous bacteria, known to trigger IL-17 cytokine production in the intestine, also induced

IL-17A-producing CD4 + T cells (Th17) in the CNS, with the subsequent development of EAE.[16]

Conversely, when orally exposed to purified Polysaccharide A (PSA), a capsular polysaccharide from the gut commensal *Bacteroides fragilis*, mice were protected against the demyelinating process in either a preventive or a therapeutic manner.[17] The protective effect of PSA was secondary to the enhancement of CD103-expressing dendritic cells that accumulated in the cervical lymph nodes. Exposure of naïve dendritic cells to PSA induced the conversion of naïve CD4+ T cells into IL-10-producing FOXP3(+) Treg cells.

Finally, additional mechanistic evidence of the role of the gut microbiome in exerting a neuroprotective effect has been provided by the CD44 knockout mouse model. This model showed a shift from proinflammatory, pathogenic Th17 cells to anti-inflammatory, tolerance-inducing Treg cells, thereby protecting the mice from EAE.[18] This shift was dependent on gut microbiota, as demonstrated by the protective role against EAE of fecal transfer from CD44KO mice to susceptible CD44 wild-type mice.

Studying MS in Humans

The role of the gut microbiome in the pathogenesis of MS generated in murine models also has been explored in human subjects, mainly focusing on microbiome composition linked to gut dysbiosis. Studies performed on Japanese patients affected by MS identified twenty-one species that exhibited significant changes in relative abundance compared to controls, with nineteen of them showing a striking depletion species in MS samples; fourteen of them belonged to Clostridia clusters XIVa and IV.[19] Clostridia clusters XIVa and IV are formed by highly diverse bacterial species, many of which are SCFA producers, including butyrate producers. Since butyrate has been shown to exert an anti-inflammatory effect related to its induction of colonic Tregs, it is conceivable that the depletion of a large subset of clostridial butyrate producers could be a contributing factor in MS pathogenesis.

Using 16S rRNA sequencing, another study showed microbiome alterations in MS characterized by increases in *Methanobrevibacter* and *Akkermansia* and decreases in *Butyricimonas*.[20] These changes correlated with variations in the expression of genes involved in DC maturation, interferon signaling, and NF-kB signaling pathways in circulating T cells and monocytes. MS patients of a second cohort showed elevated breath methane compared with controls, consistent with the observation of increased gut *Methanobrevibacter* in MS patients identified in the first cohort.

Increased *Akkermansia muciniphila* in MS patients was confirmed in another study that also showed that this microorganism induced proinflammatory responses in human peripheral blood mononuclear cells and in monocolonized mice.[21] Interestingly, the same study showed that microbiota transplants from MS patients into germ-free mice resulted in more severe symptoms of experimental autoimmune encephalomyelitis and reduced proportions of IL-10$^+$ Tregs compared with mice humanized with microbiota from healthy controls. These kinds of studies are instrumental in demonstrating that the reported gut dysbiosis plays a role in the pathogenesis of MS; it is not simply a consequence.

To summarize the central theme of findings in both animal models and human patients, the observed impact of the gut microbiome on MS pathophysiology involves both quantitative and functional changes in the composition, metabolism, gut permeability, homeostasis, and modulation of the immune system.

Rheumatoid Arthritis

Rheumatoid arthritis (RA) is an autoimmune disease in which joints are the primary target of the inflammatory process, which can systemically involve other districts, including oral mucosa, the lung, and the gastrointestinal tract. As with many other autoimmune diseases, RA pathogenesis is multifactorial, including recent evidence of the role of the gut microbiome, as demonstrated by the attenuation of the experimental arthritis-induced joint insult in germ-free mice.[22] The potential role of the gut microbiome also seems to be supported by the clinical observation that many RA patients suffer from clinical or subclinical gastrointestinal problems.

As for many other chronic inflammatory disorders, it has been hypothesized that during the preclinical phase of RA, the interplay between the gut microbiome, host factors, and environmental stimuli may lead to increased gut permeability, enhanced antigen trafficking, and an ultimate break of mucosal immune tolerance. The equilibrium between tolerance and immunity to nonself-antigens can be disrupted by this increased intestinal permeability, which in many cases may facilitate absorption of antigens and contribute to the persistence and exacerbation of some immune-mediated diseases, including RA.

Gut Mechanisms Impacting Rheumatoid Arthritis
As other chapters likewise emphasize, two major factors that may influence barrier tightness and barrier functional integrity, with an effect on intestinal permeability regulation, are

diet and intestinal microbiota. Specific to RA, gut dysbiosis has been linked to different mechanisms that may impact the host immune system and its function, including:

1. The activation of antigen-presenting cells (APCs), including dendritic cells, which can influence both cytokine mucosal micromilieu and antigen presentation, thus affecting the host immune response by influencing the differentiation and function of T cells;

2. The ability to promote the citrullination of peptides via the enzymatic action of peptidyl-arginine deiminases (PADs). The intestinal epithelium is a major producer of citrulline in the human body due to both host intestinal and microbial PAD activities. As will be discussed below, it has been suggested that peptide citrullination by the bacterial PAD enzyme expressed by *Porphyromonas gingivalis* strongly contributes to the close association between periodontitis, an inflammatory disease of the oral mucosa, and an increased susceptibility to RA;

3. Antigenic mimicry, which can result from similarities existing between foreign antigens and self-antigens, which then evoke the activation of pathogen-derived autoreactive T and B cells and thus lead to autoimmunity;

4. The impact on zonulin-mediated increased permeability, dysbiosis, and subsequent joint inflammation;[23]

5. The modulation of the host immune system by influencing T-cell differentiation and disruption of the homeostasis between Th17 cells and Treg cells. In RA mouse models, it has been shown that specific alterations in the intestinal microbiota may favor the pathophysiological action of Th17 cells in detriment to the suppressive action of Treg cells, consequently promoting Th17-mediated mucosal inflammation.

Oral Dysbiosis in RA

Among the known components of the gut microbiome, it has been shown that segmental filamentous bacteria could be responsible for the autoimmune arthritis process by inducing intestinal Th17 lymphocytes. Perturbation of the gastrointestinal microbiome also has been reported in patients affected by RA. Recently, there has been an increased interest in the role of the oral microbiome in RA pathogenesis, given the increased incidence of periodontitis reported among patients with RA and the observation that treatment of periodontitis has been shown to reduce the autoimmune arthritic insult.[24]

This comorbidity seems secondary to the oral microbiota dysbiosis characterized by the presence of periodontal pathogens, including *Porphyromonas gingivalis*, which produces enzymes capable of modifying proteins to enhance their antigenicity. This process can be detected before RA onset and has been identified as an etiologic factor in the disease process.[25] Furthermore, RA patients showed a higher bacterial load, a more diverse microbiota, an increase in bacterial species associated with periodontal disease, and increased production of inflammatory mediators including IL-17, IL-2, TNF, and IFN-γ.[26]

Besides oral dysbiosis, several studies have reported perturbation of the gut microbiome in RA characterized by a reduction in gut microbial diversity (related to disease duration and autoantibody level) and expansion of rare lineage intestinal microbes, including the pathobiont *Prevotella copri*, particularly in new-onset RA subjects. The potential pathogenetic role of this microorganism seems secondary to possible molecular mimicry between its epitopes and two host autoantigens, N-acetylglucosamine-6-sulfatase (GNS) and filamin A (FLNA). These two self-antigens, which are recognized by B and T lymphocytes in RA subjects, have been found to be highly expressed in inflamed synovial tissue. Their T-cell epitopes also present homology with *Prevotella, Parabacteroides* species, *Butyricimonas* species, and other gut microbes.[27]

More specifically, immune responses may be triggered by the presence of certain genera/species. *B. fragilis*, for example, can stimulate Th1-mediated immune responses in the initial stages of colonization by producing PSA. *Collinsella intestinalis* may contribute to RA pathogenesis by increasing gut permeability, lowering the expression of TJ proteins, and influencing the epithelial production of IL-17A.

Having shown a documented anti-inflammatory effect in the context of rheumatic diseases, including RA, butyrate-producing microbes such as Clostridia, *Faecalibacterium*, and some species of Lachnospiraceae may also play a crucial role in maintaining the integrity of the intestinal epithelial barrier. Finally, the idea that the onset of autoimmunity may be related to the ecosystem of the gastrointestinal tract is supported not only by the fact that microbiota composition in subjects with RA differs from controls, but also by the observation that an altered microbiome can be partially restored after prescribing disease-modifying, antirheumatic drugs.

Special mention should be made of the impact of specific diets on RA clinical activity as possibly secondary to microbiome modification with an impact on gut mucosal functions, particularly intestinal permeability and related antigen trafficking. While there has been no evidence of a correlation between dietary lifestyle and susceptibility to RA, there are intriguing reports showing how undertaking a Mediterranean diet based on a large consumption of fiber-rich fruits, vegetables, unrefined grains, nuts, fish, and olive oil has a positive impact on RA clinical disease characteristics, including inflammatory activity and physical function and vitality, when compared to a Western diet.[28]

Dietary regimens have a strong impact on microbiome composition and function. For example, it is well known that within the phylum Bacteroidetes, a diet rich in fiber typical of the Mediterranean diet can favor an increase in *Prevotella* genera, while high consumption of a diet rich in both animal fat (typical of the Western diet) and proteins leads to the enrichment of *Bacteroidetes* genera. Additionally, diet can highly influence SCFA production. Foods rich in fiber, such as fruits, vegetables, and legumes, which represent key components of the Mediterranean diet, favor the growth of Firmicutes and Bacteroidetes bacteria that can degrade these foodstuffs, leading to high fecal SCFA content.

This increase in SCFAs, particularly butyrate, has many beneficial outcomes, including favoring the intestinal barrier function. Reduced bacterial endotoxins and antigen trafficking can prevent the activation of effector T cells, thus hampering undesirable local as well as systemic inflammatory responses. The changes in gut microbial ecology and associated immune modulation driven by SCFAs can explain the mechanisms behind the clinical amelioration of RA in individuals following a Mediterranean diet.

Ankylosing Spondylitis

Ankylosing spondylitis (AS) is the prototype of inflammatory rheumatic diseases characterized by a male predominant, late adolescent onset with spinal involvement, characteristic extra-articular manifestations, and postinflammatory new-bone formation. Concerning extra-articular manifestations, it is interesting to point out that up to 10 percent of AS patients are reported to have IBD, and 70 percent show signs indicating subclinical intestinal inflammation. Genome-wide association studies show that more than 10 percent of the gene pathways are shared between IBD and AS.[29]

It has long been suggested that intrinsic barrier dysfunction permits nonspecific innate immune activation with the systemic translocation of bacterial endotoxins. Altered gene expression in AS can predispose some of the tight junctions between intestinal epithelial cells to increased permeability (see below). Several studies have reported specific alterations to the composition of the gut microbiota secondary to the human leucocyte antigen, namely HLA-B27, the genetic predisposition that may be responsible for gut dysbiosis, followed by a leaky gut and the subsequent systemic entrance of microbial antigens and adjuvants, which may act as a trigger for enthesitis. These adjuvants may activate entheseal stromal and immune-resident cell populations leading to the activation of the IL-23/IL-17 axis and the secretion of proinflammatory cytokines resulting in enthesitis, osteitis, and local joint inflammation.

Therefore, the pathogenesis of AS likely involves a complex interplay between genetic predisposition involving the HLA-B27 gene and environmental factors such as mechanical stress and the gut microbiome. Gut dysbiosis has been found in the terminal ileum of patients with AS together with the presence of subclinical gut inflammation. As in many other cross-sectional studies, it remains unclear, however, whether this dysbiosis is a cause or a consequence of the inflammation, and whether dysbiosis modulates immune responses in AS.

To address this issue, a mechanistic study involving fifty AS patients and twenty healthy controls was performed; researchers found adherent and invading rod-shaped bacteria in 70 percent of AS patients, compared to only 25 percent of healthy controls.[30] In AS patients, bacteria were mainly detected within the epithelium and rarely in the context of the lamina propria, while no bacteria were observed in the ileum of non-AS normal controls. Interestingly, in AS patients the bacterial scores significantly correlated with the percentages of infiltrating inflammatory cells. Identification of bacteria from the culture of ileal samples showed that the microorganisms mostly belonged to the Gram-negative bacteria *E. coli* and *Prevotella* species, while *E. coli* were the only Gram-negative species found in healthy controls.

Also of interest, these changes in the AS gut mucosa microbiome were associated with the impairment of both epithelial and endothelial barrier function. Specifically, the gene expression of intestinal tight junction CLDN1, CLDN4, occludin, and *zonula occludens 1* was downregulated in the gut of patients with AS compared with healthy controls. A similar impairment of the endothelial barrier was also detected and characterized by downregulation of *VE-cadherin* and *JAM-A* gene expression in the inflamed ileum of patients with AS compared to controls.[31]

These changes were paralleled by a significant upregulation of zonulin mRNA in the ileal samples of patients with AS that inversely correlated with the expression levels of tight junction proteins Cldn1, Cldn4, occludin, and zonula occludens 1. Zonulin overexpression was also confirmed by immunohistochemistry, with its expression identified both in epithelial cells and in infiltrating mononuclear cells.[32] Interestingly, the number of zonulin[+] cells correlated with the number of IL-8[+] cells.

Also intriguing was the finding that co-culture of Caco-2 intestinal epithelial cells with bacteria isolated from ileal biopsies of five patients with AS induced significant zonulin upregulation. Combined, these data suggest that mucosal dysbiosis causes increased zonulin expression, which leads to both epithelial and endothelial barrier dysfunction and the subsequent passage of endotoxins from the gut lumen into systemic circulation. This is demonstrated by increased levels of LPS, LPS-binding protein, and intestinal fatty-acid binding protein (IFABP) in AS patients compared to controls, with the subsequent onset of chronic inflammation.[33]

A similar mechanism of zonulin-dependent disruption of both epithelial and endothelial compartments has been described in other districts, including the lung during the process of ALI caused by viral infection associated with lung microbiome dysbiosis.[34] This mechanism is particularly pertinent to the recent COVID-19 infection.

The most severe cases are characterized by fluid and neutrophil extravasation within the airways secondary to epithelial and endothelial barrier disruption. This is followed by the disruption of endothelial barriers in other districts and the subsequent onset of vasculitis, "cytokine storm," thrombosis, and the insult of several extrapulmonary organs, including cardiovascular, renal, and cerebral districts. Interestingly enough, these effects were ameliorated by the zonulin inhibitor AT1001 in animal models of ALI,[35] and therefore this molecule is now being considered for possible trial in SARS-CoV-2-infected patients.

Atopic Dermatitis

Atopic dermatitis (AD) is a common chronic inflammatory disease of the skin clinically characterized by itching and xerosis. As in other autoimmune diseases, genetic predisposition and environmental factors, including smaller family size, urban settings, and Western diet, all seem to play a role in its pathogenesis.

Recent evidence points to an increased prevalence of the disease in Western countries, where it affects from 15 to 30 percent of children and 2 to 10 percent of adults.[36]

AD starts in early childhood and is usually the first manifestation of the atopic march that can progress to asthma, allergic rhinitis, and allergic conjunctivitis. AD has a complex pathophysiology that includes a skewed response toward Th2 immunity and defects in the innate immune system. The emergence of filaggrin as a risk allele for atopic disease emphasizes the role of the skin barrier in AD pathogenesis.[37] Filaggrin is a crucial component of the skin barrier and its loss-of-function mutation is related to AD as well as asthma. However, although AD shows signs of skin barrier defect and immunological changes, the mechanism underlying AD is not well understood, and for this reason, its treatment is often very difficult.

Emerging data suggests that AD patients have a disturbed microbial composition and lack microbial diversity in their skin and gut compared to healthy controls, which contributes to disease onset and the atopic march. However, as for many other autoimmune diseases, it is not clear whether microbial changes in AD are an outcome of barrier defect or the cause of barrier dysfunction and inflammation. Nevertheless, the general theme of ongoing cross-talk between microbial commensals and components of the immune system and their alteration affecting the maturation of innate and adaptive immunity during early life applies to AD as well.

Like the gut, under physiological circumstances the skin harbors a countless number of microbial communities that live on the tissue surface as well as the appendages, such as the sweat glands and the hair follicles. Across the skin surface, one million bacteria are found per square centimeter, with over 10^{10} bacterial cells in total. There are geographical and topographical differences in bacterial populations on the skin, depending on the local microenvironment, which includes factors such as temperature, age, amount of sebum, and sweat. Sebaceous sites are full of the lipophilic *Cutibacterium* species, while moisture-prone *Corynebacterium* and *Staphylococcus* species are present in great quantities in moist areas. The fungus *Malassezia* is abundant on the trunk and arms.

Role of the Skin Microbiome in Host Defense

When properly diverse and physiologically established, the human skin flora, possibly the body's most diverse, is instrumental in host defense. The commensal skin microbiome protects humans from pathogens and helps maintain the immune system's delicate balance between effective protection and damaging inflammation. Commensal flora such as *Staphylococcus epidermidis* produce antimicrobial substances that fight off pathogens, whereas *Cuticabacterium acnes* uses skin lipids to make SCFAs that, like in other body districts, may assist in fighting microbial threats. *Cutibacterium* and *Corynebacterium* also reduce *Staphylococcus aureus* by forming porphyrin.

The diversity of healthy skin microbiota is prominently higher in the younger population than in adults and sharply different between the two age groups as shown by beta diversity.[38] While great quantities of *Streptococcus*, *Rothia*, *Gemella*, *Granulicatella*, and *Haemophilus* are present in young children, *Cutibacterium*, *Lactobacillus*, *Anaerococcus*, *Finegoldia*, and *Corynebacterium* are more common in adults. Significant differences in skin microbiota also have been identified between AD children and adults who carry the twenty genera that are prevalent in the healthy population but not present in AD children. This might explain the much higher prevalence of AD among children compared to adults. Microbial shift may potentially contribute to the age-related reduction in AD by suppressing the growth of *S. aureus*.

Adult-associated skin commensals *Cutibacterium* and *Corynebacterium* harbor genes involved in porphyrin metabolism, which theoretically can reduce *S. aureus* infection. In addition, adult skin flora secrete metabolites with antimicrobial properties, which in turn block the growth of *S. aureus*, as shown in research in vitro and in mice. Interestingly, AD patients are reported to carry *S. aureus* on their skin at rates varying from 30 to 100 percent, whereas *S. aureus* is only found in 20 percent of healthy people.[39]

And in patients with established AD, deficiency of the skin-barrier modulator filaggrin (discussed earlier), either through genetics or derived from Th2-dominant conditions, has been shown to cause defects in corneocytes. In patients with AD, *S. aureus* has been found to bind strongly to these corneocytes in a clumping factor B–dependent manner. Filaggrin deficiency in AD is also associated with a higher pH, a condition favorable to *S. aureus* growth.

Furthermore, in AD patients the activity of serine proteases (specifically kallikreins) is increased. Hyperactive kallikreins are known to alter cathelicidin and filaggrin processing and increase the activity of Protease-activated receptor 2 (PAR2), the same receptor activated by the tight junction regulator zonulin.[40] PAR2 activation, in turn, compromises the skin barrier and increases *S. aureus* colonization.

Interestingly, besides skin dysbiosis, AD patients also show an alteration of their gut microbiome composition. Specifically, the *Bifidobacterium* counts in AD patients were significantly lower than in healthy individuals. Furthermore, *Bifidobacterium* count and percentage differed by disease state, such that lower numbers were found in those with severe AD but not in patients with mild atopic symptoms.

Conversely, *Staphylococcus* was found to be more abundant in AD patients than in healthy individuals. Enrichment of *Faecalibacterium prausnitzii* subspecies has been shown to be highly related to AD. The fecal microbiota of AD children with food allergy showed a different composition, characterized by a relatively high abundance of *Bifidobacterium pseudocatenulatum* and *E. coli* and a lower abundance of *Bifidobacterium adolescentis*, *Bifidobacterium breve*, *F. prausnitzii*, and *A. muciniphila*, than in those children with no food allergy.

Finally, there is now evidence suggesting that gut microbiota can improve the skin flora as well. As noted earlier, SCFAs (i.e., propionate, acetate, and butyrate) are end products of dietary fiber fermentation in the gut and are known to play an important role in determining the microbial composition of the skin, which is closely linked with the cutaneous immune defense mechanisms. *Cutibacterium* produces the SCFA acetate and propionic acid in the gut. Propionic acid and its esterified derivatives suppress the growth of methicillin-resistant *S. aureus*.

In the meantime, cutaneous commensals such as *S. epidermidis* and *C. acnes* tolerate wider SCFA shifts than other commensals. Together, these findings suggest that there are mutual interactions between the gut and skin. This could provide the rationale for modifying both the skin and gut microbiomes by applying moisturizers that contain nonpathogenic biomass or probiotic supplementation. This treatment could be applied during the early years of development as a preventive and as a therapeutic option in high-risk groups.

Conclusions and Lessons Learned from Autoimmune Diseases

In the past few decades, the incidence of autoimmune diseases has increased considerably in the Western world, and increasing evidence of a correlation between the presence of microbiome dysbiosis and the development of different autoimmune conditions has been reported, although the precise mechanism remains to be elucidated. Furthermore, due to the nature and the design of studies performed so far, only limited evidence is available on whether these modifications in microbiome composition could be causally related to the pathogenesis of autoimmunity or these alterations could be a consequence of an abnormal immune response.

However, there is now strong evidence that the human microbiome is able to shape the host immune response through mediator and nutrient release, and that dysbiosis of specific human bacterial species has been found to be associated with several different autoimmune conditions. Therefore, based on this evidence, the manipulation of the microbiome could represent a potential therapeutic strategy for the improvement and potentially complete restoration of the normal immune response in different autoimmune diseases, a concept unthinkable until recently, which is extremely intriguing but still not completely accepted within the scientific community.

10   The Microbiome and Neurological and Behavioral Disorders

Following the Gut-Brain Axis

As mentioned in chapter 4, the unexpected results of the Human Genome Project overturned the "one gene, one protein, one disease" paradigm, pioneering new avenues of research in many disorders. In a similar fashion, as we study how trillions of microorganisms in the human body affect our state of health or disease, one of the most tantalizing areas of interest is the role they play in neurological and behavioral disorders. The traditional view of a limited role in the development of chronic brain disease has changed dramatically with our understanding of the human microbiome in general and the cross-communication between the gut and the brain mediated by the gut microbiome in particular.

Even though they appear so different in their structure and function, developmentally speaking, the brain and the gastrointestinal system are interrelated. The ectoderm, one of the three primary germ layers in early embryo development, will lead to the formation of both the brain and some components of the gastrointestinal tract. This common origin, along with their many interconnections, including nerve networks, endocrine signaling, and hormonal interactions, are all testimony to the close, functional network between these two organs.

Therefore, it should be no surprise that this close interaction has been conceptualized in the gut-brain axis paradigm. It is also not unexpected to learn that the gut microbiome may influence the developing brain, as well as the mature brain and some of its functions. Of the many areas in which microbiome research is advancing our knowledge of how disease develops, the microbiome's role in regulating the gut-brain axis and affecting a variety of complex and multifactorial conditions presents researchers with challenging opportunities for treating and preventing these disorders.

That the brain influences gut functions, including motility, intestinal secretion (a classic example is the famous Pavlovian conditioned response stimulating gastric secretion by visual and auditory stimuli), and the release of hormones and neurotransmitters, has been known for a long time. The notion that the gut can influence brain function, however, is much more recent.

In a discussion of this "paradigm shift in neuroscience," a group of leading microbiome researchers addressed provocative and emerging topics surrounding the expanding research into the gut-brain axis. In 2014, Emeran Mayer, Rob Knight, Sarkis Mazmanian, John Cryan, and Kirsten Tillisch asserted that, although research into interactions between the gut and brain has been under way for decades, findings about linkages between the gut-associated immune system, the enteric nervous system, and the gut-based endocrine system have "largely been ignored by the psychiatric and neurological research community."[1]

This group of microbiome research pioneers asks us to consider the revolutionary nature of undertaking massive research into the human microbiome to uncover connections between bacteria and neurological conditions: "To understand the magnitude of this paradigm shift, the reader has to be reminded of the powerful grip of René Descartes's separation of mind/brain on the one side (religion, psychiatry) and body on the other side (medicine) that has dominated Western science and medicine for hundred [sic] of years."[2]

Fast forward five years, and we can safely say that the paradigm has indeed shifted. In the short span since this article appeared in the *Journal of Neuroscience*, published research on the gut microbiome and neurological/behavioral conditions has risen at an exponential rate. Following on from concepts established in early rodent experiments, human clinical trials are also shedding light on the complex interactions of the gut-brain axis, the gut microbiome, and the development of neurodevelopmental and neurodegenerative diseases.

Autism spectrum disorder (ASD), anxiety, epilepsy, depression, and Parkinson's disease are some of the brain disorders that have been linked to altered composition of the gut microbiome in humans.[3] We are still short on mechanistic answers to these interactions but long on speculation about this dynamic neurological universe, which includes ideas such as "psychobiotic" or "melancholic" microbes and the intriguing concept that humans are simply a means of transport for the one hundred trillion microorganisms that we host.[4]

Shaping the Human Brain

As discussed in chapter 3, the early formation of the human microbiome through prenatal, perinatal, and postnatal influences helps to determine the balance between health and disease in adults and throughout the life span. The development of a healthy, functional brain, with concomitant and well-functioning communications across the gut-brain axis, is an essential part of this early human growth.[5] As emerging data suggest the possibility of impaired communication along this axis in the developing pathologies of anxiety, depression, and ASD, broader and deeper research into the function of the gut microbiome in the development of the early human brain is needed.[6]

Neurodevelopment is a very complex process that is highly dependent on a variety of environmental signals, most of which come from the intestine. A typical example is the role of folic acid absorption through the intestine in maturation of the CNS. The exoderm starts as a flat sheet that will fold onto itself to generate a tube, with its top developing into the brain and the bottom into the spinal cord. The fusion of this tube depends on the availability of folic acid, which drives this process.

Folic acid deficiency, secondary to intestinal malfunctions such as chronic inflammation, which can follow from celiac disease, will cause suboptimal serum folic acid levels. This leads to incomplete fusion of the neuronal tube, and depending on the stage and the gravity, can range from malformations not compatible with fetal survival to spina bifida, which is the classic example of incomplete fusion of the neuronal tube.

Besides tube fusion, another key step of neuronal development is the maturation of the blood-brain barrier (BBB), the formation of synapses in between neuronal cells, the maturation of the microglia, and ultimately all the proper circuitry necessary for normal brain development. This extremely complex and still poorly understood process is in part shaped by environmental stimuli, including signals originating from the intestinal microbiome.

There is now growing evidence suggesting that the gut micro-biome and its metabolites participate in the formation of the BBB, and myelination, neurogenesis, and microglia maturation, from the prenatal period to senility. The microbiota can influ-ence neuroinflammation by modulating microglia and astro-cytes. Animal models suggest that the microglia, defined as "the immune system of the brain," remain in an immature status if animals develop under germ-free conditions, but they can be rescued by the administration of SCFAs.[7]

The BBB develops during the early period of intrauterine life and is composed of four key elements, namely microglia, astro-cytic end-feet, pericytes, and capillary endothelial cells. Even though there are four different cellular components, BBB selec-tivity depends on the intercellular tight junctions between endo-thelial cells in CNS vessels, which restrict the passage of solutes. At the interface between blood and brain, endothelial cells are interconnected through these tight junctions. Like the intestinal tight junctions, the BBB tight junctions were once considered static. Now we know that they are highly dynamic structures whose permeability can be modulated by several stimuli, includ-ing zonulin.[8]

The BBB is the gatekeeper regulating the exchange of mol-ecules and nutrients between the systemic circulation and the brain. The gut microbiome seems to impact the permeability of the BBB during gestation, a modulatory function that remains active throughout life and is probably mediated by the zonu-lin pathway. Germ-free mice display increased BBB permeability compared to pathogen-free mice with normal intestinal flora, increasing the risk of neuroinflammation.[9]

Furthermore, microbiome signals have been shown to affect the rate of neurogenesis during cortical development very early in life, with bacterial cell wall components inducing the proliferation of neurons in the frontal cortex.[10] The microbiome is also necessary for correct cortical myelination, which is a critical process in the development of a healthy brain.

The actual mechanisms mediating the effect of the gut microbiome on neurodevelopment are not entirely clear; however, there is evidence that some microbial-derived metabolites may play a role. For example, γ-Aminobutyric acid (GABA), the most important inhibitory neurotransmitter in the mammalian CNS, is produced by *Lactobacillus* and *Bifidobacterium* species. GABA can play a key role in brain development by facilitating postsynaptic communications and favoring the proliferation and development of neural progenitor cells.

Besides being involved in neurodevelopment, the gut microbiome also seems to influence key neurological functions of the mature brain by favoring the elaboration by the mammalian host of specific neurosignaling molecules such as serotonin. Serotonin is elaborated by the enteric nervous system of the gastrointestinal tract and CNS and regulates intestinal motility, mood, memory, learning, appetite, and sleep.

The gut microbiome controls the production of neurotransmitters by stimulating host enterochromaffin cells to produce serotonin or directly elaborate norepinephrine and dopamine, as demonstrated in *Escherichia*, *Bacillus*, and *Saccharomyces* species.[11] SCFAs including acetic, propionic, butyric, and lactic acids are produced by probiotics in the colon. Among the SCFAs, butyrate mainly elaborated by *Butyricicoccus* and *Clostridium* has been shown to induce the sprouting of dendrites, increase the number of synapses, and reinstate learning behavior and access to long-term memories.[12]

Microbiota and Complex Signaling Pathways

To add an additional layer of complexity, there is now growing evidence suggesting that gut microorganisms are connected to the CNS through multiple bidirectional signaling pathways involving neural, endocrine, and immune systems. Laura Cox and Howard Weiner elucidate tasks performed by the microbiota: besides inducing cells in the gastrointestinal tract to secrete neurotransmitters or digestive hormones that affect brain and behavior, as previously discussed, the gut microbiota can modulate immune cells that travel to the CNS and stimulate the vagus nerve with a subsequent effect on behavior. According to Cox and Weiner, "The CNS can control the gut microbiota via adrenergic nerve signaling, primarily affecting intestinal motility and by the influence of neurotransmitters on the immune mediators that shape microbiota composition and function."[13]

Gil Sharon, Sarkis Mazmanian, and colleagues describe the development of a healthy, functioning brain, largely through work done in mouse models, as dependent on "key pre- and post-natal events that integrate environmental cues, such as molecular signals from the gut." They add, "Research over the past few years reveals that the gut microbiome plays a role in basic neurogenerative processes such as the formation of the blood-brain barrier, myelination, neurogenesis, and microglia maturation and also modulates many aspects of animal behavior." In their review in *Cell*, the authors propose pathways "whereby the gut microbiome may contribute to neurodevelopment and neurodegeneration."[14]

They describe some stages in early brain development, including the long-distance migration that occurs during fetal development as cells travel to specific regions and navigate across long distances (sometimes hundreds of cell-body diameters), as the specific circuits that promulgate behavior are built. The molecular navigation of such complex physiological geography on a protracted developmental timeline makes human growth—whether it is in the gut microbiome or the brain—subject to myriad environmental factors, and chief among these is diet.

Unlike SCFAs, which are synthesized by friendly gut bacteria from dietary fiber, our bodies cannot build essential fatty acids (EFAs), which are essential to the proper development and optimum functioning of the human brain. Clinical studies show that an imbalanced dietary intake of fatty acids is related to impaired brain performance and certain diseases.[15] Dietary docosahexaenoic acid is an EFA that is critical to the functional maturation of the retina and visual cortex.

Chia-Yu Chang and colleagues discuss the important role of EFAs in the synthesis and functions of brain neurotransmitters and in the molecules of the immune system: "Neuronal members contain phospholipid pools that are the reservoirs for the synthesis of specific lipid messengers on neuronal stimulation or injury. These messengers in turn participate in signaling cascades that can either promote neuronal injury or neuroprotection."[16] Following the microbial, mechanistic progression in specific signaling pathways could lead to a new understanding of these pathologies and possible novel treatments or even preventive interventions.

There is growing evidence from research conducted in both animal models and humans showing a link between the composition of the gut microbiome and the development and functioning of the immune system. The outcome of this research, particularly through the use of germ-free mouse models, has linked specific microbes or their bioproducts or both to changes in immune functions by triggering or protecting against a variety of chronic inflammatory diseases, including T1D, asthma, and IBD.[17]

With decades of murine research related to neuroinflammation, Sarkis Mazmanian, a medical microbiologist from the California Institute of Technology (Caltech), is an early pioneer in exploring the mechanistic role of the components of the gut microbiome in neurological and behavioral disorders. We turned to him for insights into the role of mouse and human research and thoughts on this shifting paradigm of the gut-brain-microbiome axis in the specific conditions of ASD and Parkinson's disease.

ASD and the Gut Microbiome

ASD is the fastest growing developmental disability in the United States, according to the CDC. The prevalence rate continues to climb from one in five thousand in the mid-1970s to the CDC's current estimate of one in fifty-nine. Boys are disproportionately affected, with a rate of one in twenty-five.

The prevalence rate is also affected by ethnicity and race, with more non-Hispanic white children identified with ASD than non-Hispanic Black and Hispanic children. Children across the autism spectrum are being diagnosed at younger ages, leading to a substantial and growing early childhood group of children being treated for ASD.[18]

Children and adults affected by ASD suffer from a higher number of gastrointestinal disorders than the general population and exhibit altered composition of the gut microbiome as well as increased intestinal permeability.[19] Mayer and colleagues contend that ASD is an exception to the traditional and increasingly outdated view that microorganisms play a limited role in the development and function of the CNS. They describe this "brain disease" as having "long been suspected to be related to altered gut microbiota," a concept that has recently been revisited both in rodent models and in human subjects.[20] Teasing out the mechanistic factors that lead to these conditions through the study of the microbiome could lead to novel treatments and possible prevention.

Mazmanian and colleagues are progressing in the hunt for these "mechanisms that mediate the interdependent and complex interactions between the microbiome and animals, as well as their influences on human health."[21] His most recent findings, published in *Cell* in collaboration with Rob Knight and twenty-two other scientists from leading West Coast institutions, include inducing hallmark autistic behaviors in germ-free mice engrafted with gut microbiota from human donors with ASD. This phenotype seems to be secondary to epigenetic changes induced by the microbiome and involving key genes controlling behavior. According to the authors:

> The brains of mice colonized with ASD microbiota display alternative splicing of ASD-relevant genes. Microbiome and metabolome profiles of mice harboring human microbiota predict that specific bacterial taxa and their metabolites modulate ASD behaviors. Treatment of an ASD mouse model with candidate microbial metabolites improves behavioral abnormalities and modulates neuronal excitability in the brain. We propose that the gut microbiota regulates behaviors in mice via production of neuroactive metabolites, suggesting that gut-brain connections contribute to the pathophysiology of ASD.[22]

This *Cell* paper is the culmination of decades of work using murine models to show how gut microorganisms might affect neurological and behavioral disorders (figure 10.1).

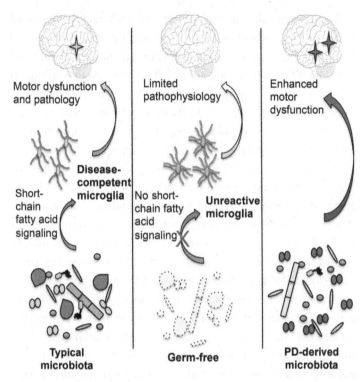

Figure 10.1

Effect of gut microbiota on neurofunctions. Signals from gut microbes are required for neuroinflammatory responses as well as hallmark gastro-intestinal and α-synuclein-dependent motor deficits in a model of Parkinson's disease. Adapted from T. R. Sampson, J. W. Debelius, T. Thron, S. Janssen, G. G. Shastri, Z. E. Ilhan, C. Challis, et al., "Gut Microbiota Regulate Motor Deficits and Neuroinflammation in a Model of Parkinson's Disease," *Cell* 167, no. 6 (December 1, 2016): 1469–1480, https://doi.org/10.1016/j.cell.2016.11.018.

Surprisingly, Mazmanian did not start with data that emerged from the lab, but with a hunch. "We and other researchers had been working on the relationship between the gut microbiome and IBD. Many others, including Jeff Gordon, were working on interactions with metabolic systems. There's a lot of neuroscience research at Caltech, and I thought, why not study the nervous system? Why would the nervous system be 'immune' from education and regulation by the microbiome?"[23]

As sometimes happens with an imaginative and curious scientist like Mazmanian, he found an interesting clue to his future research by investigating old research literature. He found evidence from electron micrograph studies showing that mice and humans share a robust enteric nervous network, with myriad connections between the nervous system and other key components of the gut mucosa, including immune and epithelial cells. One 1967 image of a neuron synapsing with a white blood cell triggered an "aha" moment for Mazmanian. He thought about how, as part of our metazoan evolution, animals evolved in a microbial world. Microbes existed two billion years before eukaryotes, and they inhabit all terrestrial, aquatic, and biological ecosystems. This led to his question, "Why wouldn't they [microbes] have a role in shaping the nervous system?"[24]

Mazmanian noted that when he began his research into whether the gut microbiome could influence key brain functions, there was little to no rigorous and reproduced evidence for the biological mechanisms linking gut-brain connections to behavior, but he and other researchers have since found "similar and different connections between the microbiome and the nervous system." Just as coauthor Alessio Fasano began by studying *V. cholerae* and ended up in autoimmune research, Mazmanian started looking at how microbes affect enteric infections. "My original intention was to understand the biological basis for how gut bacteria affected the course of infections in the gut," he said. "I never published a single paper on that topic."[25]

Instead, while he was studying the immune profile of germ-free mice, he found a proportional decrease in CD4 positive immune cells in the spleen, which led him to go beyond thinking about microbe-immune interactions in terms of enteric infections. "I started thinking about how microbes educate the immune system and are involved in immunological development. We started to see if we could identify an organism and then a molecule that recapitulates this phenotype of immune development, and the story went on from there."[26]

Learning from *Bacteroides fragilis*

During his years of training at Harvard, Mazmanian was working in Dennis Kasper's lab on one specific microbiome species, which would turn out to be instrumental in shedding light on how the gut communicates with the brain. Kasper and Mazmanian worked on *B. fragilis* and its influence on the immune system through its sugar component, PSA (polysaccharide A). This led to the foundation for Mazmanian's future work on how *B. fragilis* may affect behavioral functions in the host.

His team found that oral administration of *B. fragilis* in mice that were offspring of mothers exposed to the viral mimetic polyinosinc-polycitidylicacid (polyI:C) during key developmental time periods led to an ASD phenotype. Taking this research a step further, Mazmanian used this ASD model to explore the role of *B. fragilis* in ameliorating inflammation.

Specifically, he noted that offspring born from pregnant mice exposed to polyI:C during key neurodevelopmental time periods showed increased gut permeability, alteration of microbiome composition and function, changes in metabolomic profiles, and ultimately an ASD-like behavior characterized by defects in communication, stereotypical ASD behavior, and signs of anxiety as well as sensorimotor changes. Interestingly enough, when these mice were orally treated with B. fragilis, both intestinal and behavioral disturbances were corrected. Furthermore, B. fragilis was also capable of decreasing the level of several metabolites that were increased in the ASD-type mice.

"The lesson we learned from B. fragilis is similar to that of microbes that drive the evolution of other environments, whether it's a deep-sea vent or a savannah ecosystem," said Mazmanian. "It's like in the movie, The Wizard of Oz, with the little man behind the curtain who was turning all the knobs—the microbes are controlling everything. They helped build us, and they know how we work. What we want to learn is what the microbes already know, and by doing that I think we can get some insight into human biology and hopefully help people affected by a variety of diseases in which the microbiome may play a role."[27]

What Motivates a Microbe?

As a pediatric gastroenterologist, coauthor Alessio Fasano has developed tremendous respect for microbes from a research and clinical perspective. Mazmanian shares this deep respect but reminds us that bugs are selfish, as they try to shape an optimum environment for their reproduction. "Microbes could have evolved ways to ensure the longevity and robustness of the human body," he said. "They evolve with activities that result in a benefit to the microbe by regulating behaviors to ensure their propagation."[28] Some of these interactions have already been identified, but as Mazmanian noted, there are many, many, many interactions between the gut microbiome and the nervous system of the human that we have yet to discover.

"Imagine a scenario where there's a metabolic web or metabolic network in the gut because one microbe is producing a metabolite and another microbe figures out it can use that metabolite as a nutrient source, and the microbes build their own little ecosystem," said Mazmanian. "Now that you have all those chemicals floating around in the gut, perhaps we also just piggybacked or highjacked those processes, and somehow we adapted to utilize those molecules for our own benefit, whether it be an energy source or some sort of biological, immunologic, or neurological function."[29]

Following this argument, it seems clear that to achieve its overall goal, namely propagation of its species, a microbe may aim at establishing a symbiotic relationship with its host, not an antagonistic one. Mazmanian lends an interesting perspective to the evolution of pathogenic microbes. He calls pathogenesis a dead-end event. "Someone is going to get hurt, and it's not a good, long-term strategy for anybody. I believe that the organisms we call pathogens are very early in their coevolution with humans, and they just haven't figured out their symbiotic relationships—they haven't figured out how to coexist with us." Mazmanian adds, "The ones that have coevolved the longest have figured out how to not be harmful, but ultimately how to benefit human health."[30]

We fully share his vision, since we have several examples supporting the notion that pathogenic traits are not necessarily part of the evolutionary plan of microorganisms. While working at the Center for Vaccine Development in Baltimore, coauthor Fasano had the opportunity to experience this concept firsthand while working on enteric pathogens. Let's take the classical example of *E. coli*. The vast majority of *E. coli* strains are in a symbiotic relationship with their human host, and they represent one of the most abundant species in our gut microbiome. But a few *E. coli* species, including enterotoxigenic *E. coli*, entero-aggregative *E. coli*, enterohemorrhagic *E. coli*, enteroinvasive *E. coli*, and enteropathogenic *E. coli*, are pathogens that can cause severe gastrointestinal or systemic disorders—or both.

What makes the difference between these pathogens and the symbiotic *E. coli*? All pathogenic *E. coli* acquired their pathogenic genes later in their evolutionary history, through the infection of phages carrying these genes that have been integrated into the bacterial chromosome or their plasmids, or both, in what are defined as "pathogenic islands." Pushing this concept to a more extreme example, *Shigella*, the deadly pathogen that causes severe morbidity and mortality particularly in children, is nothing more than an *E. coli* that acquired specific pathogenic traits to make it a totally different microorganism, even if it shares more than 95 percent of its genome with *E. coli*.

The acquisition of these pathogenic traits gave these pathogens an edge in competing with other microbiome species in terms of nutrient procurement, finding a privileged nutritional niche within the host, or temporarily escaping the host immune defense. This may appear to be a good strategy in the short term but, as pointed out by Mazmanian, in the long term, harming your own host would be detrimental. The virulence approach will eventually impact a microbe's overall goal of conservation and propagation of the species. Learning how to establish a symbiotic relationship with the host is a much more efficient strategy for long-term survival.

Alzheimer's Disease

Alzheimer's disease is the most common neurodegenerative disease affecting humankind. As with any other chronic inflammatory disease discussed in this book, the number of Alzheimer's disease patients continues to rise; it has doubled from approximately twenty-two million to forty-six million in only twenty-five years.[31] The disease is characterized by the atrophy of the brain's cerebral cortex and loss of neurons and synapses. It manifests clinically with short-term memory loss, mood swings, decline of verbal memory, and loss of motivation, planning, and intellectual coordination skills.

As with other chronic inflammatory diseases, there is a genetic component in Alzheimer's disease, as testified by genome-wide association studies that have identified twenty genetic-risk loci. What is interesting about this finding is that none of these twenty loci are located within coding regions, suggesting a strong epigenetic modulation by environmental factors, pointing to the gut microbiome as a possible trigger.

A small number of studies have been published showing dysbiosis in subjects affected by Alzheimer's disease. Some of these studies performed at the phylum level showed a decreased level of Firmicutes and Actinobacteria and an increase in Bacteroidetes compared to healthy controls. These findings have been replicated in animal models of Alzheimer's disease, showing similar changes in microbiome composition. This dysbiosis triggers neuroinflammation through the elaboration of proinflammatory cytokines causing increased permeability of the BBB activation of immune cells, reactive gliosis (inflammation of the glia), and ultimately, neurodegeneration. Since Bacteroidetes belong

to the Gram-negative family of bacteria, they elaborate endotoxins like LPS that trigger inflammation through the release of
proinflammatory cytokines.

Furthermore, injection of LPS during brain development
increases microglial activity and results in increased levels of proinflammatory cytokines in murine models. Animal studies also
show that *Akkermansia* may affect the levels of cerebral amyloid
beta (Aβ) 42 peptide, a biomarker of neurodegeneration and
senile plaques.[32] As noted earlier, *Akkermansia* has several beneficial effects on gastrointestinal biology. It increases gut barrier
integrity, intestinal remodeling, and control of intestinal absorption capacity. Therefore, the finding of an inverse correlation
between an abundance of *Akkermansia* in the gut microbiome and
the amount of pathogenic Aβ 42 in the brain is intriguing.

Taken together, it is possible to hypothesize that the concomitant decrease of *Akkermansia* (causing loss of intestinal barrier
function) and increase in Bacteroidetes may lead to the enhanced
translocation of LPS from the intestinal lumen into systemic circulation and ultimately the brain, culminating in Alzheimer's
disease pathology in genetically predisposed individuals. This
degeneration seems to be mitigated by the activation of the innate
immune system as a consequence of exposure to LPS, as also
suggested by several risk factors for Alzheimer's disease, including aging, systemic infection, inflammation, obesity, and brain
trauma, all involving activation of the innate immune system.

Parkinson's Disease

In his research into both ASD and the neurodegenerative amyloid
disorder known as Parkinson's disease, Mazmanian uses humanized models in mice. The rapidly rising rate of ASD diagnosed in

children is paralleled at the end of life by an increasing trajectory of neurodegenerative diseases seen in the elderly. In the United States, Alzheimer's disease affects about five million people and Parkinson's disease affects approximately half a million, with fifty thousand new cases diagnosed each year.[33]

In this condition, there is an abnormal aggregation of certain neuronal proteins that results in interference with many cellular functions. The affected neural cells eventually die, leading to classical motor symptoms including bradykinesia, resting tremor, and rigidity. Nonmotor symptoms affect the olfactory, gastrointestinal, cardiovascular, and urogenital systems. Although some symptoms can be treated, there is no way to slow the progression of the disease.

The rapidly growing elderly population in the United States is expected to double by 2050, leading to an increased economic and social burden for families faced with these diseases. Although the exact cause is not known, a prevailing theory is that dopamine decreases when clusters of neurons found in the midbrain die, resulting in the tremors or shaking and difficulties with balance, speech, and coordination associated with Parkinson's disease. Like ASD, Parkinson's disease is a multifactorial disease, and researchers are investigating genetic and environmental risk factors, including diet, exercise, and pesticides.

Bring the gut microbiome and the gut-brain axis into the picture of Parkinson's disease pathology, and a broader picture emerges of the development of this devastating condition. Researchers are connecting the dots involving the gut-brain microbiota axis to show that these interactions are significantly modulated by the gut microbiota via immunological, neuroendocrine, and direct neural mechanisms.[34] Gastrointestinal symptoms, including constipation, dysphagia, and hypersalivation, can precede the pathogenesis of Parkinson's disease by several years or even a decade, which supports the hypothesis that, to paraphrase Hippocrates, Parkinson's disease begins in the gut.

Once again, we find ourselves looking at the leading role that intestinal permeability, inflammation, and cytokine-induced toxicity play in these complex conditions, and more specifically, in the neuronal damage associated with Parkinson's disease. Pro-inflammatory cytokines and T-cell infiltration have been shown in the brain parenchyma of Parkinson's disease patients, and evidence of inflammatory changes has been reported in the enteric nervous system, the vagus nerve and its branches, and glial cells.

The role of inflammation in Parkinson's disease has been strengthened by findings that show the presence of alpha-synuclein deposits in brain biopsies from autopsies of patients with the disease. Research suggests that misfolding of alpha-synuclein in patients with Parkinson's disease has further substantiated the role of inflammation in this disease. It has been suggested that misfolding of alpha-synuclein could begin in the gut and "spread 'prion-like' via the vagus nerve into the lower brainstem and ultimately the midbrain." This concept has become known as the Braak hypothesis.[35]

Similar to ASD, Parkinson's disease presents a compelling research challenge for the role of the gut-brain-microbiome axis. Mazmanian's group tested the hypothesis that gut bacteria regulate motor defects and the development of Parkinson's disease. After a 2016 paper was published in *Cell*, Mazmanian said, "We have discovered for the first time a biological link between the gut microbiome and Parkinson's disease. More generally, this research reveals that a neurodegenerative disease may have its origins in the gut, and not only in the brain as had been previously thought."[36]

In related research in 2016, a Finnish group led by Timo Myöhänen corrected motor symptoms associated with Parkinson's disease in mice. After earlier research established the role that the PREP enzyme plays in increasing the formation of alpha-synuclein aggregates in the brain, the researchers were looking to uncover the importance of the connection between PREP and alpha-synuclein by blocking PREP. In the mouse model, large amounts of alpha-synuclein were produced, leading to misfolded proteins and the motor symptoms associated with Parkinson's disease. Myöhänen's group reported that the PREP blocker eliminated added damage to the motor area and had "cleared the brain of nearly all accumulations of alpha-synuclein in as little as two weeks after treatment."[37]

Realizing the Promise of the Microbiome

As we have emphasized, promising research exists showing that particular components of the microbiome are associated with certain clinical conditions. But we are still unsure which comes first—dysbiosis or the disease—and whether the microbial differences cause disease or are the consequence of the clinical condition, or whether they are unrelated. We agree with Mazmanian and his colleagues, who say that rather than continue to catalog bacterial populations "we must extend this foundational research approach to test the functional and ecological roles that a given microbial population plays, as well as decipher the physiological effects individual bacteria or consortia of bacteria have on their animal hosts."[38]

These are expensive and time-consuming studies, but they are essential to unlocking the full potential and promise of the human microbiome in the treatment and prevention of neurodegenerative and neuropsychiatric disorders, as well as many other diseases. What Mazmanian calls "basic rules" are emerging from the gut-brain-microbiome research using murine models, humanized mouse models, and human clinical trials. One hypothesis is that specific neurological pathways have evolved in response to certain microbial populations, while other pathways are unaffected by this training from the microbiome but are subject only to genomic or environmental clues, and that the nervous system gains information through microbial metabolites directly or through one or more of the immune, metabolic, and endocrine systems.

Finding answers to these complex questions will require extensive collaboration across many disciplines in both basic science and applied medicine. As much promise as mouse models hold, and taking into consideration their limitations outlined in chapter 13, we agreed with Mazmanian when he said, "We need successes in the clinic, and we need successes in people, or else the microbiome community is going to experience a backlash from society, as has been the case in other scientific areas in the past. Unless we have evidence from the clinic that the microbiome modulates, impacts, maybe even ameliorates a disease state or condition, the research is reduced to intellectual curiosities."[39]

Of course, his statement is right on target. It demonstrates that, up to this point in time, the entire field of microbiome research has been focused on positioning the foundations for in-depth study of how the microbial communities colonizing our body interact and establish symbiotic relationships with their host. Now it is time to move to the next level: understanding what can go wrong in this symbiotic relationship and finding possible therapeutic targets to ameliorate inflammation and, therefore, treat diseases. This is our next necessary task, and developing strategies to exploit the gut microbiome to treat neuroinflammatory diseases will be a phenomenal achievement.

11 The Microbiome and Environmental Enteropathy

Dysbiosis in Developing Countries

In this book, we have covered a variety of diseases in which the microbiome may play a pathogenetic role that mainly affects people from Western, industrialized countries. We also have pointed out that the departure from a lifestyle typical of developing countries increases the risk of dysbiosis and, therefore, of the onset of noninfectious, chronic inflammatory diseases. Based on this premise, it could be inferred that microbiome-related diseases almost exclusively affect people living the privileged life of the well-to-do in the Western world.

While this is generally true, it would be dismissive not to pay attention to environmental enteropathy, which is one of the most consequential conditions affecting children living in developing countries. In this condition, dysbiosis is more related to the inappropriate distribution and abundance of microorganisms along the gastrointestinal tract, rather than to the unbalanced ecosystem in the colon, and this uneven distribution seems to play a pivotal pathogenetic role.

Environmental enteropathy/environmental enteric dysfunction (EE/EED) is a chronic disease mainly affecting the proximal intestine. It is characterized by the loss of barrier function, bacterial overgrowth in the small intestine, and low-grade intestinal inflammation leading to small intestinal villous atrophy that in some aspects resembles celiac disease enteropathy. It is also characterized by malabsorption and systemic inflammation in the absence of diarrhea. While the cause or causes of EE/EED are still an object of discussion, there is some agreement with the proposition that the ingestion of fecally contaminated food and water by children from low-income countries plays a role.

The clinical outcome of EE/EED is physical stunting and, most likely, low neurocognitive development as possible consequences of an imbalanced gut-brain axis. Because these conditions do not cause overt symptoms leading to an increased morbidity and mortality of affected children, who in a superficial clinical evaluation may be considered healthy individuals, these clinical traits have not generally been recognized as a priority health issue. To properly appreciate the impact of this condition on the health of entire populations, it should be pointed out that one in every three children in developing areas becomes stunted in their critically formative first two years of life. This is related, at least in part, to increased exposure to environmental pathogens, as the children are often weaned in settings of inadequate water and sanitation.

The potential developmental consequences of EE/EED can be devastating for the full physical and neurocognitive development of one-third of the world's children growing up in impoverished areas. Furthermore, now that there is a much better appreciation of the potential long-term consequences that may affect not just the single individual but entire nations in their journeys to independence and prosperity, it is imperative to consider EE/EED not only as a health issue but also as a social and political problem that needs to be properly addressed by capitalizing on novel information about its pathogenesis, which offers several possible therapeutic targets.

Historical Perspective

The discovery of EE/EED happened by serendipity in the early 1960s, when a series of studies aimed at investigating the causes of symptomatic diarrhea and malabsorption in tropical regions uniformly identified a high prevalence of abnormal intestinal permeability and intestinal histological abnormalities. These included villous blunting and mucosal inflammation in apparently asymptomatic, healthy, and well-nourished adults and children from the same population who were used as control subjects. Studies in Peace Corps volunteers stationed in Thailand and Bangladesh showed that these features were acquired and were similar to those observed in the Indigenous population.

Furthermore, postmortem histological examinations of neonatal intestines also showed that these abnormalities were not present at birth but rather occurred after six months of age. Interestingly, these changes were reversible; Peace Corps volunteers living in India or Pakistan who returned to the United States showed resolution of their histological damage and absorptive impairment, usually within two years after leaving tropical areas. Based on the initial studies, which were exclusively conducted in the tropics, the condition was labeled "tropical enteropathy."[1]

As mentioned previously, this problem of environmental enteropathogens inducing enteric dysfunction has been recognized for a long time in developing areas. Coauthor Alessio Fasano had firsthand experience of this phenomenon more than twenty years ago, even though he was not aware of it at the time. As a member of the University of Maryland Center for Vaccine Development (CVD), he was working with a group trying to solve an apparent dichotomy concerning a live, attenuated, oral cholera vaccine candidate. It was working extremely well in North American volunteers, but it was poorly effective in protecting children in developing countries. This conundrum was not unique to this vaccine candidate, since similar differences had been reported for other oral vaccines, including polio and bovine rotavirus.

Mike Levine, the director of the CVD, called Fasano into his office to ask his opinion as a pediatric gastroenterologist about an attenuated oral cholera vaccine named CVD 103-HgR. It performed extremely well in terms of tolerability and protection when tested on urban populations in Baltimore, Maryland, but offered very little protection against cholera when tested in children from impoverished areas of developing countries. Fasano told him that several possible explanations could be considered, but the most plausible one was that children from developing countries had a different ecosystem in their small intestine, perhaps due to the ingestion of contaminated food and water, which could lead to colonization by a large number of bacteria that might compete with vibrio colonization when ingested as oral vaccine.

Levine paused for a moment before asking if this change in the gut ecosystem had ever been described before and, if so, how it would be diagnosed. Fasano explained that there is a well-defined clinical condition called small intestinal bacterial overgrowth (SIBO) characterized by an excessive presence of bacteria in the small intestine that can be diagnosed by a hydrogen breath test. In this test, the subject orally ingests a sugar probe (glucose or lactulose), and the breath exhalation of hydrogen produced by the fermentation of the sugar by bacteria in the small intestine is a sign of SIBO.

With his typical pragmatic attitude, Levine asked Fasano to buy a hydrogen-breath testing machine and set up an operation in Santiago, where CVD Chile was located, to test schoolchildren already scheduled to receive the vaccine for possible SIBO. Fasano flew to Santiago and enlisted the help of his colleague, the director of CVD Chile, Rosanna Lagos. In a few weeks, they had set up a lab to perform hydrogen breath tests on more than two hundred children scheduled to receive the cholera vaccine before the end of their school year.

Sure enough, it turned out that the subgroup of children who were diagnosed with SIBO showed a decreased vibriocidal seroconversion and, therefore, showed fewer protective antibodies against *V. cholerae*.[2] This provided a possible explanation as to why several live oral vaccines, including polio, rotavirus, and the CVD 103-HgR developed by Fasano and colleagues, could be less immunogenic and, therefore, less protective in developing countries than in industrialized countries.

It took an additional twenty years to appreciate a possible link between SIBO and what is now officially recognized as EE/EED as an immediate causal factor connecting poor sanitation and stunting. According to WHO, approximately 155 million children under the age of five were stunted worldwide in 2016.[3] With slow progress in stunting reduction in many regions, and the realization that a large proportion of stunting is not due merely to insufficient diet or diarrhea alone, the evidence of the potential role of an unbalanced microbiome in its pathogenesis is rising to center-stage attention.

More recently, EE/EED has been associated with the concept of the disruption of gut barrier function, the passage of microorganisms or their bioproducts from the intestinal lumen to the lamina propria (or both), and mucosal inflammation— does this chain of events sound familiar? This ultimately leads to damage to the villous architecture and further impairment of digestive and absorptive functions and barrier disruption, thereby generating a vicious circle of impaired gut functions. A large number of well-designed studies have suggested that an impaired intestinal barrier coupled with local and systemic inflammatory responses that can lead to impaired growth are also linked with poor household environments that likely predispose subjects to enteric infections and EE/EED.

However, specific biomarkers to diagnose EE/EED or, even better, predict its occurrence in at-risk children in order to ameliorate these mechanisms and improve clinical outcomes have not been identified and validated. In collaboration with colleagues from the University of Virginia under the leadership of Richard Guerrant, a pioneer in the study of EE/EED, Fasano lab members have examined potential biomarkers that might associate functional and structural "enteropathy" with malnutrition or with subsequent growth impairment in children enrolled in a malnutrition study at a nutrition clinic serving several impoverished communities in and near Fortaleza, Ceará, in Northeast Brazil.[4]

These studies have identified key, noninvasive biomarkers of intestinal barrier disruption (LPS translocation) and of intestinal and systemic inflammation as potential biomarkers for early detection of children at risk of stunting. Subsequent developments open the possibility of implementing early interventions for enteropathy and its growth and developmental consequences in children in impoverished settings.

More than Caloric Intake

During the same time as our early vaccine studies, several public and private agencies were implementing an ongoing and aggressive campaign to stop hunger and tackle childhood malnutrition. Interestingly enough, while this campaign had a tangible effect in decreasing malnutrition-associated mortality, it had little impact on improving stunted growth in these children.

It appeared that a shortage in calories was not the only element at play in jeopardizing their growth, but rather that some impairment of the intestinal function in charge of digesting and absorbing foodstuffs was also part of the equation. We now know that the enteropathy that characterized EE/EED is responsible for the poor growth, with the caveat that these children living in poor countries are not only at a disadvantage because they do not have enough food to eat, but also because the little nutrition they have available is not optimally utilized due to intestinal damage causing malabsorption.

Therefore, the combination of our observation of SIBO in children living in poor socioeconomic and hygiene conditions and the lack of full growth recovery despite targeted feeding provided some mechanistic understanding of tropical enteropathy and its clinical outcome. As mentioned previously, the term "tropical enteropathy" was adopted because most of the people affected lived in tropical regions whose climate facilitated the environmental conditions for its onset.

However, epidemiological data from 1999 comparing intestinal permeability among asymptomatic volunteers from fourteen different countries showed tropical enteropathy was present in many but not all areas across the tropics. It was absent in countries with high socioeconomic standards such as Singapore and Qatar, a finding that supports the idea that these abnormal changes are dependent on socioeconomic status and not on tropical climate.[5] Therefore, "tropical enteropathy" has been renamed "environmental enteropathy (EE)" or "environmental enteric dysfunction (EED)."

Pathogenesis of EE/EED

Based on the considerations outlined earlier, the hypothesized pathogenesis of EE/EED can be summarized as follows: starting soon after weaning, children living in poor sanitary conditions are continuously exposed to stool-contaminated food and water containing intestinal bacteria, viruses, and parasites. This increased load of ingested microorganisms changes the ecosystem of the small intestine, allowing local colonization causing SIBO, which is defined as a subclinical quantitative abnormality ($> 10^5$ CFU/mL) of bacteria in the upper gastrointestinal tract. SIBO was also shown to be associated with a high relative risk of faltered growth, since the presence of SIBO was negatively correlated with the growth rate from birth to two years of age.[6]

Interestingly, in this study, intestinal permeability and systemic inflammation, which are usually elevated in patients with EE/EED, were not associated with the presence of SIBO, suggesting that SIBO can be an early step leading to the subsequent onset of EE/EED. In line with this hypothesis are the results we generated from our study in collaboration with Guerrant's group from the University of Virginia, as already described.

With the goal of identifying noninvasive biomarkers predictive of EE/EED onset, we assessed fecal, urinary, and systemic biomarkers of functional and structural enteropathy and growth predictors in 375 children from six to twenty-six months of age with varying degrees of malnutrition (stunting or wasting) in Northeast Brazil. Biomarkers that correlated with stunting included plasma IgA anti-LPS and anti-FliC, zonulin, and intestinal fatty acid–binding protein (I-FABP), suggesting early functional gut-barrier impairment; and citrulline, tryptophan, and lower serum amyloid A (SAA), suggesting impaired defenses.[7]

In contrast, subsequent growth was predicted in children with higher fecal myeloperoxidase or alpha 1 anti-trypsin (A1AT), or higher lactulose/mannitol (L/M) values, and plasma LPS, I-FABP, and SAA, showing intestinal barrier disruption and inflammation. Interestingly, biomarkers clustered into markers of (1) functional intestinal barrier disruption and translocation (IgA anti-LPS and anti-FliC, zonulin, and I-FABP); (2) structural intestinal barrier disruption and inflammation (A1AT, L/M, Reg1, and MPO); (3) systemic inflammation (kynurenine, cSD14, SAA, and LBP); and (4) impaired growth (tryptophan and citrulline).[8]

Cluster dendrogram analysis showed that the biomarkers themselves tend to cluster into three main groups. These groups reflect intestinal translocation, intestinal mucosal barrier disruption and inflammation, and systemic inflammatory responses that likely contribute to the potential long-term growth, developmental, and metabolic consequences that have been attributed to repeated intestinal infections and "environmental enteropathy" in early childhood. Those responses that indicate early tight-junction effects (zonulin) cluster with the indicators of recent LPS translocation, while those that indicate intestinal cell and structural barrier disruption (I-FABP, L/M, %L, A1AT, and Reg1) cluster with the marker of intestinal inflammation, MPO.

Finally, markers of systemic acute phase or proinflammatory responses (SAA, kynurenine, K/T, sCD14, and LBP) cluster to indicate the systemic responses to intestinal barrier disruption and inflammation that lead to troubling lasting growth and to developmental and metabolic consequences. Furthermore, multivariate pathway analyses also show that barrier function, intestinal inflammation, and systemic markers are linearly associated with each other as well as with either previous stunting or subsequent growth.[9]

Taken together, these data suggest a temporal EE/EED "march" from the initial ingestion of contaminated water and food with enteric microorganisms to the onset of SIBO with subsequent activation of the zonulin pathway. The zonulin-dependent functional impairment of the gut barrier allows passage of LPS and other endotoxins derived from the abundant microflora from the intestinal lumen into the lamina propria, thereby causing low-grade, local inflammation. Intestinal inflammation causes the shift from functional to structural disruption of the barrier function in a vicious cycle that leads to the enteropathy characteristic of EE/EED and systemic inflammation. As a clinical consequence, affected children experience growth faltering, low efficacy of oral vaccines, and poor neurocognitive development.

Dysbiosis Associated with EE/EED

As mentioned elsewhere, there is a reciprocal influence between intestinal barrier function and microbiome composition and function. EE/EED is no exception to this rule, because additional key elements influencing this interaction, including nutrition and inflammation, are in play in this condition. It should also be reiterated that dysbiosis can be secondary to inappropriate distribution of the microflora along the gastrointestinal tract (SIBO) or changes in the composition of the fecal microbiome.

Small Intestinal Bacterial Overgrowth

While the ideal test for SIBO would be the analysis of duodenal aspirates, its invasive nature and technical complexity make this test impractical. The hydrogen breath test described earlier is much less invasive and cheaper, and for this reason it is favored and widely used to diagnose SIBO in EE/EED studies. In Brazil, three distinct studies used the hydrogen breath test to investigate bacterial overgrowth in school-aged children living in a slum in either the countryside of the state of São Paulo or the city of São Paulo, Brazil.[10] The results were compared with those obtained from the control group, which was comprised of children from a private school in the same municipality.

The breath test using lactulose as a substrate showed that SIBO was more prevalent in the children living in slums (ranging between 30.9 and 61 percent) compared to controls (ranging between 2.1 and 2.4 percent).[11] There were discrepancies among the three studies concerning differences in stunting in children living in slums with and without SIBO. Two studies showed no differences, while a third study showed worse stunting in children with SIBO compared to those without it. However, all three studies showed a significantly lower mean height-for-age and Z-score in the group of children with SIBO living in slums when compared to matched controls living in more affluent areas.

Changes in Fecal Microbiome Composition

Researchers in a different study report that Bangladeshi children and adults living in an impoverished urban environment, where EE/EED is highly prevalent, have a different microbiome composition compared to healthy and affluent children in the United States. Specifically, Bangladeshi children showed enhanced levels of *Prevotella*, *Butyrivibrio*, and *Oscillospira* and depleted levels of *Bacteroides* relative to U.S. children. The Bangladeshi children also showed a more unstable microbiome composition, which fluctuated greatly when assessed on a monthly basis for up to six months.[12]

Additionally, Bangladeshi subjects had a reduced Bacteroidetes/Firmicutes phyla ratio compared to U.S. children, and no enteric pathogens were present in their stools, which mainly harbored commensals. Of all possible elements in play to explain the dysbiosis reported in EE/EED children, continuous exposure to fecally contaminated food and water and malnutrition can both have a big impact in inducing compositional changes in gut microbiota.

A study from Bangladesh showed an intriguing association between fecal microbiota and severe acute malnutrition in monthly assessments. In this study, gut microbiota maturity was measured by "relative microbiota maturity index" and "microbiota-for-age Z-score," which were calculated from the child's fecal microbiota relative to healthy children of similar chronological age. Gut microbiota immaturity correlated with growth stunting as well as malnutrition.[13]

A study from the metropolitan region of São Paulo compared the microbiota of one hundred school-aged children living in slums with that of thirty children living under more favorable socioeconomic conditions recruited from private schools. Children living in slums had a higher number of bacteria, organisms from the Firmicutes and Bacteroidetes phyla, the *Escherichia* and *Lactobacillus* genus, and lower counts of *Salmonella*. A lower prevalence and counts of *C. difficile* were also observed.[14]

Interestingly, a larger number of archaea *Methanobrevibacter smithii* were observed in children living in a slum, accompanied by a larger production of methane in the exhaled air, indicating a different pattern of bacterial metabolism. In children diagnosed with SIBO who were living in an impoverished urban area, the fecal microbiota showed a lower count of Bacteroidetes and Firmicutes and a higher count of bacteria of the *Salmonella* genus.

Considered together, these studies suggest that SIBO is not limited to the abnormal regional intestinal presence of microorganisms, but that qualitative (compositional) changes in microbiota in the small intestine also play an important role in the development of EE/EED. Nevertheless, while there is a mechanistic explanation of how SIBO may lead to EE/EED, to conclude from the findings outlined in this chapter that this condition may also be caused by changes in microbiome composition is an interpretative stretch.

Indeed, in populations where EE/EED is prevalent, people tend to have a diet low in animal fat and protein and high in starch, fibers, and plant polysaccharides. Therefore, they would have a very different microbiome composition compared to subjects embracing a Western diet. These dietary differences result in significant differences in the microbiome. In order to move from the descriptive data mentioned previously to more mechanistically oriented studies that could link gut dysbiosis (SIBO and/or changes in stool microbiome composition and function) to EE/EED, more mechanistic and better-designed prospective studies are necessary.

An example of a well-designed study trying to achieve this goal is the Study of Environmental Enteropathy and Malnutrition (SEEM) in Pakistan. This study aims to establish a cohort of 350 malnourished and 50 well-nourished children in Matiari, Pakistan, from birth to six months. Other SEEM goals are to assemble serum, fecal, and urine samples for assessment as biomarkers of EE/EED; to provide educational and nutritional interventions according to the level of malnutrition of the child; to evaluate the subset of malnourished children who fail to respond to educational and nutritional interventions by upper gastrointestinal endoscopy to identify treatable causes of malnutrition; and to use the upper gastrointestinal endoscopy biopsy specimens for a detailed assessment of histopathology, gene expression, and immune profiling to better characterize the pathophysiology of EE/EED, validate current candidate biomarkers, and discover novel biomarker candidates.[15]

Importantly, this study provides a unique opportunity to examine whether there are identifiable relationships between histologically diagnosed EE/EED and the configuration of the proximal small intestinal and fecal microbiota. Therefore, SEEM has been designed to include two primary substudies: 1) longitudinal analyses of growth in birth cohort members and 2) correlations of multiomic phenotyping with biopsy analysis, including correlating gut microbial community features with features of the duodenal mucosal gene expression profile and immune phenotypes, with the ultimate goals of gaining a better understanding of disease pathogenesis and identifying mechanism-based biomarkers.[16]

To conclude, while recent work has provided important clues to help unravel the complex pathogenesis of EE/EED and suggest possible strategies for controlling this condition, effective diagnostic methods and, therefore, potential therapeutic targets remain elusive. The outcome of the SEEM study and similar ongoing studies can mitigate the growing concern about EE/EED associated with its impact on longitudinal public health issues, such as growth faltering, low efficacy of oral vaccines, and poor neurocognitive development.

12 The Microbiome and Cancer

A New Principle of Cancer Therapy

Of all the possible connections of the role of the microbiome in the pathogenesis of chronic inflammatory diseases, the one that was definitely not on coauthor Alessio Fasano's radar screen when he started to focus his research efforts in this field was a link to tumoral diseases. Cancer is the second leading cause of death in the United States, with approximately 1.7 million new cases diagnosed in 2017.[1] In a healthy person, noncancer cells keep a tight control on cell reproduction. When cancer cells develop, this mechanism is disrupted, and the cells target abnormal metabolic pathways to generate energy.[2]

Nearly two decades ago, Douglas Hanahan and Robert Weinberg proposed six "Hallmarks of Cancer" as an organizing principle for the diversity of neoplastic disease. These biological hallmarks are sustained proliferative signaling, evading growth suppressors, resisting cell death, enabling replicative immortality, inducing angiogenesis, and activating invasion and metastasis.[3] Within that framework, they argued that genome instability and inflammation foster these functions that permit cancer cells to "survive, proliferate and disseminate." In 2011, they added reprogramming of energy metabolism and evading immune destruction.[4]

Since then, the field of cancer biology has experienced some of the most remarkable advances in knowledge, resulting in innovative and successful treatments unimaginable until the recent past. Many cancers that were terminal a few decades ago are now fully treatable with complete remission. People with nontreatable cancers, once they are diagnosed, have life expectancies that continue to climb with treatment breakthroughs.

Seminal discoveries that mechanistically link the loss of control of cell reproduction to specific pathways, which can be targeted to slow down the progression of many tumoral diseases, have been responsible for much of this progress. But the most remarkable discovery in cancer biology has arguably been the appreciation of the role that the immune system plays in the pathogenesis of tumoral diseases. Demonstrating the impact of these discoveries, in 2018 the Nobel Prize in Physiology or Medicine was awarded to Tasuku Honjo and James Allison for the discovery that the body's immune system can be harnessed to attack cancer cells.[5]

Targeting the Immune System

Specifically, Allison's portion of the Nobel Prize was awarded for his studies on a known protein called CTLA-4, which functions as a brake on the immune system. His research team discovered that releasing the brake, and therefore unleashing our immune system to attack tumors, could have therapeutic potential. He further developed this hypothesis into a therapeutic approach for cancer patients based on a new principle that shifted the paradigm of current treatment.

Allison's intuition in targeting this protein was not to put a brake on the immune system, as his colleagues interested in the treatment of autoimmune diseases tried to exploit. Rather, he came up with a 180-degree different approach: he employed antibodies against these CTLA-4 proteins to release the brake on the immune system, which would then be able to attack the highly proliferative cancer cells.

Allison and his team first tried this approach in 1996 with spectacular results.[6] They observed that mice in which they induced cancer were completely cured by treatment with antibodies that block the CTLA-4 protein, disengaging the T-cell brake and unleashing the immune system to attack the cancer cells. Largely unsupported by the pharmaceutical industry, which showed little interest in this technology, Allison and his group continued their studies to develop strategies applicable to human cancer. In 2011, the fifth clinical trial in humans was performed, which showed striking results in patients affected by melanoma, a skin cancer. Allison's team observed the cancer disappear in a manner that had never been seen before in any other therapy with this patient group.[7]

While Allison was developing his line of research, in 1992 Honjo discovered PD1, a protein that also operated as a brake on the immune system, but with a different mechanism of action than CTLA-4. In animal testing, PD1 inhibition also showed a powerful anticancer effect. Like Allison's path, these preclinical studies were validated in 2012 in the first human clinical trials, showing efficacy of PD1 inhibitors in the treatment of patients affected by a variety of cancers.

What was most remarkable about these clinical trials was the observation that even in untreatable metastatic cancer, the researchers saw long-term remission and even a possible cure. Since the publication of these two studies exploiting checkpoint inhibitors to unleash the immune response against cancer, many other studies and subsequent treatments have followed. This has fundamentally changed the destiny of a multitude of patients affected by advanced cancer that was previously considered untreatable.

Overturning Paradigms

Given the importance of the immune cells in cancer biology and the treatment explained earlier, it should be no surprise that the microbiome and its mutually influential relationship with the immune system may also be in play in cancer. The fact that some microbes can induce cancer has been an accepted concept in scientific literature since the early 2000s, with the first evidence provided by *Helicobacter pylori*, a microorganism linked to gastric cancer that causes ulcers.[8] Its discovery is linked to another Nobel Prize in Physiology or Medicine, awarded in 2005 to Barry Marshall and J. Robin Warren.

Early in his career as a physician and clinical microbiologist at Royal Perth Hospital, in Perth, Australia, Marshall linked up with pathologist Warren to study the association of spiral bacteria with gastritis. According to the 2005 Nobel Prize press release, they "made the remarkable and unexpected discovery that inflammation in the stomach (gastritis) as well as ulceration of the stomach or duodenum (peptic ulcer disease) is the result of an infection of the stomach caused by the bacterium *Helicobacter pylori*." The two Australians were lauded by the committee for having "tenacity and a prepared mind (that) challenged prevailing dogmas."[9]

When Warren and Marshall discovered *H. pylori* in 1982, peptic ulcer disease was thought to be caused by stress and other lifestyle factors. Following on from two decades of research, it is now established that spiral bacteria cause the vast majority of duodenal ulcers and gastric ulcers. The discovery of *H. pylori* accelerated the examination of the interactions among chronic infection, inflammation, microbial agents, and cancer.

After this early evidence linking specific microbes to cancer, many other examples followed with a variety of microorganisms. These include toxigenic Gram-negative bacteria *B. fragilis* and *Fusobacterium*; *Cutibacterium acnes* leading to sustained proliferative signaling; and some *E. coli* strains, leading to genome instability and mutation. We also see mounting evidence suggesting that modulating the gut microbiome may affect responses to a variety of cancer therapies. This concept goes hand in hand with the novel concept of personalized medicine in which, as can be seen throughout this book, the microbiome will play a pivotal role.

Checkpoint Inhibitors in Mice

Three pivotal studies from *Science* link the microbiome composition with the effectiveness of the checkpoint inhibitors previously described, further highlighting the connection between the microbiome, the immune system, and cancer. In the first of these three studies, dysbiosis induced by broad-spectrum antibiotics was associated with the failure of PD1 inhibitors to effectively treat cancer in both patients and mice. Moreover, in all three studies, fecal microbiome analysis performed in independent patient populations led to the identification of specific bacterial entities that positively correlate with the clinical response.

When fecal samples from responder and nonresponder humans were transplanted into germ-free mice, even more remarkable results transpired. The mice then received cancer implants and checkpoint inhibitors, and they reproduced the responder and nonresponder phenotypes when they underwent fecal transplant. In other words, mice that were engrafted with the microbiota from nonresponder patients did not respond to the checkpoint inhibitor, while mice that were engrafted with the microbiota from responder patients did.

Among the many similarities of these three independent studies, it is important to report that one specific strain, namely *Akkermansia muciniphila*, was overrepresented in responder patients in two of the three studies. Also, the genus of bacteria to which *Akkermansia* belongs, the *Verrucomicrobiaceae* family, was also enriched among responder patients in all three studies. These findings are encouraging, but nevertheless, we are still lacking in the deep knowledge needed to provide effective, personalized anticancer treatment based on the patient's gut microbiome.

If the vision of the French group headed by breast cancer physician and researcher Laurence Zitvogel comes to fruition, this will be the successful outcome of her current research. A professor of immunology and biology at the University of Paris-Saclay Medical School and scientific director of the Department of Immuno-Oncology at the Gustave Roussy Institute in Villejuif, France, Zitvogel pioneered the concept of immunogenic cell death, a paradigm shift that takes advantage of the immunogenic abilities of apoptosis.

A leader in the advancement of cancer immunology and immunotherapy, she has focused her research on the emerging role of the commensal microbiome in different cancers. We turned to Zitvogel to provide insight into recent findings that support the hypothesis that the gut microbiome acts as an imperative determinant of the immune setpoint of cancer, and what this might mean for future treatments.[10] Through her research, which has provided us with a deeper understanding of the links between the microbiome and cancer, Zitvogel calls for deeper collaborations among various groups to accelerate progress toward what she calls "codified" treatment.[11]

A Pioneer in Oncoimmunology

"Over the past few years, a multidimensional network of functional connections has emerged between the gut microbiome, intestinal physiology, whole-body metabolism, the immune system and neoplastic disease."[12] Zitvogel and her colleague Guido Kroemer provide this excellent description of the broad overview needed to tackle the complex and challenging questions in the collective quest to develop new cancer treatments. The overarching goal of the Zitvogel lab is to develop new tools to demonstrate how specific cancers in diverse body locations are associated with the gut microbiome, and whether the composition and function of the gut microbiome can be linked to the pathogenic onset of cancer.

At the "heart of the matter" of all or most of the current issues related to the microbiome, and perhaps especially in cancer, is the relationship between gut dysbiosis and cancer, or autoimmunity or allergy—whatever the chronic inflammatory condition is that affects the person's health. Zitvogel's research follows on from that seminal premise to address the following specific questions: Does cancer modify the composition of the gut microbiome, and is there a single core stressor that activates the mechanism that eventually causes cancer? Is there a particular type of dysbiosis involved in specific types of cancer? Finally, how do particular therapies affect the composition and function of the gut microbiome in the cancer patient?[13]

Through a project funded by the European Commission as part of its Horizon 2020 initiative, Zitvogel's organization is coordinating a group of nineteen partners from Europe and North America to develop accurate methods to answer these questions (see below). At this point in oncoimmunology research, she said, "Everything is possible. And knowing that there is a link between the diet and the microbiome—if we can unravel that, little by little, we can learn which mechanisms will be the ones activated on the journey of developing cancer."[14] This dedicated scientist embraces a broad vision as she seeks complex answers to the questions of how the gut microbiome acts as the "conductor" of the "musical score" that results in cancer.

We will highlight two aspects of this complex conversation: the link between the gut microbiome and cancer and the role the microbiome plays in the effectiveness of the patient's response to canonical anticancer therapies. The fundamental question is: Why does the response of some patients to anticancer drugs seem to depend on the gut microbiome, and what role does it play in cancer therapy and, in particular, in immunotherapy?

The Big Picture

Before we drill down into some of the complexities of cancer immunotherapy, let's look at the broad picture of cancer rates in the developed world today and what might drive those numbers. A startling statistic shows that although Europe makes up

9 percent of the world's population, its population represents 25 percent of the global cancer burden. According to Jacques Ferlay and colleagues, there were nearly four million new cases of cancer (excluding nonmelanoma skin cancer) and almost two million deaths from cancer in Europe in 2018.[15]

Cancer is the second leading cause of death worldwide, and the most common cancers are female breast, colorectal, lung, and prostate. In the United States, skin cancer cases outnumber all other cancer cases combined.[16] According to Brazilian researchers, basal cell carcinoma, which accounts for 75 percent of primary skin tumors of epithelial origins, has jumped from 20 to 80 percent over the last thirty years.[17]

According to Zitvogel, this increased prevalence in the developed world is most likely due to a combination of factors. Two of those are an increase in cancer diagnoses, due to improved diagnostic capabilities, and an increased negative environmental effect, leading to an accelerated rate of cancer in the Western world. Environmental factors include microenvironmental pollution, or what is called the "exposome," a term created by Christopher Wild in a groundbreaking 2005 paper that explored the role of environmental risk factors in parallel with the human genome.[18] The CDC defines the exposome as "the measure of all the exposures from environmental and occupational sources to an individual from before birth across a life span and how those exposures relate to health."[19]

According to the CDC, "Understanding how exposures from our environment, diet, lifestyle, etc., interact with our own unique characteristics such as genetics, physiology, and epigenetics impact our health is how the 'exposome' will be articulated."[20] This includes exposure to noxious and toxic materials, nutritional deficiencies, and poor habits in the Western diet. In accordance with Zitvogel's point of view, it is surely plausible that the factors that make up the exposome are probably key causative elements linked to the cancer epidemics we are witnessing, especially in the Western world. Examples include exposure to cigarette smoke or asbestos and the connection to increased rates of lung cancer and the rising rate of apical basal carcinoma.

Cells Gone "Haywire"

Cancer is an abnormal situation in which cells lose their inhibitory capability to stop growing. Our healthy cells are programmed to grow until they reach their final maturation. If you become injured and lose a piece of tissue—a wound or the blunting of villi in celiac disease are classic examples—the cell turns on its programming to regenerate lost tissue. We know little about the signaling that turns this machinery on to create new tissue.

Cancer is the inappropriate turning on of this machinery when it is not needed. There are cells that are somehow programmed to reproduce when it is not appropriate; you could say they've gone "haywire." What is this wiring that is telling the cells to reproduce? We know that there are specific signals that have been linked to cancer, that is, specific genetic, epigenetic, and metabolic pathways linked to this response that causes cell proliferation to go out of whack.

What we do understand is that the reproduction of these cells is the balancing act of physiology. On one hand, you can have an activation of specific transduction pathways that are turned "on" all the time (a typical example is the stem cells of the intestine, which regenerates itself every week). Pluripotent stem cells are extremely active all the time; they don't require a specific microenvironment or immunomodulation but are programmed to stay on for their entire life span.

These same stem cells also have the ability to accelerate their reproductive pace if, as in the case of celiac disease, epithelial cells are destroyed more rapidly due to the inflammatory insult. Once this insult is removed (in the case of celiac disease, through elimination of gluten from the diet), the stem cell compartment goes back to its routine business of weekly renewal of the epithelial layer. This plastic capability of the stem cell compartment to accelerate or decelerate its reproductive pace is linked to specific pathways involved in the cell's cycle, including the Wnt and Hedgehog (Hh) signaling pathways.

In studying these pathways, our researchers at the MIBRC found that in patients with active celiac disease (when stem cells are forced to step up to compensate for the accelerated disruption of mature epithelial cells), there is an increase in the number of stem cells and an expansion of the Wnt responding compartment. Further, we observed an alteration in the number and distribution of mesenchymal cells, which is predicted to be part of the intestinal stem cells niche. At the molecular level, we found downregulation of Indian Hh and other components of the Hh pathway. Combined, these changes are a sign of appropriate compensation by stem cells for the inflammatory insult typical of celiac disease.[21]

On the other hand, sometimes in these tissues, because of changes in the microenvironment, these pathways are unnecessarily turned on. This may eventually lead to tumoral transformation. Fighting cancer is most likely just an everyday task for our immune system. According to Alan Tan of the Cancer Treatment Centers of America, it's likely that your immune system regularly fights off cancer or precancer with a mechanism to "filter out a small amount of cancer cells to prevent us from having visible cancer in the body." He noted that the balance can become lost over time with overt senescence, leading to precancer and cancer.[22]

Cancer development results from co-occurrence of the dysregulation of cell intrinsic pathways and escape from immunosurveillance. On one end, cells proliferate when they should not, and on the other end, the immune system cannot get rid of them, with interconnection between intrinsic and extrinsic cues. This is the perfect storm that creates carcinogenesis.

Stratifying Cancer Patients

Other factors to consider as we tease out the various components of cancer and the gut microbiome include the fact that the microbiome is different at different stages of cancer and for different treatment types (chemotherapy versus checkpoint inhibitors). Zitvogel noted that the checkpoint inhibitors used for lung cancer patients can result in more severe side effects, including checkpoint inhibitor pneumonitis in patients with non-small-cell lung cancer.[23] Conversely, the side effects of chemotherapy, associated, for example, with breast cancer treatment, are toxic but not as severe.

For effective treatment, Zitvogel said we must classify these individuals into subsets by the type of cancer, the stage of cancer, the location of the cancer, and what is typically associated with this type of cancer. There are certain cancers, like prostate cancer, that tend to occur in older individuals. Breast cancer commonly occurs in women ages twenty-eight to forty-five, and this illustrates another challenge. What is the role of sexual hormones in linking specific cancers with certain ages and specific locations in the body that affect this link between the human microbiome and cancer?

There is so much to study in the next few years in this field of research, and the main question is: Does the gut microbiome "cause" the cancer, or does the cancer "cause" the dysbiosis in the gut microbiome? This chicken-or-egg question is at the heart of our discussion, whether it concerns obesity, cancer, or any of the other conditions we've examined in part II of this book.

When we examine that question more deeply, certain patients seem to respond to treatment differently depending on the composition of their gut microbiome. It stands to reason that a resilient microbiome would be a more effective tool in fighting cancer and maximizing treatment, but this remains to be proved. Once we answer this question, using the gut microbiome as a potential avenue to treat or even prevent cancer becomes possible.

As the coordinating institution for the Oncobiome project with Horizon 2020, Zitvogel's group is tackling this conundrum. They aim to recruit approximately nine thousand cancer patients from ten European countries to study microbiome signatures and treatment response for four types of cancer as part of the European Union's Horizon 2020 Programme. The European Commission website describes the project's aims:

ONCOBIOME will pursue the following aims: 1/ identify and validate core or cancer-specific Gut OncoMicrobiome Signatures (GOMS) associated with cancer occurrence, prognosis, response to, or progression on, therapy (polychemotherapy, immune checkpoint inhibitors, dendritic cell vaccines) or adverse effects, 2/ decipher the functional relevance of these cancer-associated gut commensal ecosystems in the regulation of host metabolism, immunity and oncogenesis, 3/ integrate these GOMS with other oncology hallmarks (clinics, genomics, immunomics, metabolomics) 4/ design optimal companion tests, based on those integrated signatures to predict cancer occurrence and progression. With high carat interdisciplinary experts, ONCOBIOME expects to validate cancer or therapy-specific Gut OncoMicrobiome Signatures (GOMS) across breast, colorectal, melanoma and lung cancers adjusting for covariates, to unravel the mode of action of these GOMS in innovative platforms, thus lending support to the design of cancer preventive campaigns using well characterized pre- and pro-biotics.[24]

Unraveling Immune Communications

Zitvogel noted that some species in a patient's microbiome make chemotherapy more efficient and some make it less efficient. As we look at these results, can we begin to talk about personalized therapy? Zitvogel prefers the term "codified" for customizing treatment to specific species and exploiting the entire ecosystem of bacteria, viruses, parasites, archaea, and genetic background to predict who will respond more or less favorably to cancer treatment. Zitvogel thinks that eventually people will be classified based on this profiling of the microbial ecosystem.[25]

What is essential, and something that Zitvogel's lab is pursuing, is to uncover the mechanism that connects the gut immune system and the systemic immune system—how do they eventually communicate? The researchers hope to understand how immune cells migrate from the gut to the site of the cancer and how this affects the immunology of cancer. What kind of "language"— with factors like endocrine activity, cytokines, neurokines, signals throughout the nerves, adipose tissue involvement, bacteria, and increased gut permeability—changes the immune cells and is involved in cancer development and behavior?

Advances in microbiome research have quite literally moved us into unexplored territory, or what Zitvogel calls the "dark matter." She is advocating for three things as we move into this uncharted territory: to identify new bacterial strains and put them in repositories with access for researchers; to map the level at which these new strains are functioning, not only locally but at other sites in the body, such as bone marrow, adipose tissue, thymus, or tumor lesions; and to determine the proper growth conditions for these new strains as they optimize their relationships with the other microorganisms. Finally, another imperative is finding which strains are what she calls "happy" and with whom they can cohabit ("co-occurrence networks") to help understand the dialogue between the microbiome and organs and tissues.

Whether it is chemotherapy or immunosuppressants, trying to understand what the medications used in cancer treatments are doing to commensal bacteria in the gut is another important part of the picture. The ecosystem of the cancer site (for example, the colon and the ileum have different immune systems) is also important. The progression of cancer in the small intestine and the colon follows different trajectories. "We need to understand the difference between the ileum and the colon, and we are looking at these immune systems in my lab," said Zitvogel.[26]

When it comes to drilling down into the trajectories of intestinal cancer, animal models cannot fully recapitulate all the features of cancer that are present in the human system. Therefore, we need to develop new models to help understand the dialogue between the organs and tissues and the microbiome and then test potential medications or other treatments for cancer.

At the MIBRC at MGH, we have developed "mini-guts" or intestinal organoids, which replicate the small intestine. Celiac disease and necrotizing enterocolitis (see chapter 9) are two conditions we are studying using the new model. Zitvogel views intestinal organoids as a promising model to study how microbes can circulate through the gut by way of the lamina propria to the cancer site.

The "five pillars" of autoimmunity described in chapter 6 can be applied to cancer pathogenesis: genetics, environmental factors, increased intestinal permeability, immune system dysregulation, and an unbalanced microbiome. Zitvogel agrees that multiomic modeling, mathematical models, mucosal immune and host genomics, metabolomic and proteomic profiling, and looking at environmental factors are the way forward in studying the role of the human microbiome in cancer research.[27] It will require scientists who have been isolated in their research to come together to cross-fertilize ideas and share subsequent findings. In both medicine and biomedical research, for far too long—nearly two centuries!—we have been narrowly focused on the disease; now it is time to return our focus to the individual patient.

Cancer Dysbiosis

As Zitvogel points out, we are starting to see fairly convincing research evidence that people with cancer also have dysbiosis in their gut microbiomes. An immune system primed wrongly, because of gut ecosystem dysfunction, is less effective in fighting cancer—or celiac disease, or HIV, or obesity, and so on. This interconnected system brings back uncontrolled antigen trafficking and is inundated by enemies, which again affect the immune system. The bottom line is that there are two interconnected conditions—cancer and gut dysbiosis. If a person mounts an inappropriate immune response that is unable to fight cancer, the immune response depends on the function of the gut.

But returning to the concept of personalized or "codified" treatment, Zitvogel said there are some people in cancer treatment who do not have a problem with an unbalanced microbiome; they do not suffer from the "gut disease," but only the cancer.[28] Studying these patients is important, as well as working together to try to understand how the loss of barrier function makes our immune system less effective in fighting cancer. Examining this link will give us insights into the possible connection between dysbiosis and the loss of barrier function in cancer pathogenesis.

Is there a "forest behind the trees": a metabolic disorder caused by cancer that is impairing the gut barrier? Determining the priming factor that is driving this biological phenomenon is crucial for targeting individual interventions. But in Zitvogel's words, "We have to not jump on the microbiome as a business opportunity, but we must learn to walk first before we try to run. We have to be very careful in making statements and conclusions. We take this responsibility very, very seriously."[29]

A Patient Revolution

As microbiome research studies and potential interventions for the prevention and treatment of cancer evolve, there is another essential part of the puzzle that Zitvogel emphasized: patients participating in their cure. Almost universal access to current scientific research combined with robust social media platforms has changed the landscape from the 1980s, when the primary care provider and oncologist made most of the decisions for cancer patients.

Now many patients want to be active in their treatment beyond standard medical treatments such as chemotherapy, surgery, radiation, and immunotherapy. They enact lifestyle changes by making dietary modifications, taking prebiotics or probiotics or herbal supplements, participating in acupuncture, taking exercise and yoga classes, and pursuing other adjunctive therapies. Zitvogel rightly sees this patient component as essential to developing new therapies and understanding the "oncobiome."

"We believe that with the participation of the patients, and their general practitioners, we will witness a social movement that will change the way we target future cancer therapy," said Zitvogel. "And without their help, we would go nowhere."[30]

With appreciation for the complexity of the interplay between the human microbiome and its host and the very different clinical outcomes of this interaction, we completely share her assessment that rather than passively assisting with this scientific revolution, patients should play a much more active role in their contributions to bringing microbiome discoveries to therapeutic fruition. To achieve this goal, it is necessary to move microbiome science from its mainly descriptive nature to more mechanistic studies linking microbiome composition and function to specific pathways proven to be involved in the pathogenesis of a variety of chronic inflammatory diseases. Part III of this book considers how to capitalize on the massive amount of information that has been generated about the human microbiome so that we can implement personalized intervention and primary prevention for a variety of conditions in which the pathogenetic role of the microbiome has been hypothesized.

III Manipulating the Microbiome to Maintain Health

13 From Association to Causation: A New Approach to Microbiome Composition and Function in Disease Development

In Search of a "Healthy" Microbiome

If you have had the perseverance to read parts I and II of this book, you might think that the wealth of research on the human microbiome published to date has been able to link dysbiosis to almost every disease affecting humankind, leading to therapeutic interventions never explored before and with practically unlimited applications. Compared to classic drug-based therapies, manipulation of the human microbiome does offer a myriad of advantages, including a more natural and, therefore, more acceptable therapeutic approach to chronic inflammatory disease. It also gives us the ability to tackle the causes of diseases rather than the consequences and the possibility of fixing multiple problems caused by dysbiosis with a single intervention.

However, although research focused on the role of the human microbiome in health and disease is growing at a phenomenal rate, the clinical applicability of these findings is under scrutiny, given the descriptive nature of most of the research literature on the topic. Current studies evaluate the microbiota at different taxonomic levels and at different time points, using different collection and storage approaches at different sites and on different platforms, and with different computational strategies. Exponential growth in this field is challenged by the narrow focus of individual studies, small sample sizes, lack of standardization, and, most important, cross-sectional study design comparing patients affected by any given disease to matched healthy controls. This approach assumes that healthy subjects, by default, harbor a "normal microbiome" considered as an ideal target to maintain health. However, there is no clear understanding of a "normal" microbiome.

To specifically tackle this issue, a group of academic, government, and industry experts focused on the gut microbiome, the most studied human microbiome in relation to human diseases. The North American branch of the International Life Sciences Institute convened the workshop "Can We Begin to Define a Healthy Gut Microbiome through Quantifiable Characteristics?" in Washington, D.C., in December 2018. Their objectives were to (1) develop a collective expert assessment of the state of the evidence on the human gut microbiome and associated human health benefits; (2) see if there was sufficient evidence to establish measurable gut microbiome characteristics that could serve as indicators of health; (3) identify short- and long-term research needs to fully characterize healthy gut microbiome–host relationships, and (4) publish the findings.

Interestingly and perhaps not surprisingly, they reached the following conclusions:

1) mechanistic links of specific changes in gut microbiome structure with function or markers of human health are not yet established; 2) it is not established if dysbiosis is a cause, consequence, or both of changes in human gut epithelial function and disease; 3) microbiome communities are highly individualized, show a high degree of

interindividual variation to perturbation, and tend to be stable over years; 4) the complexity of microbiome-host interactions requires a comprehensive, multidisciplinary research agenda to elucidate relationships between gut microbiome and host health; 5) biomarkers and/or surrogate indicators of host function and pathogenic processes based on the microbiome need to be determined and validated, along with normal ranges, using approaches similar to those used to establish biomarkers and/or surrogate indicators based on host metabolic phenotypes; 6) future studies measuring responses to an exposure or intervention need to combine validated microbiome-related biomarkers and/or surrogate indicators with multiomics characterization of the microbiome; and 7) because static genetic sampling misses important short- and long-term microbiome-related dynamic changes to host health, future studies must be powered to account for inter- and intraindividual variation and should use repeated measures within individuals.[1]

The last point about attempts to measure the dynamic changes in the gut microbiome is particularly important. While it is hard to disagree with all the conclusions listed, is finding the "normal microbiome" a logical and attainable goal? The engraftment, development, and dynamic nature of gut microbiome composition and function is so personalized that defining a "normal" microbiome would be like defining the "normal" length and color of someone's hair.

Let's say that the overall goal is to "select" a gut microbiome that can establish an ideal symbiotic relationship with its human host as a distinct genetic individual in a particular environment with a particular lifestyle. If the ultimate goal is to maintain a physiological profile of our metabolic processes that dictate the balance between health and disease, then we need to accept the notion that each of us has our own "healthy" microbiome. This microbiome might not necessarily be healthy for a different person. This concept poses real challenges in finding specific targets to manipulate the human microbiome in patients affected by a variety of chronic inflammatory diseases with the ultimate goal of reestablishing health.

If we add these challenges to finding an "ideal" microbiome to the exponential increase in literature published on microbiome research in the last decade (based mainly on murine models), should we conclude that we have wasted time and money so far? Absolutely not, since this decade of deep scientific inquiry into the human microbiome is vital if we are to move beyond current approaches and ultimately capitalize on this foundational knowledge. So, what roadmap should we follow to apply the human microbiome to therapeutic and preventive interventions? Let's look at the path we have taken thus far and learn from the experience we've already accumulated.

Animal Studies

Strengths and Limitations of Animal Models Mechanistically Linking the Microbiome to Human Diseases

Murine models have been widely used in biomedical research, due to many similarities with human anatomy, physiology, and genetics and the possibility of generating genetically modified mouse models to perform functional studies. Low management costs, high reproductive rates, and short life cycles are additional advantages of the mouse model, which has been increasingly used to study the role and functioning of the gut microbiota and its association with diseases.

Specific experimental approaches like murine humanized models (transfer of fecal human microbiome into the mouse host) allow functional and mechanistic research on host-microbe interactions, thus helping to assess causality in disease-associated alterations in gut microbiota composition. Manipulations that are instrumental in gaining mechanistic insights into gut microbiota research include modifications of the host genetic background (gene knockouts), of the gut microbiota composition (microbiome inoculation in germ-free or gnotobiotic mice), and of the environment, including dietary interventions, antibiotic treatment, and fecal transplantations.

Thanks to these approaches, we have acquired key information about the pathological mechanisms of several diseases, including the role of gut microbiota in the pathogenesis of IBD, in controlling the energy balance of the host in obesity, and in the cross-talk between the gut and the brain pertinent to neuroinflammatory diseases. Although results from such experiments have yielded important breakthroughs in understanding the dynamic and complex relationship between the gut microbiota and its host, translating such results from murine models to humans remains challenging due to important differences between the human and murine host.

Anatomical Differences in Mouse and Human Intestines

Mice and humans do share some similarity in physiology and anatomical structures, thus providing the rationale for the frequent use of mouse models in biomedical studies. However, if we closely scrutinize the gastrointestinal tract, which harbors most of the host microbiome, there are substantial anatomofunctional differences due to the divergent diets, feeding patterns, body sizes, and metabolic requirements of humans and mice. Specifically, the small-intestine-to-colon length ratio is almost three times higher in humans than in mice and, more important, our small-intestine-to-colon surface ratio is more than twenty times higher compared to mice.

Furthermore, mice have a large cecum to ferment plant materials and produce vitamins K and B, which are reabsorbed through coprophagy, the practice of eating their own feces. In contrast, the human cecum, which absorbs salt and electrolytes and breaks down cellulose from plant matter, is quite small. Combined, these structural differences reflect the dietary need of mice to use the large intestine to extract nutrients and energy from largely indigestible components in their diet, while humans almost exclusively digest and absorb nutrients in the small intestine.

These regional anatomical and functional differences along the gastrointestinal tract in humans and mice, which are related to a greater fermentation capacity of mice, translate into major differences in the diversity and composition of their gut microbial communities in the colon. In the murine host, these microbiome communities are responsible not only for the fermentation of indigestible food components but also for the production of essential vitamins and SCFAs.

Different Composition and Function of Gut Microbiota in Humans and Mice

Even if the two major phyla Bacteroidetes and Firmicutes are equally dominant in both human and mouse hosts, a deeper taxonomic analysis has demonstrated that 85 percent of the bacterial genera found in the gut microbiomes of mice are not detected in the human intestine.[2] We find that a comparative analysis of both microbiome composition and function between humans and mice is hampered by several differences in research techniques.

These include the selection site of the samples used for analysis (stool samples are used for human gut microbiome studies, while cecal contents are usually used in mouse gut microbiome studies) and the sequencing analysis itself. In humans both 16S rRNA and metagenomic sequencing are used, particularly in more recent studies, while in mice it is mostly the 16S rRNA approach. Furthermore, other variables, including diet, lifestyle, and exposure to pathogens, may impact the reported differences in microbiome composition between mice and humans.

Based on the preceding discussion, it is possible to reach the wrong conclusion that murine models have a limited role in linking the host microbiome to clinical outcomes. Despite some limitations, there are indisputable advantages in using murine models to gain insights into microbiome composition and function in disease pathogenesis.

Overcoming Limitations of Mouse Models to Move Microbiome Research from Association to Causation

Microbe–Phenotype Triangulation

A typical example of how the wise use of mouse models can still be relevant to dissect human and microbiome interaction in health and disease comes from researchers Neeraj Surana and Dennis Kasper of Harvard Medical School. One of the major conundrums in this field is moving from association to mechanistic causation. Human microbiome-wide association studies often generate long lists of microorganisms potentially implicated in health or disease, but moving beyond these correlations to address causation remains a challenging proposition. Studying all the identified microbes, or even just the top candidates, in mouse models is a daunting task.

Surana and Kasper attempted to overcome the problem by employing a new method for studying cause-and-effect relationships within the microbiome. They observed that mice colonized with a mouse microbiome were more sensitive to DSS-induced colitis than mice colonized with a human microbiome. So, instead of focusing only on microbes that differed between the two groups, they cohoused the mice to generate hybrid-microbiota animals with intermediate phenotypes. Performing four pairwise comparisons between mice with less or more severe disease led to the identification of only one taxon, *Lachnospiraceae*, which was present in all comparisons.[3]

The researchers managed to isolate a single cultivable *Lachnospriaceae* isolate from the human microbiome, called *Clostridium immunis* (a previously unknown bacterial species from this family), that, when inoculated into colitis-prone mice, protected them against colitis-associated death.[4] Thus, using microbe-phenotype triangulation, the authors were able to move beyond the typical correlative microbiome study and identify causal microbes dictating different phenotypes. Identification of disease-modulating commensals by microbe-phenotype triangulation, particularly single organisms, may open new paradigms relevant to the mechanism of action from the perspective of both the host and the bacterium.

Elucidating the relevant host pathways and cell types involved in disease onset or protection will be critical for patient stratification and targeted interventions involving microbiome manipulation. Having a single organism with a well-defined phenotype will also allow the identification and characterization of bioactive postbiotics that can be exploited for therapeutic use. However, we need to be mindful of the genetic plasticity of bacteria and the much more complex microbial ecosystem including viruses, parasites, and fungi as important members of this large, coevolving ecosystem that lives on and within us. Additional mechanistic study to understand these human-microbe interactions is needed.

Monocolonization

Engraftment of germ-free mice with a single strain (monocolonization) has recently been used to establish the impact of a single human microbiome species on immune system development. This has been shown with specific *Clostridium* and *Bacteroides* strains capable of stimulating Treg differentiation[5] and with their bioproduct, as in the case of PSA from *B. fragilis*, which is capable of exploiting the toll-like receptor 2 pathway to favor intestinal colonization.[6] Using the same monocolonization strategies, a multitude of microbiome strains have been shown to differentially impact various aspects of the immune response.

However, despite their multiple useful aspects, mono-association approaches have many limitations. The presence of a single microorganism does not reflect the complex ecosystem in the gastrointestinal tract, instead allowing expansion of the strain into gut niches not typically colonized by the specific species, and it can lead to an aberrant impact on the host immune response. Additionally, natural microbial communities are characterized by critical interactions with other bacterial, fungal, or viral species that affect the dynamic interaction of a specific strain with the host.

Finally, it has been clearly shown that germ-free mice have inappropriate immune development, making the baseline readout necessary to interpret the impact of a single strain on the immune system extremely challenging. Ultimately, the monocolonization approach does not allow for the establishment of the collective epigenetic effect of complex microbial communities on the host immune system.

Human Microbiota-Associated Mice

To overcome the limitations of the monocolonization experiments previously outlined, human microbiota-associated (HMA) mice (germ-free mice colonized with a complete human microbiome isolated from a single subject) have been used to model the pathogenetic role of complete and complex human microbial communities in disease. One of the first applications of HMA mice aimed to establish the role of the gut microbiome in obesity by transplanting the gut microbiome from twin pairs discordant for obesity into two groups of germ-free mice.[7]

This study showed that mice engrafted with the "lean" microbiota remained lean, whereas recipients of the "obese" microbiota became obese. Following this first proof-of-concept study, many HMA models have been used to establish the pathogenetic role of the microbiota in a variety of diseases, including allergic diseases (asthma), functional disorders (IBD), chronic inflammatory diseases, and neuroinflammation (Parkinson's disease, multiple sclerosis, etc.). This growing body of research increases our scientific confidence in using this combined human-murine approach to link specific microbiome communities to disease pathogenesis.

However, many of these studies have not been reproduced, casting doubts on their validity, and the HMA model suffers from other limitations. Along with the inherent differences between mice and humans in intestinal anatomy, physiology, and gut ecology outlined earlier, the HMA model does not consider the impact of the host, in connection with the microbiome, on disease pathogenesis. Environmental, behavioral, nutritional, and genetic factors are intrinsically different in HMA mice and their human donors.

Since these host factors influence microbial communities in health and disease, they are impossible to replicate in mouse models. Thus, it is not surprising that microbial communities in HMA mice do not perfectly reflect the host-microbiome interactions present in human donors. Some human species do not engraft in the germ-free mouse host.[8] Furthermore, the relative abundance of the various species and strains in the gut microbiome of HMA mice is substantially altered when compared with the gut microbiome of human donors after transplantation; we lose the original dysbiosis profile from the human donor hypothetically responsible for disease onset. Finally, there are species-specific microbiome strains present in the human intestine that do not engraft or do not exert their effect on the immune system in a different host (for example, mice), or both.

Final Thoughts on Murine Models

The advantages of murine models are (1) manipulations that are not feasible in humans to study the causal role of gut microbiota in health and disease; (2) a deep knowledge of and ability to manipulate the mouse genome, which allows for isolating a specific gene or pathway to study complex gut microbiota–host interactions by using genetic knockout, knock-in manipulations; (3) relatively low maintenance costs, a high reproductive rate, and a short life cycle; (4) gut physiology and anatomy comparable to its human counterpart, with both humans and mice being omnivorous mammals; (5) a homogenous genetic background, which can minimize "noise" variables that can affect metabolic pathways related to gut bacteria–host interactions; and (6) good control of other variables that can affect microbiome composition and function, including diet and housing conditions.

Limitations of these models include substantial differences in anatomy, genetics, and physiology that hamper the possibility of fully recapitulating human host-microbiome interactions. Gut microbiota–host interaction is highly host-specific, and therefore, murine data related to this cross-talk may not apply to the human host. Key environmental factors affecting gut microbiome composition and function in play in the human host, including genetic background, mode of delivery (caesarean or vaginal), feeding regimen (breast or bottle), infections, exposure to antibiotics, diet, and social activities, are not present in mice, and therefore, murine microbiota cannot reflect a "real-life" human gut microbiota.

The use of murine models to study host-microbiome interaction can be useful, but only if we do not overinterpret the results obtained and are cautious in extrapolating findings to human clinical applications. Based on these considerations, there has been a surge of interest in exploring alternatives to murine models to link specific microbiome composition and function to disease pathogenesis.

Murine-Independent Approaches to Studying Host-Microbiota Interactions in Human Diseases

Using the Immune System to Identify Microbiome Components Linked to Disease Pathogenesis

The major limitations of murine models suggest that novel approaches are needed to more efficiently establish the causal role of the microbiome in human disease. Given the intimate relationship between the microbiome community and the host immune system, interest has shifted toward immunologically important gut microbial species that probably play a more direct role in human disease pathogenesis. Among other immune responses, deploying secretory immunoglobulin A (SIgA) antibodies in the gut lumen has been shown to coat many components of the gut microbiome, thus playing an important role in protecting mucosal surfaces against pathogens and maintaining homeostasis with commensal microbiota.

Based on this rationale, an approach referred to as SIgA sequencing (SIgA-seq) has been used to focus on microbiome components targeted by mucosal antibody responses to identify immunologically relevant microbes that may play causal roles in disease. The IgA-seq approach involves isolation of IgA-coated bacteria and noncoated bacteria by cell sorting, after which the coated bacteria are subjected to 16S rRNA gene sequencing to determine which specific organisms are coated by IgA. Using this technique, putative disease-driving bacteria were identified in IBD and were used in animal models to show their pathogenetic role in gut inflammation.[9]

However, the "flip of the coin" of the microbiome-SIgA interaction also suggests that free SIgA may have a differential effect on mucosal homeostasis as compared to the SIgA-microbiome complex. In line with this consideration, we investigated the relationship between microbiota-SIgA complexes and host inflammatory epithelial cell responses. We used a multicellular, three-dimensional (3D) organo-typical model of the human intestinal mucosa composed of an intestinal epithelial cell line and primary human lymphocytes/monocytes, endothelial cells, and fibroblasts. We also used human SIgA from human colostrum, and we used a prominent bacterial member of the first colonizers, E. coli, as a surrogate commensal.[10]

We found that free and microbiota-complexed SIgA triggered different epithelial responses. Free SIgA upregulated mucus production, the expression of polymeric immunoglobulin receptor (pIgR), and the secretion of IL-8 and tumor necrosis factor α; microbiota-complexed SIgA mitigated these responses. These results suggest that free and complexed SIgA have different functions as immunoregulatory agents in the gut and that an imbalance between the two may affect gut homeostasis.

Culture-Based Studies of the Gut Microbiota

The most important breakthrough in the field of microbiome studies is indisputably our ability to study its complexity and composition using culture-based techniques. It was previously supposed that the vast majority of gut microbes were unculturable, since only a very few gut microbiota species, which are primarily obligate anaerobes, grow on standard media under aerobic conditions. However, recent high-throughput, anaerobic, microbial culture approaches using next-generation sequencing to classify large numbers of bacterial strains have revealed that a much greater proportion of the human gut microbiota can be captured in monoculture than previously anticipated.

These new approaches have revealed that, contrary to our previous assumption, the vast majority of gut microbiota species are culturable, often under relatively standard anaerobic culture conditions. Specifically, a broader effort to capture increased gut microbial diversity in culture, combined with a meta-analysis of previously isolated species (culturomics), concluded that more than 75 percent of all known species inhabiting the human gut have been captured in monoculture at some point in time.[11] Supporting these findings, an additional large-scale culture study using diverse media has suggested that up to 95 percent of species present at less than 0.1% relative abundance in fecal samples are theoretically culturable.[12]

These studies also revealed major shortfalls of culture-independent, sequencing-based methods, including missing some specific strains identified with culture-based studies and, most important, relatively shallow limits of detection based on sequencing depth. This last limit is particularly worrisome, since many studies have shown that different strains of the same species with distinct, and sometimes even opposite, effects on the host cannot be differentiated by 16S rRNA culture-independent sequencing.[13] Species-level classifications of the microbiota are therefore often insufficient to capture critical functional differences between bacterial strains. For this reason, use of a previously cultured strain to test the role of a particular species in disease is a poor alternative to direct isolation of that species from the gut microbial community of interest. Thus, although culture-independent methods have revolutionized microbiota research, the resurgence of culture-based studies of the microbiota will be critical in developing a complete picture of microbial contributions to human disease.

Multiomic Profiling

The application of recently developed "omics"-based profiling techniques—including shotgun metagenomics, proteomics, epigenomics, glycomics, and metabolomics, as well as the development of new functional profiling approaches such as IgA-seq—have the potential to allow the identification of causal microbiome species mechanistically linked to the onset of a variety of chronic inflammatory diseases. For example, functional profiling–based approaches theoretically can be used to identify specific microorganisms that affect a variety of host physiological pathways regulated by the microbiota, such as the regulation of epithelial permeability or the epigenetic switching of genes controlling the immune response, either directly or through the production of specific bioactive metabolites.

These kinds of studies will lead to a multitude of new hypotheses regarding potentially causal roles for specific microbiota components in disease pathogenesis. However, the large number of postulations generated will inevitably exceed our capacity to test them one by one using currently available experimental models.

One emerging solution to this shortfall is to integrate omics and experimental data by using advanced computational approaches (such as machine learning) to reduce the number of experimental tests necessary to establish robust and predictive models. Integration of various datasets to create more sophisticated algorithmic models may eventually allow for many emerging hypotheses to be tested in silico rather than in vivo or in vitro. Given the magnitude of the task, creating synergistic and self-reinforcing interactions between experimental data and omics data is critical in developing a complete picture of the causal role of the microbiota in human health and disease. We will review these computational approaches in depth in chapter 17.

Human Gut Organoid Model

The intestinal micromilieu is instrumental in controlling mucosal homeostasis through complex bidirectional interactions with the epithelial cells covering the gastrointestinal tract. The gut microenvironment includes 3D tissue architecture, multiple cell types, extracellular matrix, innate immunity mediators, and, most important, indigenous colonizing microbiota.

The intestinal mucosal epithelium contains a multitude of specialized epithelial and immune cells. These cells coordinate their functions to protect against infection or exposure to other noxious environmental elements by serving as a barrier against luminal toxins, commensals, and pathogens. They also sample microbial antigens and recruit innate and adaptive immune effectors.[14]

The relative composition and function of these cells vary by gastrointestinal region and consist of integrated cross-communication networks of different cell types and effectors critical for protection against pathogens. Epithelial cell polarity establishes barrier function, regulates uptake/transport of nutrients, and maintains epithelial architecture, which are all critical features in the interface with the luminal microbiota.

As already discussed, the intestinal tract contains prokaryotes, viruses, archaea, and eukaryotes, some of which protect the host against pathogen colonization by a variety of mechanisms, including epithelial cell turnover, mucin synthesis, and triggering bacterial sensors on host cells. The reciprocal interaction between a specific host and its coevolving microbiota contributes to tissue function and homeostasis, which then determine microbiota composition and function, which, in turn, may affect the host epithelial function.

This constant exposure to a vast amount of complex microbiota highlights the critical interface that this single-cell layer forms between the host and our environment. The well-documented contribution of environmental factors to the functional development of the human intestinal epithelium directly implies epigenetic mechanisms in orchestrating this complex interplay. These mechanisms may lead to the switch from genetic predisposition to clinical outcome related to a variety of chronic inflammatory diseases. The development of intestinal epithelial organoid culture systems generated from human tissue provides researchers with unprecedented opportunities to study functional aspects of host-microbiome interactions by faithfully recreating these interactions to properly model the impact of microbiome composition and function on disease pathogenesis.

We have already exploited this powerful tool in several settings: (1) to study the transcriptional changes triggered by an enteric pathogen (*Salmonella*) in the gut mucosa of the host to evade host immune defense;[15] (2) to study potential evolutionary changes in the microbiome–gut mucosa interaction in the pathogenesis of necrotizing enteropathy;[16] and (3) to study the response to an environmental trigger of inflammation—for example, gluten in celiac disease.[17] With additional knowledge about specific components of the gut microbiome involved in disease pathogenesis, along with the ability to perform cocultures with immune cells, gut organoid technology could play an important role in dissecting the complex host-microbiome interactions instrumental in the health-disease balance.

Human Studies

Limitations of Cross-Sectional Human Study Designs to Link the Microbiome with Disease Pathogenesis

The vast majority of reports on the microbiome's role in the pathogenesis of a variety of human diseases are based on cross-sectional studies, which describe alterations in the host microbiota composition after the disease has become apparent. Because of this, the technical approaches used to study microbiome composition and, more important, function are limited in indisputably supporting a cause-and-effect relationship linking these microbiome shifts mechanistically with disease pathogenesis.

To overcome this limitation, a prospective cohort design is required to capture changes that precede or coincide with disease and symptom onset. These prospective studies also must integrate microbiome, metagenomic, metatranscriptomic, and metabolomic data with comprehensive clinical and environmental metadata to build a systems-level model of interactions between the host and the development of disease. The creation of novel biological computational models and a pathway to move from association to causation are essential in providing a mechanistic approach to exploring the development of chronic inflammatory diseases. This can only be done when these diseases are investigated as complex biological networks.

We are much more likely to discover clinically meaningful and successful interventions when study design is based on proven premises and established mechanistic understanding of the paths leading to a break of tolerance and the onset of chronic inflammatory diseases. Therefore, we need to transition from descriptive to mechanistic studies of the microbiome for promising translational medicine (moving research discoveries from the lab bench to clinical applications for patients) to be possible. Given the hypothesis that the development of the microbiome during the first one thousand days of life has a lasting effect on an individual's future health and risk of disease, pediatric researchers are well positioned to lead this transition. Large-scale, collaborative, prospective, longitudinal, and multiomic studies are needed to create targeted, personalized interventions or primary prevention strategies.

Longitudinal data collection must begin before the onset of disease to understand the contribution of multiomic covariates to the development of disease. It is now clear that knowing the functionality of the microbiota is central to understanding the mechanistic role of the microbiota in the development of chronic inflammatory disease. Metatranscriptomics, metaproteomics, and metabolomic analyses all are necessary, because gene expression is consistent with DNA abundance for less than 50 percent of genes in the microbiome.

Optimal computational models to analyze longitudinal multiomic covariates are under investigation, but large datasets, of which there are few, are needed to evaluate them. When properly designed, these pediatric, longitudinal multiomic studies have the potential to dramatically impact our understanding of and approach to these complex chronic inflammatory diseases. Only then will breakthrough treatment and prevention strategies likely emerge.

In the short-term study of COVID-19, using multiomic data and a machine learning model, Wanglong Gou, Ju-Sheng Zheng, and colleagues identified a set of proteomic biomarkers to predict progression to severe disease among infected patients. They identified core gut microbiota features and created a blood proteomic risk score for the prediction of the progression of the disease. The core gut microbial features and related metabolites could provide an interventional target among vulnerable populations.[18]

In chapter 14 we will discuss in depth our current knowledge regarding the microbiome's contribution to chronic inflammatory diseases in childhood by focusing on four important pediatric examples: obesity, food allergy, autoimmune disease (celiac disease), and ASD. We will discuss how to move research in these fields from descriptive to mechanistic approaches, with an ultimate goal of being able to predict and even prevent disease.

14 Preventive Medicine: Monitoring the Microbiome for Disease Prediction and Interception

Methods of Microbiome Study

As mentioned at the end of chapter 13, we stand at a crucial crossroads in the field of human microbiome studies. To link microbiome shifts in composition and function mechanistically with disease pathogenesis, we must move away from cross-sectional, case-control studies and move toward a longitudinal cohort design to capture changes that precede or coincide with disease and symptom onset. These prospective studies must also integrate microbiome, metagenomic, metatranscriptomic, and metabolomic data with comprehensive clinical and environmental data to build a systems-level model of interactions between the host and the development of disease.

The creation of novel biological computational models and a pathway to move from association to causation are both essential in providing a mechanistic approach to explore the development of chronic inflammatory diseases. These diseases must be investigated as complex and highly integrated biological networks, rather than as a puzzle of disconnected information that needs to be examined and then arbitrarily glued together. Studying the microbiome as a "stand-alone" research focus without the context of its symbiotic host—genetic makeup, metabolic and proteomic profiles, the surrounding environment, lifestyle, dietary habits, exposure to stressors, and more—is like trying to understand the mechanics of a car by only looking at the car's tires.

To add another layer of complexity, we must acknowledge that, even if we approach the study of the human microbiome in an integrated fashion with due consideration of diverse variables, we still need to deal with a dynamic and unstable "interactome." Together with the microbiome, these other variables are changing all the time in a mutually influencing fashion. It's like looking at a kaleidoscope. If you rotate the tube just a little bit, you put into motion all the pieces of glass, and they are rearranged into different positions leading to different mutual interactions, which results in an ever-changing view.

To capture these variables in a dynamic and integrated fashion, designing longitudinal birth cohort studies surveying the incidence of early childhood diseases seems to be the most logical approach. Through this method, we can mechanistically link the epigenetic pressure of microbiome shifts on the host genome causing the switch from genetic predisposition to clinical outcome. For insights into ongoing birth cohort studies, we turned to three of our colleagues from the Division of Pediatric Gastroenterology and Nutrition at MGHfC.

Obesity and the Microbiome

As discussed in chapter 8, there is great interest in understanding whether shifts in microbiome composition and resulting dysbiosis are contributing to the recent rise in obesity numbers. Lauren Fiechtner, director of the Center for Feeding and Nutrition at MGHfC, and her colleague Elsie Taveras, chief of the Division of General Academic Pediatrics at MGHfC, have been working on the complex etiology of childhood obesity. They examine the issue from various points of view, including social, economic, and macro-environmental. Fiechtner's recent focus on the microbiome represents a departure from this broad perspective to study the role of the microbiome as it relates to obesity in the first one thousand days of life.

"It's becoming increasingly clear that the factors that put the risk of developing obesity into motion seem to be at play very early in life," said Fiechtner.[1] To understand which mechanisms are involved in driving development of obesity in early childhood, Fiechtner and colleagues designed a longitudinal study of the microbiome. They collected samples from a prospective cohort of children to study differences in the gut microorganisms of those who eventually develop obesity and those who do not. Following on from this evidence, manipulating the composition and function of the microbiome could represent a significant public health intervention target with the potential to radically change the landscape of the obesity epidemic.

Researchers like Fiechtner and Taveras may discover a novel target for intervention in the microbiome domain to improve public health at the intersection of nutrition, the microbiome, and obesity. One of the critical questions in this public health puzzle is: What role does quality versus quantity of food play in caloric intake and nutritional impact on the composition of the microbiome in children with obesity? Highly processed foods and "fast food" or "junk food" favor hypercaloric intake and increase the risk for cardiovascular disease, obesity, and cancer.

And as MGH hepatologist Lee Kaplan informed us in chapter 8, obesity is an inflammatory disease.

As an expert in nutrition and the social factors associated with obesity, Fiechtner can provide insight into nutritional interventions. These include improving the quality of food for a better metabolic outcome and potentially manipulating the composition and function of the microbiome to effectively slow down—if not arrest—the obesity epidemic. But first we must understand the impact of nutritional interventions on the microbiome and its metabolic profile in terms of calorie balance. Once we have that piece of the puzzle, the outcome of longitudinal studies like the one Fiechtner and Taveras are conducting could help us to better tailor nutrition to manipulate the microbiome to treat obesity.

Fiechtner is looking for microbial drivers of obesity in infants with early, rapid weight gain, which is related to later onset of childhood obesity. We asked her what we might expect twenty years from now in terms of personalized intervention for children with obesity or, even better, primary prevention for at-risk children. Her vision is that, with a comprehensive, personalized profile from stool analysis and other metabolic data, the pediatrician could guide the family to a specialized diet in tandem with a probiotic or prebiotic to change the microbiome's function to favor a better metabolism that can ameliorate obesity.[2]

Nutritional Role of the Microbiome

Fiechtner emphasized that obesity cannot be treated in a vacuum or with a single intervention, as we learned from Lee Kaplan. But she called manipulation of the microbiome "an amazing tool" that may also work for children at the other end of the spectrum—children who are not growing well.[3] For them, the focus is on high-calorie foods, which could result in negative consequences in the long term. Understanding the role of the microbiome in children diagnosed with failure to thrive gives us another tool to tackle this problem.

Let us suppose that Fiechtner's work and similar research leads to the understanding that the combination of a high-quality diet in conjunction with a probiotic, prebiotic, or synbiotic will supply whatever has been lost because of poor diet and mitigate the inflammation that leads to obesity. This line of research in understanding the nutritional role of the microbiome could lead to potential collaborations with food industry partners to develop products that are friendlier to a balanced microbiome. However, this scenario raises the complex issue of human behavior. If someone can eat unhealthy food and take a pill to balance their microbiome, what problems does that pose on a societal as well as a scientific level?

Scientists like Fiechtner could be asked by legislative bodies to guide policymakers' decisions concerning nutrition and obesity, raising difficult issues. On the one hand, scientists hope to advance scientific knowledge to ameliorate the obesity epidemic through microbiome manipulation. On the other hand, policymakers may be tempted to use this knowledge as an "easy way out" to avoid a comprehensive approach to the problem of obesity, including addressing health disparities and advocating for healthier lifestyles. The best policy and foundation for future decisions would be to ensure that all members of society have access to inexpensive, healthy food, including plenty of fresh produce, whole grains, and lean protein.

However, even if these changes were enacted, a certain number of people might only be effectively treated by also manipulating the microbiome through probiotics. Under the same American "roof" we have people with no means to eat enough or eat well, and others who can afford to eat well but spend billions of dollars on weight-loss supplements and diet plans. If it became possible to redistribute some of this "food wealth" in both quality and quantity, people would be healthier. This food redistribution would also avoid wasting an average of approximately 30 percent of the food produced, thereby contributing to the sustainability of food procurement for an ever-growing worldwide population.[4]

At the moment, this would be a desirable but not attainable goal, while there are now other possible strategies to mitigate the unprecedented challenges the obesity epidemic poses in our pediatric population. For example, as a pediatrician, coauthor Alessio Fasano was not trained to deal with heart attacks, fatty liver, or joint pain due to overweight and stress on bones, which are now increasingly frequent in pediatric practice.

Making this situation even more worrisome, Fiechtner stated that there is a huge disparity concerning obesity epidemics when related to race and income. Racial and ethnic minorities and low-income populations are experiencing rising rates of obesity, while the prevalence of childhood obesity has plateaued for white and higher-income populations.[5] She is convinced that systemic change is needed to provide underprivileged families with access to healthy food. Fiechtner also advocates nutritional counseling, as well as creative ideas for healthier living within the current constraints of society.[6]

Providing Healthy Food to Every Community

Food justice and access are easy to talk about but much harder to implement. Fiechtner is convinced, however, that step by step, whether imposing a tax on sugary beverages or making other changes to food policies, nutrition and food access will improve. Building on interventions such as the "My Way Café," a pilot program launched by the Shah Family Foundation to cook healthy, nutritious meals in school kitchens in Boston public schools, may help naturally reshape a more symbiotic microbiome to mitigate the risk of obesity. Michelle Obama's efforts to champion exercise and better nutrition for children through local food sourcing in the federal school lunch program (partially overturned by the Trump administration[7]) will aid the progress that Fiechtner hopes to see throughout the United States. However, these changes will only become possible and continue if school lunchrooms and other consumer settings cease to be used as political pawns for the interests of big business and food lobbies.

However, the situation is more complicated than simply changing school lunch policies. "Food deserts," those city neighborhoods, towns, and other places in this country lacking adequate food sources, are another challenge. When Fiechtner lived in rural Mississippi, she noticed that there was only one grocery store in the community. To be precise, she said it was not even a grocery store; it was a Walmart that would often run out of food. The closest grocery store was ten miles away, and many of the very-low-income residents did not own cars.[8]

The only alternative for feeding a family was fast-food restaurants and convenience stores, and "there were tons of them," said Fiechtner. "They could feed their children Slim Jims, but not fresh fruits and vegetables." She noted that the school had fresh fruits and vegetables, so many families relied on the school lunch program for their children's access to some healthy foods.[9]

But there have been incremental and positive changes in underserved communities. As part of a growing trend, United

Supermarkets in Texas began offering free health screenings every weekend, with blood pressure and glucose checks and other tests.[10] Some supermarkets have added color-coded nutrition labels to make nutritional choices easier for people with chronic conditions, and some stores also offer nutritional counseling with dietitians. In California, Northgate González Market, a Latino-owned family supermarket chain, is working to improve community health and quality of life through its Viva la Salud Program, with bilingual healthy-food labels and a program to help customers make healthy food choices.[11]

Imagine implementing this model after microbiome studies provide us with solid evidence on the type of diet needed to modulate the microbiome's function to change metabolic profiling to ameliorate obesity and its complications. For now, according to Fiechtner, the United Supermarkets in Texas model is slowly expanding to the rest of the United States with a lot of good community work guiding people toward healthy options and away from fast and processed food. She adds that food banks around the country are following this approach to move families toward healthy food choices.[12]

These and similar strategies can obtain favorable results not only for children at risk of obesity but also for future generations. As we addressed in chapter 3, healthy nutrition is a critical issue for pregnant women and their infants, since maternal gestational weight gain during pregnancy can strongly influence the risk of developing obesity in offspring. Fiechtner and her colleagues have published research showing that gestational weight gain is associated with differences in the microbiome.[13] Additional research from her group on pregnant women shows the beneficial effects of fish consumption on the microbiome of their offspring.[14]

These examples from a growing body of literature show how maternal lifestyle and nutritional habits can affect the microbiome composition of the infants, which can lead to negative clinical outcomes including obesity. As we learn more about the complex etiology of obesity and the role of the maternal and infant microbiome, sharing this information may be another strategy to mitigate the complications of obesity in future generations. The projects that Fiechtner and Taveras are pursuing will contribute to efforts to implement effective nutritional strategies and other approaches to improve the health of future generations.

Food Allergies and the Microbiome

Food allergies are common diseases affecting approximately 8 percent of children and 5 percent of adults. In societies incorporating Western lifestyles, their prevalence has increased over the last two decades, suggesting an important environmental contribution to susceptibility. Some studies have identified birth delivery mode, pet exposure, and older siblings as significant risk-modifying factors in the development of food allergies. These aspects and other factors significantly impact the composition of the intestinal microbiome and its role in shaping the immune system, leading to research into the gut microbiome and food allergic manifestations.

Studies in cross-sectional human cohorts support a role of dysbiosis in the pathogenesis of food allergies, and limited data suggest that this dysbiosis occurs early in life, preceding the onset of sensitization. Studies from animal models have shown that the composition of the intestinal microbiome confers susceptibility to food allergy. However, even though our understanding of microbial regulation of food allergy is in its infancy, the state of the field supports an important contribution of intestinal microbes to susceptibility to these diseases.

The research challenge is to identify commensal-derived micro-organisms for therapeutic use to prevent or perhaps treat food allergies. Achieving this goal and answering the many still-unanswered questions requires novel approaches to mechanistically link microbiome composition and function to the pathogenesis of food allergies. Above all, prospective studies on human cohorts using well-defined clinical outcomes of individuals with food allergies are needed to definitively establish if changes in microbial composition precede allergy development.

The dynamic composition of the gut microbiome in early life demands multiple sampling to assess its role in a window of susceptibility. A better understanding of the general features of a protective microbiome and environmental factors such as diet, which could regulate its protective development, are needed to realize our goal of manipulating the intestinal microbiome for prevention or treatment of food allergy. To learn more, we reached out to Victoria Martin, pediatric gastroenterologist at MGHfC. Martin and MGHfC pediatric gastroenterologist Qian Yuan lead a birth-cohort, prospective study called the Gastro-intestinal, Microbiome and Allergic Proctocolitis (GMAP) Study. She shared her thoughts on the pathogenesis of food allergic manifestations and the scientific roadblocks that need to be overcome in this emerging field of research.

Shifting the Research Focus

Martin emphasizes the need for broad, large-scale studies performed in diverse populations. By now, it is apparent that there is a strong personalized component in different diseases, which requires a mechanistic understanding applicable to different types of people in different parts of the world. One of the first hurdles, noted Martin, is conducting these studies to determine how to properly analyze data for insights into the role of the gut microbiome in the break of immune tolerance that leads to food allergies.[15] With this information, we will be able to identify therapeutic targets tailored to specific subpopulations that are stratified based on the results obtained.

This aligns with the concept of personalized medicine, meaning that one size is certainly not going to fit everyone. Our colleague therefore suggests that study designs must be carefully constructed and aimed at personalized interventions. Although individuals may suffer from the same conditions, certain interventions to manipulate microbiome composition may work in one subgroup of patients but not in another. These meaningful results can be achieved only if studies are carefully designed and populations carefully selected to move away from the "educated guess" that currently is applied to interpret the role of specific microbiome species supposedly linked to disease pathogenesis.

It should come as no surprise that interventional studies aimed at manipulating the microbiome, based on what we have discovered so far in cross-sectional studies, have led to less than satisfactory results. Martin noted that many of these disappointing results aimed at manipulating the microbiome have been performed by "putting the cart before the horse"—in other words, carrying out interventions on people before having any sort of mechanistic understanding.[16]

Continuing in this direction may lead to patients' distrust in some of these ideas and approaches, said Martin, since the lack of specific personalized targets to manipulate the microbiome may not only generate negative results but, even worse, may also cause harmful outcomes. To mitigate this risk, Martin became involved in the GMAP Study that began in 2015.[17] The study's primary focus is food protein–induced allergic proctocolitis, also referred to as cow's milk protein allergy. This is an entity that pediatricians commonly see in clinical practice—in general pediatrics, in gastroenterology, and in allergy practices.

There is not much consensus on how to diagnose or manage it, and there is no understanding about its epidemiology or pathophysiology. There has been a dramatic rise in general in food allergies in children, and food protein–induced allergic proctocolitis is one of the earliest food allergic manifestations in childhood. Martin and her colleagues are studying how this disease develops by prospectively following infants from birth. They recruited healthy newborns from a suburban pediatric practice in Massachusetts and have been following them for six years. The focus of GMAP is to survey these children as they grow and develop and focus on the composition of their gut microbiomes as it relates to their clinical outcome, namely staying healthy versus developing early food allergic manifestations.

"The real strength of this study is that we collect microbiome samples essentially every time they visit their pediatrician, which as any parent knows, is an awful lot in the first couple of years," said Martin.[18] Subsequent samples are collected at least once a year. Due to the frequent surveillance of the microbiome and the unbiased recruitment of infants with no particular risk factor, this study will allow the observation of any microbiome shift before, during, and after onset of any disease of interest, with the major focus on food allergies. Meanwhile, it also allows researchers to study and compare, on a large scale, the dynamic development of the healthy infant microbiome.

High Rates of Diagnosis

Although this is an ongoing study, Martin shared a staggering preliminary finding: in the general, healthy pediatric population, 17 percent of children were diagnosed with food protein–induced allergic proctocolitis, a dramatically higher rate than what has been reported. This is partly due to permissive diagnostic inclusion criteria used by Martin and colleagues. Since there are no validated biomarkers for diagnosis, the researchers relied on diagnostic criteria used by pediatricians and gastroenterologists in clinical practice.[19]

While this approach may lead to a substantial overdiagnosis, it allows clinicians to understand the consequences of making this diagnosis in real clinical settings while researchers search for microbiome signatures that can be objectively linked to real loss of tolerance to food antigens, said Martin. Based on GMAP's preliminary results, it is likely that many children are diagnosed with food protein–induced allergic proctocolitis and placed on an elimination diet in the first few months of life. Martin noted that it is likely that only a subgroup needs this strict dietary intervention, while for the remaining healthy children this substantial diet restriction may be harmful.[20]

We now have fairly convincing evidence that antigen exposure early in life is the best strategy to prevent IgE-mediated food allergy, as already reported for peanuts and other foodstuffs.[21] The concern of Martin and her team is, if close to 20 percent of the population undergo dietary restrictions when they are babies, what is that doing to their microbiome and to their immune system's ability to learn oral tolerance?[22]

The next step for Martin's group is to look at rates of IgE-mediated food allergies in this population to try to understand what role diet elimination and the microbiome might be playing in that disease process. Is the high percentage of food protein–induced allergic proctocolitis in this cohort due to (a) the unique study design of their prospective cohort, or (b) the result of a true increase in incidence compared to twenty years ago, or (c) perhaps a combination of both?

Martin believes that it is probably the result of a true increase in incidence. She suspects that this is a consequence of earlier thinking—that there was not much harm in diet elimination implemented as an empiric treatment option without a sigmoid-oscopy with biopsies or an open-milk challenge.[23] There has been an unequivocal rise in other food allergic diseases in the United States, as demonstrated by the doubling of IgE-mediated food allergy in the last decade. So, it should not be surprising that one of the earliest manifestations of food allergies is on the rise too.

Targeted Probiotics to Prevent Food Allergy

Martin hopes the outcome of GMAP and studies like it will be the identification of microbiome targets for both personalized treatment and primary prevention. On the preventive side, the ideal and plausibly achievable result is demonstrating that at-risk populations can be identified through these types of studies. For example, with outcomes from the GMAP Study, in the future we might be able to identify a pregnant woman with certain characteristics that indicate her baby will be at increased risk of developing a food allergic manifestation early in infancy. We could then enact primary prevention and disease intercep-tion with a targeted probiotic given to the baby or even to the mother prenatally. The design of the probiotic would be based on the findings of the disease-associated dysbioses from GMAP.

Alternatively, this study could lead to personalized interventions to manipulate the microbiome of children who develop food allergic manifestations. If we can identify key microbiome shifts that activate the pathophysiology of disease development, said Martin, we can target specific proinflammatory strains linked to the onset of the disease or supply protective strains to counteract them.[24] Doing this kind of personalized, sophisticated microbiome analysis in a noninvasive way (and truly understanding how a patient's microbiome functions) to develop preventive or therapeutic strategies would provide a remarkable alternative to the only tool we currently have available to treat food allergy, namely, elimination diets.

How far are we from this goal, and what roadblocks must we remove to get there? Martin thinks there are several large, complicated hurdles to overcome. First and foremost, as already noted, developing a true mechanistic understanding of the role of the microbiome in the pathogenesis of chronic inflammatory disease is a must. And this goal most likely will not come from only one study or one population. Rather, it will evolve from many integrative analyses from diverse studies designed in a prospective way. Harmonization among several prospective studies like those described in this chapter would be highly advantageous. Mechanistically linking specific strains to the protection or instigation of inflammation is another necessary step to identify targets for microbiome manipulation to ameliorate or prevent food allergy.[25]

Once these targets are identified, novel human-relevant models will be needed to validate findings. The typical mouse model may not be the best agent for microbiome studies. Innovative tools to test therapeutic interventions, including ex vivo models like human gut organoids, should be pursued. Only then can we move to the last hurdle: testing those interventions that pass the scrutiny of preclinical tests in terms of safety and efficacy in well-designed clinical trials. We are optimistic that in the near future Martin's generation of clinical investigators will be able to use the outcome of these studies in their routine clinical practice.

Celiac Disease and the Microbiome

Celiac disease is an autoimmune enteropathy triggered by the ingestion of gluten-containing grains like wheat, barley, and rye in genetically susceptible individuals. With the knowledge that gluten is the environmental trigger, it represents a unique model of autoimmune disease. Other factors in its development are a close genetic association with HLA genes (DQ2 or DQ8) and a highly specific humoral autoimmune response resulting in auto-antibodies to tissue transglutaminase. However, the early steps following from the exposure of intestinal mucosa to gluten that lead to the loss of tolerance and the development of the autoim-mune process are still largely unknown.

Studies now suggest that this loss of gluten tolerance does not always occur at the time of gluten introduction into the diet of genetically susceptible individuals; rather, it can occur at any time in life as a consequence of other, unknown environmental stimuli. A proof-of-concept study published by our group shows that a unique interplay between a particular microbiome and host may lead to alterations in metabolic pathways, resulting in the production of specific metabolites prior to the onset of autoimmune disease.[26] Many environmental factors known to influence the composition and function of the intestinal micro-biome are also thought to play a role in the development of celiac disease. These include birthing delivery mode, infant feeding type, history of infection, and antibiotic use.

We have previously reported that infants at risk of celiac disease had a decreased representation of Bacteriodetes and a higher abundance of Firmicutes compared to control infants with a nonselected genetic background. Their microbiome showed a delay in maturation at two years of age, although maturation was complete in not-at-risk infants at one year.[27] In the same study, infants who developed autoimmunity had decreased lactate signals in their stools. This coincided with a diminished representation of *Lactobacillus* species, and this shift preceded the first detection of positive antibodies representing the biomarkers of celiac disease autoimmunity.

Our results have been confirmed by colleagues who compared microbial communities in infants carrying the celiac disease–linked DQ2 haplotype with infants without a compatible haplotype. They observed distinct differences in microbiome composition at one month of age; infants carrying DQ2 showed a higher abundance of Firmicutes and Proteobacteria compared to infants without a genetic predisposition.[28]

Two large prospective cohort studies of infants at risk of celiac disease found that it develops quite early in life in this group, which further supports the notion that early environmental factors may be pivotal in the development of the disease. These studies showed that 16 percent of infants with a first-degree relative with celiac disease and who carry HLA DQ2 or DQ8 or both will develop celiac disease by age five; most will be diagnosed by age three. These studies further demonstrated that 38 percent of infants who are first-degree relatives of celiac disease patients and who carry two copies of DQ2 will develop the condition by age five.[29]

As mentioned earlier and in chapter 13, the major limitation in performing preclinical studies on human gastrointestinal inflammatory diseases is that animal models do not completely recapitulate the complexity of host-microbiome interactions. These interactions influence the activation of specific metabolic pathways dictating the tolerance–immune response balance in humans. To date, no large-scale, longitudinal studies have defined if and how gut microbial composition and metabolomic profiles may influence the loss of gluten tolerance and subsequent onset of celiac disease in genetically susceptible subjects.

To overcome this deficit, Maureen Leonard, clinical director of the Center for Celiac Research and Treatment at MGHfC, together with a multidisciplinary team of MGHfC colleagues and Italian institutions, have embarked on a ten-year birth cohort study. They are investigating the role of the developing intestinal microbiome and the resulting metabolome as additional factors that may play a key role in the onset of and predisposition to celiac disease autoimmunity. The Celiac Disease Genomic, Environmental, Microbiome, and Metabolomic (CDGEMM) Study is looking at the natural history of celiac disease and other autoimmune disorders.

Leonard and colleagues hypothesize that the introduction of gluten into the diet combined with the particular gut microbial composition of infants genetically at risk for celiac disease activates specific metabolic pathways. These pathways then contribute to the loss of gluten tolerance and to the onset of autoimmunity, as reflected by specific metabolomic phenotypes. They hope to identify and validate specific microbiome and metabolomic signatures to predict loss of tolerance in children genetically at risk for autoimmunity. "Our ultimate goal is to generate the body of knowledge that will help us to eventually implement early interventions to prevent loss of tolerance to gluten and the development of celiac disease," said Leonard.[30]

In addition to repeated serological screening for celiac disease in children until age ten, detailed environmental information is obtained frequently, and stool is collected every three months for the first three years of life and every six months thereafter. Infants' microbiomes and metabolomes will be compared longitudinally, with attention to differences before and after the introduction of gluten, and before and after the development of celiac disease when it occurs, as well as other environmental factors that may be at play in the development of celiac disease.

Leonard explained that, within the longitudinal study, researchers will perform a nested case-control analysis. Infants who develop celiac disease will be matched with control infants with a genetic predisposition to the condition who have not developed celiac disease. A second analysis will match infants who develop celiac disease with control infants who do not carry the HLA-predisposing genes. These analyses will help researchers address environmental factors that may contribute to alterations in the microbiome and predispose infants to the development of celiac disease.[31]

The long-term and more ambitious goal of CDGEMM is to understand how these environmental factors or combinations of environmental factors, like mode of delivery, antibiotic use, viral illnesses, and introduction to certain foods in the diet, cause changes in the childhood microbiome or shifts in biological functions, contributing to the instigation of celiac disease. With this knowledge, Leonard thinks clinicians will be able to intervene with diet or personalized prebiotics or probiotics before the loss of tolerance to gluten occurs that leads to disease onset.[32]

In clinical practices of the future, pediatricians and family physicians would perform early genetic testing and frequent stool analysis. By combining these data with family history and known environmental exposure, clinicians could then monitor the potential risk of developing celiac disease and intervene to prevent its occurrence.

Prevention Instead of Treatment

Finally, we asked Leonard how she would invest resources to achieve these goals if given unlimited funds. She said the top priority is to attract more computational scientists to help create tools to build these models to predict and prevent disease.

"We'll have the data, but we need to be able to analyze it and then create something that will be utilized in the clinic, and quickly," said Leonard. "We need to be able to validate our findings in other cohorts. So, we need to have other scientists who are willing to take this super long but hopefully fruitful approach to try to discover how to prevent disease." Another roadblock she identified is shifting the focus of research funding to longitudinal studies and scientists who follow infants at risk of disease, with the understanding that it may take longer than the three-to-five-year funding cycles the National Institutes of Health currently provides to study outcomes.[33]

She predicts that these data will be translated into apps on patients' smartphones, with the ability to monitor factors like environmental exposure to toxins, antibiotics, temperature, diet, activity, weight, blood pressure, and heart rate. "This accurate, real-time data collection can improve our predictive model for disease," said Leonard. "Once we have direct health and environmental data from patients and computational researchers to develop algorithms for routine clinical practice for health care professionals, we will be able to radically change the way we conduct patient visits. We will implement true personalized medicine, with many more preventive visits, rather than visits to treat disease."[34]

The outcome of studies like CDGEMM will allow us to implement primary prevention strategies for celiac disease and other chronic inflammatory diseases through manipulation of the microbiome. This represents a complete shift of paradigm in chronic inflammatory disease pathogenesis and the treatment of these lifelong disorders. The identification of metabolomic phenotypes specific to celiac disease can also help to define additional diagnostic tools and therapeutic interventions. CDGEMM's biorepository will allow for future epigenetic studies and validation of biomarkers. These findings may have a far-reaching impact on other chronic inflammatory diseases in which diet-genome-microbiome interaction in the pathogenesis of the disease has been hypothesized.

In turn, these developments could help to establish strategies aimed at reexamining the process of oral tolerance to gluten and other environmental antigens, opening the way to novel approaches to the prevention and treatment of chronic inflammatory and other autoimmune diseases. Three million people in the United States are affected by celiac disease, and approximately seventeen million people suffer from other autoimmune diseases. With no effective strategies to prevent these conditions, the findings from studies like CDGEMM could have a tremendous impact on public health.

Autism Spectrum Disorder and the Microbiome

First introduced in chapter 10, ASD is not a single disorder but a spectrum of related disorders with a shared core of symptoms defined by deficits in communication, social reciprocity, and repetitive, stereotypical behaviors. Identifying specific biomarkers for patient stratification and possible common disease pathways are of special interest to researchers and clinicians.

Although the underlying reasons for this pandemic are difficult to study empirically, selected studies suggest that part of the recent rise in prevalence is likely due to extrinsic factors such as improved awareness and recognition and changes in diagnostic practice or treatment availability.[35] Nevertheless, these changes are insufficient to explain this phenomenon, and other environmental factors are likely at play. Based on the gene-environment interaction theory, several therapeutic approaches have been proposed with conflicting, sometimes opposite results.

These suboptimal results are probably due to the fact that, like many other multifactorial diseases, ASD is a final pathological destination that can be reached through different pathways. With these considerations, it is imperative to stratify the ASD population based on the identification of specific biomarkers that can assist in personalizing interventions for the most effective preventive or therapeutic results. The magnitude of the disease and its extreme sex/gender bias (it is four times more common among males than females) also suggest a multifactorial cause of ASD that could affect the gender structure of entire populations. Many individuals with ASD have symptoms of associated comorbidities, including seizures, sleep problems, metabolic conditions, and gastrointestinal disorders, which have significant health, developmental, social, and educational impacts.

The neuroanatomical and biochemical characteristics associated with ASD's pathogenesis involve mechanisms that are direct consequences of the effects of low-grade, feverless, systemic inflammatory events, while the protective mechanisms against ASD's pathogenesis have strong anti-inflammatory components.[36] The gut microbiome drives immunoregulation during the first three years of life. Faulty immunoregulation, as well as inflammation, predisposes an individual to psychiatric disorders, including ASD, while psychological stress drives further inflammation via pathways that involve the gut microbiome through the gut-brain axis network.[37]

Although there is limited epidemiological information, a recent meta-analysis confirms that children with ASD experience four times more gastrointestinal symptoms than control groups, and that these symptoms may identify a unique subgroup of children with ASD.[38] Other studies suggest aberrant immune system activation and alterations in gut microbial composition that correlate with ASD's severity.[39] Finally, we have reported a link between the gut microbiome and immune function in ASD[40] and improvement of both gastrointestinal and behavioral symptoms in ASD subjects in whom gut dysbiosis was corrected by microbiota transfer therapy.[41]

While numerous cross-sectional studies have shown a correlation between microbiome changes and ASD, they suffer from the same limitations as other pathologies mentioned in this chapter. They fail to mechanistically link microbiome composition and function to disease pathogenesis. To establish causality in ASD by identifying alterations in gut bacteria composition mechanistically linked to the onset of the disease, we launched a prospective birth cohort study in 2019 like the one outlined earlier for celiac disease.

GEMMA Study of Autism Spectrum Disorder

Launched in 2019, GEMMA, which stands for Genome, Environment, Microbiome, and Metabolome in Autism, will use a combination of patient cohorts, animal models, and prospective patient samples. Using diverse methods, our GEMMA research team will investigate whether specific patterns of gut dysbiosis influence epigenetic modification in the host, modify metabolic pathways, alter gut permeability and immune response, and, ultimately, trigger and/or increase the severity of ASD and related gastrointestinal comorbidities.

Moreover, using specific biomarkers to stratify patient populations for targeted interventions, the project will attempt to validate the hypothesized mechanisms of disease by correcting gut dysbiosis or eliminating alimentary triggers and observing whether changes in proposed disease-specific pathophysiology occur. Based on the in-depth evaluation of six hundred at-risk infants observed from birth, GEMMA will provide solid mechanistic evidence of ASD onset and progression in relation to dynamic changes in abnormal gut microbiota, causing epigenetic modifications that control gut barrier and immune functions.

GEMMA will support novel customized prediction (personalized medicine) and disease interception (intervention) approaches that attempt to modulate gut microbiota to reestablish and maintain immune homeostasis. The biomarkers identified in this project will contribute to a better understanding of the pathogenesis of ASD in at-risk children and advance the possibility of manipulating the microbiome through the administration of prebiotics, probiotics, or synbiotics along with dietary changes for prevention and treatment. This approach represents a paradigm shift in ASD pathogenesis and early intervention.

The identification of specific ASD metabolic phenotypes also will help to define biomarkers for diagnostic tools and patient stratification models for other conditions in which interplay among the genome, microbiome, and metabolome has been suspected or proved. Finally, the project will generate a unique biobank of more than sixteen thousand blood, stool, urine, and saliva samples prospectively collected, which can be exploited in future multiomic studies. These samples will provide substantial value to the research community beyond the completion of the GEMMA project.

Risk Factors for Autism Spectrum Disorder

To achieve these goals, the GEMMA research team will rely on data mining and biostatistical analysis of multiomic data collected from preclinical and clinical studies of animal models and prospective at-risk patient cohorts to identify mechanisms common to ASD and gastrointestinal disorders. As mentioned earlier, the project's aim is to link epithelial gut barrier disruption resulting in local immune cell infiltration and activation of the onset of ASD, with or without gastrointestinal symptoms, to a specific host genetic-epigenetic profile, microbiome compositions, and metabolic signatures, using both human observational studies and mouse model studies.

Preclinical work on mouse models will provide valuable information on the validation of specific biomarkers mechanistically linked to the onset of ASD to rationalize future patient stratification for primary intervention. The project has an interventional arm to test the assumption that probiotics, prebiotics, and synbiotics (plus or minus dietary interventions) in a specific, targeted population stratified by validated biomarkers can attenuate the expression of ASD and its comorbidities.

The consortium assembled for this five-year project employs the multidisciplinary skills of seven industrial partners, four universities, three patient networks, and two research institutes, including our team of pediatric gastroenterologists from MGHfC and our European Biomedical Research Institute of Salerno (EBRIS) in Italy. These institutions have members with expertise in behavioral disorders, genetics, gut mucosal biology and immunology, microbiome studies, immunology, metabolomics, statistical analysis of multiomics, animal models, infant and clinical nutrition, and clinical trials.

Based on clinical outcomes, and after assessment using recognized scoring systems for ASD diagnosis, the population will be divided into four groups: neurocompetent children with no gastrointestinal symptoms; neurocompetent children with gastrointestinal symptoms; ASD children with no gastrointestinal symptoms; and ASD children with gastrointestinal symptoms. In each group, twenty to thirty individuals will be selected, and a series of omics platforms will be used to characterize their phenotype. For the selected infants, we will also collect genomic samples from family members to distinguish between inherited and de novo properties and to better detect the variants related to ASD. This project represents a departure from classic ASD research focused on genetic, behavioral, and neurological aspects of the disease.

However, the primary roles of environmental risk factors—gut barrier dysfunction, immune dysregulation, and additional metabolic disruptions—in the pathogenesis of ASD have recently gained significant attention. The striking heterogeneity among individuals who share the same diagnosis is consistent with the prevailing notion that there are a variety of etiologies for ASD. Moreover, the spectrum of ASD symptoms and the challenges in identifying specific causes, treatments, and molecular biomarkers underscore the need to better define the clinical subtypes of ASD and provide tailored treatment to subclasses of ASD individuals. In observational trials, GEMMA will validate a mechanistic chain of interactive events—affected by the host's genome, microbiome, and metabolome, along with gut mucosal biology and immune changes—that will lead to the clinical outcome of ASD.

Developing a Multiomic Analytics Platform for Autism Spectrum Disorder

Modern high-throughput measurement technologies have given us powerful tools to measure the properties of cells and organisms, and they are constantly producing masses of new omics data. Next-generation sequencing methods have been used to measure the genomics and transcriptomics data from different types of cells, tissues, and organisms. The metagenomics of gut microbiota also have recently been measured using these methods. Simultaneously, we need to better understand the system-level functions of these domains, and thus association analyses of such data are gaining importance.[42]

Several analysis algorithms and association analysis methods currently exist that can be applied to these measurements. Most of the association analyses have focused on the association of two data types, for example, studying the role of genetics associated with other types of data. Recent studies have shown that alterations in single nucleotide polymorphisms (SNPs) can affect gene expression levels, alternative splicing, DNA methylation, and miRNA-mediated gene expression levels in different types of cells, and SNPs also have been associated with transcript isoform variation and alternative splicing. Copy number changes in genomes have been associated with changes in gene expression values in various cell types, and metabolomic data have been associated with transcriptomics data. Recent studies also focus on multiomic analysis of the metagenomic data as well.

Given the anticipated complexity and high-dimensionality of merging all omics data-types in the GEMMA project, our research team will initially determine the structure of the merged data to optimize downstream analyses. GEMMA will use dimensionality reduction techniques and modeling techniques discussed in chapter 17. This project will develop a first-of-its-kind multiomics analytics platform that will allow third-party researchers to share and integrate data from various omics data repositories.

If successful in this particular aim, GEMMA will provide an opportunity to create a high-quality, multiomic tool that could help in hypothesis generation, hypothesis validation, and patient stratification studies. Moreover, such a tool could continue to evolve via the contributions of the wider research community to expand the breadth and depth of quality data available in the public domain, potentially leading to the discovery of new mechanisms and corresponding biomarkers of ASD and related disorders. Although the initial focus of the platform will be ASD, the platform can be extrapolated to other chronic inflammatory diseases, particularly autoimmune disorders, in which the pathogenic role of dysbiosis and metabolic unbalance has been either proved or hypothesized.

The Power of Diet in Modulating the Gut Microbiome

As we have seen again and again—no matter what the disorder— one of the most important lessons to be more deeply explored in longitudinal studies aimed at disease interception is the role of diet. The negative impact of a poor diet on health and its role in chronic disease is now taking center stage at the global level.

In a paper from Ashkan Afshin and colleagues, the burden of a suboptimal diet as an important preventable risk factor for noncommunicable chronic inflammatory disease was systematically evaluated in adults.[43] The study evaluated the consumption of major foods and nutrients across 195 countries and quantified the impact of their suboptimal intake on mortality and morbidity associated with chronic inflammatory disease. The main inputs to the researchers' analysis included the intake of each dietary factor, the effect size of the dietary factor on disease endpoint, and the level of intake associated with the lowest risk of mortality. Then, by using disease-specific population-attributable fractions, mortality, and disability-adjusted life-years (DALYs), the researchers calculated the number of deaths and DALYs attributable to diet for each disease outcome.

Based on this approach, the authors of this paper found that in 2017, 11 million deaths and 255 million DALYs were attributable to dietary risk factors. Besides a high intake of sodium, all the other listed risk factors linked to deaths and DALYs globally (low intake of whole grains, fruits, nuts and seeds, vegetables, fibers, and legumes) have a proved negative impact on gut microbiome composition and function and favor dysbiosis and inflammation.

The corollary of these findings is that nutrition is driving many noncommunicable diseases, and dietary interventions would represent an effective, cheap way to improve health. We agree with our colleague Lauren Fiechtner that the cost of implementing healthy nutrition habits would be much cheaper than the cost of all the drugs we currently prescribe. It would be cheaper than any drug we can develop to fight inflammation, so why not improve nutrition early in life to prevent DALYs and early mortality? Following the nutritional route would also avoid the potential side effects of drugs that are fighting the consequence (inflammation) rather than the cause (poor eating habits).

Unfortunately, the vast majority of people still do not appreciate that poor eating habits may lead to a certain chronic inflammatory disease, which then requires a specific drug to treat it. Following that, patients often have to take a second drug to treat the side effects caused by the first drug. And, because these are chronic conditions that require chronic treatments, patients must deal with the complications of prolonged treatments and inefficient control of inflammation, ultimately causing a poor quality of life.

So, of all the likely elements in this chain of events—eating, developing a chronic inflammatory disease, treating it, treating any complications of the treatment, inefficient control of inflammation, and experiencing a poor quality of life—only one is mandatory. We must eat at least three times per day; the other events are avoidable.

Eating well serves multiple purposes at once, which operate like dominoes. If we eat well, we don't develop chronic inflammatory diseases, so we don't need to take medications, we don't develop complications, we don't have a decreased quality of life, and our life expectancy is longer. We would vastly improve public health and eliminate many chronic conditions by simply doing *one* thing well that we must do every day—eat! But developing and sustaining healthy eating habits is a challenging concept to implement in the minds and behaviors of many people.

Fighting a Sweet Battle

As parents, many of us have experienced these challenges with our own children and the addictive attraction of sugar and food industry products. Grandparents too often enjoy giving their grandchildren special treats—and the kids really enjoy eating them! Even Fiechtner humorously acknowledges the power of a sweet treat over her own children. "It gives them instant joy," she said. Some researchers have linked a child's "sweet tooth" to evolutionary development during a child's growth spurt.[44] Harvard University evolutionary biologist Daniel Lieberman calls sugar a "deep, deep ancient craving."[45]

But Fiechtner points out that it is the long-term consequences of these items as a regular part of a child's diet that can have detrimental consequences. She is hopeful that the food industry will take some responsibility to improve what we call "junk food." While it is impossible to negate the effects of continually eating food that is nutritionally substandard but that tastes good—like chips and ice cream—we must find a happy and healthy balance.

Nevertheless, it is amazing how much joy children experience eating unhealthy food, even if they have not been exposed to food advertisements, have never been in a supermarket, and know nothing about the nutritional difference between a bag of chips and an apple. When it comes to older children and adults, the food industry approaches the advertisement of their products with the rigors of science. Just as nicotine will make you addicted to smoking, some food companies know that if you activate the sweet-tasting center in a specific part of our brain, we get immediate gratification that shortcuts our rational reasoning about the health consequences of eating junk food.

And unhealthy food is so tasty that we tend to eat more of it than of its healthy counterpart, losing out on both quantity and quality and setting the stage for an unbalanced microbiome in genetically predisposed individuals. This brings us to the verge of what has been previously unthinkable. For the first time in human history, the life expectancy of the next generation is projected to be shorter than that of our current generation. We have the power to stop these epidemics of noninfectious, chronic inflammatory diseases, if we have the common sense to put our scientific discoveries in the microbiome domain at the service of public health policies.

15   Treatments for Disease: Prebiotics, Probiotics, Synbiotics, and Postbiotics

Rebalancing the Microbiome as a Therapeutic Intervention

The plethora of research focused on the human microbiome in recent years has fueled a parallel interest in developing interventions to ameliorate dysbiosis. The ultimate goal of this research is treating noninfectious, chronic inflammatory diseases in which the pathogenetic role of the microbiome has been postulated. The premise for these interventions is that a balanced microbiome is needed to maintain a state of health.

Therefore, in the case of dysbiosis, the therapeutic goal is to reestablish a balanced microbiome. This assumption is supported by the concept that the first forms of life on Earth were prokaryotic, and that the evolution of all eukaryotic life, including humans, was shaped by a symbiotic relationship based on a well-established ecosystem in equilibrium with the surrounding environment and its inhabitants.

In humans, all surfaces and cavities are colonized by a microbiome that shapes the physiological functions and metabolic pathways vital to the state of health of each organ and tissue. This health equilibrium occurs through a carefully orchestrated balance between the host response and its colonizers. These microbiomes supply essential "services" to the ecosystem that benefits health through homeostasis. Changes in microbiota composition and function can lead to dysbiosis, which can have significant consequences for disease development or progression.

Based on this premise, in this chapter we review key strategies to reestablish a balanced microbiome to regain health, taking into account the limitations related to the therapeutic applicability of current microbiome knowledge outlined in chapter 13. Most studies aimed at linking dysbiosis to disease pathogenesis are focused on the gut microbiome, and the major focus of this chapter is on therapeutic strategies aimed at treating gut dysbiosis, including prebiotics, probiotics, synbiotics, and postbiotics. Nevertheless, it is worthwhile to remember that changes in the gut ecosystem may lead to consequences affecting the function and metabolic profile of many organs and tissues in our body.

Prebiotics

Definition

Prebiotics were first described in the mid-1990s as "non-digestible food ingredients that beneficially influence the health of the host by stimulating the activity of one or more commensal colon bacteria."[1] They exert a beneficial effect on the health of the host by several nonmutually exclusive mechanisms. These include (1) acting as a fermentable substrate for specific symbiotic bacteria that lead to the release of SCFAs in the gut intestinal tract, which influences many molecular and cellular processes; and (2) acting directly on several cellular functions, including epithelial barrier function and immune response from cells belonging to the GALT.

There is growing interest in nutrients with prebiotic properties for their potential to promote a more tolerogenic environment by decreasing antigen trafficking, which affects gut permeability and immune response. This action then epigenetically influences the immune response toward an anti-inflammatory tone. By directly affecting these functions, or by selecting commensal bacteria capable of fermenting prebiotics, or both, this environment favors a symbiotic microbiome.

New knowledge about the mode of action and specificity of prebiotics led to their redefinition in 2017 by the International Scientific Association for Probiotics and Prebiotics as "substrates selectively used by micro-organisms of the host conferring benefits for his/her health." The three criteria prebiotics must meet are to (1) be resistant to digestion in the stomach and upper intestine; (2) be fermentable by the gut microbiota; and (3) specifically stimulate the growth and/or activity of intestinal bacteria beneficial to our health.[2]

As outlined earlier, since gut microbiome functions can affect many organs in our body, the benefits of prebiotics are not limited to gut functions but can affect many biological processes throughout the body.[3] Prebiotics are usually composed of linked sugars such as oligosaccharides and short-chain polysaccharides, which are not digestible by intestinal enzymes. This gives them the ability to serve as nutritional substrates for microorganisms considered beneficial.

Structural Characteristics and Sources

The first prebiotics that humans are exposed to are the HMOs present in breast milk (see chapter 3). They have presented a nutritional puzzle for many years. Abundant in human milk, they are not digestible by the breast-fed baby. Both human milk and colostrum contain large concentrations of oligosaccharides characterized by a lactose-reducing end elongated with fucosylated or sialylated N-acetyl-lactosamine units, or both. More than 150 HMOs have been identified that differ in size, charge, and sequence; the most frequent HMOs are the neutral fucosylated and non-fucosylated oligosaccharides.[4]

Since HMOs provide no direct nutritional value to the infant, it is postulated that like other prebiotics, they may represent preferred substrates for several species of gut bacteria. This promotes the growth of beneficial intestinal flora and helps to shape the gut microbiome. When HMOs ferment in the microbiome, SCFAs are generated. SCFAs contribute to general intestinal health by stimulating the growth of gut commensal bacteria, providing direct nourishment for epithelial cells lining the gut, and modulating host-epithelial responses. These combined effects also prevent colonization of pathogens in the gut epithelial barrier by acting as decoy receptors and providing a selective advantage for colonization by commensal bacteria.

With the introduction of solid foods into the diet, fructans like fructo-oligosaccharides (FOS) and inulin become the most abundant prebiotics ingested. Due to their modulatory effects on the gut microbiome, these are the most-studied food-derived prebiotics. They are found in many plant-based foods, including leeks, onions, garlic, artichokes, chicory, asparagus, and bananas as well as cereals like rye and corn. The recommended daily fiber intake is typically achieved with a balanced European diet characterized by the ingestion of three to eleven grams of natural prebiotics.

In comparison, the typical U.S. diet provides only one to four grams of daily prebiotic intake.[5] Because of this dietary shortfall, "second-generation" prebiotics have recently been designed to

enhance functionality and improve protective effects by changing their chemical structure through digestion with a specific
enzyme produced by a probiotic. Nevertheless, although there are
many commercially available foods and food ingredients claiming prebiotic activities, currently only lactulose, FOS, and galacto-
oligosaccharides (GOS) have a proven prebiotic effect and status.

Mechanisms of Action

As discussed, prebiotics can influence host health by two distinct
mechanisms: indirectly, by acting as a fermentable substrate for
beneficial bacteria and contributing to the establishment and
maintenance of a symbiotic microbiome, or directly, by affecting
key gut functions needed to maintain homeostasis. The indirect
effects of prebiotics have been the most-studied mechanism of
their beneficial effect on the gut microbiome.

In one example, the consumption of prebiotics like inulin
has been shown to increase the abundance of beneficial bacteria,
including *Bifidobacterium* and *Lactobacillus*, which contribute to
gut health by preventing colonization of enteric pathogens.[6] And
the inulin-induced increase of *Bifidobacteria* has been correlated
with an increase of acetate production by these microorganisms, with a subsequent decrease in the abundance of *C. difficile* pathogens in the intestinal tract and an inhibition of their
translocation from the gut lumen to the blood.[7] Also, new prebiotics, including apple pectin and 1-kestose, seem to favor the
proliferation of *F. prausnitzii*, which is known to possess an anti-
inflammatory effect, and *Eubacterium eligens* DSM3376, which is
capable of inducing in vitro secretion of the anti-inflammatory
cytokine IL-10.[8]

But the most-studied indirect effect of prebiotics leading to
gut health is the elaboration of SCFAs by enteric commensals

following their fermentation. SCFAs are used by gut microbiota for their own metabolism or are released in the lumen, where they can interact with different cells. These include intestinal epithelial cells (IECs) and innate/adaptive immune cells that affect several cellular processes, including gene expression, differentiation, proliferation, and apoptosis. While their multiple mechanisms of action remain the object of active research, it has been demonstrated that SCFAs can activate G protein–coupled receptors (GPRcs) to modulate cell development, function, and survival.[9]

These effects are the consequence of the activation of several signaling pathways mediated by a variety of protein kinases (including the adenosine monophosphate (AMP) activated protein kinase and mitogen-activated protein kinases), the mammalian target of rapamycin (mTOR), the signal transducer and activator of transcription 3 (STAT3), and the nuclear factor kappa-light-chain-enhancer of activated B cells. The outcome of these activations depends on the targeted cell.

In IECs, SCFAs trigger the secretion of cytokines and the production of chemokines.[10] GPRc-dependent activation of SCFAs also seems to regulate the expression of an antimicrobial peptide on IEC-activating mTOR and STAT3 signaling, thus preventing the colonization of pathogens on the gut mucosa.[11] Given the close interaction between IECs and immune cells present in the gut lamina propria, SCFAs can indirectly affect the function of immune cells like dendritic cells, Tregs, and FOXP3+ immune cells. While FOXP3 full-length (FL) is capable of properly downregulating the Th17-driven immune response, the alternatively spliced isoform FOXP3 Δ2 is not.

Prebiotics and Immune Cells in Celiac Disease

Gloria Serena and other colleagues from our group at the MIBRC at MGH have studied the role of both forms of FoxP3 in the autoimmune disease of celiac disease. The active state of celiac disease has been associated with impairments in Treg cell function rather than with their reduction, which has been described in other autoimmune diseases. Our researchers explored whether imbalances between FoxP3 isoforms may be pathogenetically linked to the disease. They showed that intestinal biopsies from patients with active celiac disease showed increased expression of FOXP3 Δ2 isoform over FL, while both isoforms were expressed similarly in nonceliac control subjects.[12]

Conversely to evidence from the intestine, peripheral blood mononuclear cells from healthy subjects did not show the same balance between isoforms. We therefore hypothesized that the intestinal microenvironment may play a role in modulating alternative splicing. The proinflammatory intestinal microenvironment of active celiac disease patients has been reported to be enriched in butyrate-producing bacteria, while high concentrations of lactate have been shown to characterize the preclinical stage of the disease.

Serena and colleagues showed that the combination of interferon alpha and the SCFA butyrate triggers the balance between FoxP3 isoforms in healthy subjects, while the same does not occur in celiac disease patients. This was one of the first demonstrations mechanistically linking the FoxP3 alternative splicing process with a microbial-derived SCFA, thereby providing additional evidence of the epigenetic role of the gut microbiome on the immune cells mediated by the elaboration of specific metabolites.

SCFAs also promote the gut barrier function itself through different mechanisms, including the production of IL-18, a cytokine that contributes to intestinal epithelium homeostasis. Using intestinal organoids developed from duodenal biopsies from both healthy subjects and patients affected by celiac disease, the MIBRC researchers explored the role of microbiota-derived SCFAs in modulating the epithelium's response to gluten. When compared to healthy controls, RNA sequencing of celiac disease organoids revealed significantly altered expression of genes associated with gut barrier, innate immune response, and stem cell functions.[13] Monolayers derived from celiac disease organoids exposed to gliadin showed increased intestinal permeability and enhanced secretion of proinflammatory cytokines compared to healthy controls, an effect that was mitigated by the SCFA butyrate, lactate, and PSA.

These data confirmed the finding from another group that showed a direct effect of prebiotics on the maintenance of epithelial barrier function.[14] Specifically, the investigators showed that the application of prebiotics onto immortalized, gut-derived epithelial cell lines and human intestinal organoids directly promoted barrier integrity, thus preventing pathogen-induced barrier disruptions involving the induction of select tight junction proteins through a protein kinase–dependent mechanism.

Probiotics

Definition

The definition of "probiotics" has changed from early descriptions more than a century ago, due to an increasing knowledge of their characteristics and effects on the human host. Probiotics have been defined reductively as live microorganisms with health-promoting properties. Although the term "probiotics" was introduced by Daniel Lilly and Rosalie Stillwell in 1965, the father of the entire "probiotic movement" is clearly Elie Metchnikoff, the Russian comparative zoologist and pioneering immunologist. Corecipient with Paul Ehrlich of the Nobel Prize in Physiology or Medicine in 1908, he became the first biologist to describe the beneficial effects of probiotics.[15]

With a visionary intuition that remains an object of active debate, Metchnikoff linked regular consumption of yogurt to the delay of the aging process and the promotion of longevity. His theory, presented in his book *The Prolongation of Life*, translated into English from the original Russian in 1908, postulated that putrefactive bacteria present in the gut release toxins and other harmful substances to the host, so accelerating the process of aging. Based on Metchnikoff's theory, the administration of host-friendly bacteria present in yogurt could counteract this action, restore intestinal balance, and enhance human health.[16] Compare his theory to the definition of probiotics by the Food and Agriculture Organization of the United Nations and WHO as "live microorganisms that, when administered in adequate amounts, confer a health benefit on the host."[17]

Mechanisms of Action

The use of probiotics to implement human health traveled underground for many decades, only to resurface to the attention of clinical and scientific communities with the advent and popularity of studies on the human microbiome and its role in health and disease. Based on renewed attention to this topic, a large body of literature has been published that shows the positive outcomes of probiotic supplementation on several gastrointestinal disorders, including antibiotics-associated diarrhea, IBS, necrotizing enterocolitis, ulcerative colitis, lactose intolerance, and colorectal cancer.[18]

Parallel to these reports, and in an attempt to make sense of the multiple beneficial effects of probiotics reported, many articles have endeavored to review their mechanisms of action. These mechanisms include modulating the immune system, inducing anti-inflammatory and antioxidant responses, competing with pathogens, and subtracting key nutrients and/or preventing their colonization, as well as producing antimicrobial substances. Additionally, probiotics could exert antiproliferative effects in colon cancer cells by inducing apoptosis and cell-cycle arrest.

Nevertheless, these reports are sometimes contradictory, not replicated by subsequent studies, and based on suboptimal study designs. To bring some order to the matter, we reached out to Jon Vanderhoof of Boston Children's Hospital and Boys Town National Research Hospital. A good friend and superb pediatric gastroenterologist and clinical investigator, he has been involved in the use of probiotics to treat a variety of gastrointestinal disorders since their "rediscovery." We asked him for a historical perspective on the general use of probiotics to treat gastrointestinal disorders and, more specifically, food allergies. What follows is his account of the some of the highlights of the days before "probiotics" became a household word.

Of course, he first returned to Metchnikoff's hypothesis and exploitation of fermented foods, based on Metchnikoff's observation that people who ate fermented foods seemed to be healthier. The next logical step was to consider taking the bacteria involved in the fermentation of food and feeding patients with these live bacteria. This method assumed that these bacteria were what made the patients healthier, and somehow the bacterial microorganisms would positively influence the health and well-being of the human host. This was an important part of where the "probiotics revolution" began.[19]

Based on this rationale, several investigators started exploring the use of different live probiotics—like *Lactobacillus bulgaricus* and *acidophilus* and other strains—by feeding them to patients, even though there was not much scientific evidence to support their therapeutic use. According to Vanderhoof, the scientific foundations for the use of probiotics in specific diseases were proposed later on by researchers like Sherwood Gorbach at Tufts University and Stig Bengmark in Sweden. These researchers were trying to isolate organisms that appeared to be more specifically beneficial than others—thinking that there might be some strains that were more effective at colonizing and growing in the gastrointestinal tract. This might make these microorganisms more effective at killing enteric pathogens by mechanisms that were not entirely clear.[20]

In the late 1990s, Gorbach realized that the positive health benefits of probiotics, and specifically those related to Lactobacilli, apply to only a few strains among those that were commercially available at that time. He reasoned that these differences were probably related to a key feature, namely the ability of a probiotic to colonize the intestinal tract to influence human health. Vanderhoof noted that this requirement disqualified many of the strains used in fermented dairy products in the 1990s.[21]

The Success Story of *Lactobacillus* GG

Lactobacillus GG (LGG), a variant of *L. rhamnosus* and extensively studied in adults and children, would play an important role in the development of probiotics and in Vanderhoof's career. When consumed as a dairy product or as a lyophilized powder, LGG showed good colonization of the gastrointestinal tract for several days. For this reason, LGG has been successfully used to treat traveler's diarrhea, antibiotic-associated diarrhea, and relapsing *C. difficile* colitis. In infantile diarrhea, duration of the attack is primarily reduced, and LGG-fermented milk lessens intestinal permeability defects caused by exposure to cow's milk or rotavirus infection. It confers subsequent beneficial effects on intestinal immunity as measured by increased numbers of IgA and other immunoglobulin-secreting cells in the intestinal mucosa, stimulation of local release of interferon, and increased antigen transport to underlying lymphoid cells, which serves to increase antigen uptake in Peyer's patches.

And in an animal model of colon cancer, LGG showed a reduced incidence of chemically induced tumors in the large bowel of rodents, paralleled by extensive safety signals in both animal models and humans. Vanderhoof pointed out that at the same time, similar efforts in Scandinavia and Japan using other probiotic strains, including *L. plantarum*, reached similar conclusions.[22]

Microbiome researcher Erika Isolauri, a colleague of Vanderhoof, is head of the Nutrition, Allergy, Mucosal Immunology and Intestinal Microbiota Research Program of the Department of Pediatrics at the University of Turku in Finland. In her early research in the laboratory of Jehan-François Desjeux at the Conservatoire National des Arts et Métiers in Paris (where coauthor Alessio Fasano spent a few months), she raised the science of probiotics to the next level in her work with Ussing chamber assays. The Ussing chamber is what Fasano used to discover many toxins from enteric pathogens in his early research. This led him to the discovery of the protein zonulin (introduced in chapter 6), which plays such a prominent role in the regulation of gut permeability.

Isolauri hypothesized that the LGG strain could influence gut permeability, which she proved through Ussing chamber experiments that showed reduced gut permeability. The general idea was that gut permeability was important in the development of food allergies. She postulated that the risk of allergy might be ameliorated by reducing gut permeability with LGG.[23]

Isolauri proceeded to design clinical studies using LGG, and the results seemed to support her hypothesis. Vanderhoof recalled that researchers were then becoming interested in the concept of gut bacteria influencing, besides barrier function, immune cell differentiation and Th1-Th2 immune response balance by affecting cytokine production and, ultimately, reducing allergic responses. At the same time, clinical investigators were performing clinical studies looking at illnesses in day-care settings, including diarrheal illnesses and respiratory illnesses. All these studies provided the rationale to use probiotics to redirect the immune system to ameliorate inflammation.

The outcome of these studies led to the realization that favoring LGG engraftment also changed the gut ecosystem, proving that in the world of the microbiome nothing works in isolation. When organisms like LGG are administered, they tend to change the neighborhood by favoring the growth of some microorganisms and retarding the growth of others, thereby changing the metabolic profiles in the gut micromilieu.

For example, LGG appears to favor the growth of some butyrate producers and possibly other microorganisms, producing yet undescribed effects that influence the probiotics' mechanism of action. These findings suggest that probiotics may exert their action by nonmutually exclusive mechanisms, such as direct interaction with the immune system due to colonization of the gut mucosa, or indirectly when residual, luminal, and nonadhering microorganisms secrete molecules (postbiotics) that in turn influence barrier function and immune parameters.

Moving Probiotics from the Lab to Clinical Applications

To put into perspective the history of the development of probiotics from preclinical proof of concept to their therapeutic use, let's take a closer look at the role Vanderhoof played in this process. In the mid-1990s, a food company with a huge market share in oats in his hometown of Omaha, Nebraska, approached Vanderhoof for advice. One of the company's executives was investigating a company in Sweden working on fermented oats as a remedy for severe postsurgery complications.

To exploit this opportunity, the company contacted Vanderhoof to review some of the data they had generated using fermented oats–derived probiotics and, specifically, the *L. plantarum* that the Swedish company had isolated from their preparations. He was so intrigued by the data and by what he read in the literature on the topic that he agreed to fly to Sweden. There he spent what he described as "one of the most enlightening afternoons of my life" at the home of Swedish surgeon Stig Bengmark.[24]

Bengmark used his slide projector to help narrate a fascinating story. Concerned about the effects of preoperative antibiotics used prophylactically on his patients undergoing liver surgery, Bengmark asked his surgery resident to review the medical records. What they learned was that many of these patients experienced a high incidence of multiorgan failure, something that was not detected as frequently in patients who did not receive the antibiotic treatment.

Bengmark was an innovative thinker interested in nutrition. He developed a gruel of fermented oats that he started feeding to his patients, with the remarkable result that the incidence of multiorgan failure basically dropped to zero. From this fermented oats formulation, Bengmark and his collaborators isolated this *L. plantarum* organism. Using his slides, Bengmark showed Vanderhoof the data he generated with this probiotic strain in terms of ameliorating inflammation with mechanisms that closely resembled LGG's mechanism of action.[25]

Returning to Omaha, Vanderhoof delved deeper into the story. He was surprised to learn that LGG was patented by Sherwood Gorbach in Boston, a scientist whom he knew well. His next stop was in Boston, where he met Gorbach, who showed him his data and told Vanderhoof that he had sold the LGG patent to a Finnish company. With this information in hand, the Nebraska company sent a team that included Vanderhoof to Finland to negotiate the acquisition of the U.S. rights for the LGG patent.[26]

As a scientist and not a businessman, Vanderhoof was somewhat reluctant to negotiate rights to a Finnish patent, but he and the team were successful. With the rights to use LGG as a probiotic, the company intended to put it into food, but Vanderhoof convinced them it would be better to put the probiotic into a pill. The company had never done anything like that, so they made this the focus of a small subsidiary, which eventually led to the creation of Culturelle®.[27]

Vanderhoof subsequently launched pilot clinical studies using Culturelle® with very promising results, which provided the impetus to perform additional, larger, multicenter studies, most of them in adults. The results showed remarkable efficacy in treating antibiotic-associated diarrhea and, thanks to studies performed by Isolauri and colleagues, atopic dermatitis.[28] Since then, Culturelle® has become one of the most prescribed probiotics worldwide.

Based on these very promising results, the company wanted to use LGG for other clinical indications and in different formulations. Vanderhoof suggested adding it to baby formulas used to treat food allergies to see if it would increase efficacy in treating these diseases.[29]

This idea spawned another business venture, which accelerated when Roberto Berni Canani from Fasano's alma mater, Università Federico II in Naples, approached Vanderhoof with data from his pediatric patients. These data showed that children who received the formula with LGG in it seemed to outgrow their allergies sooner or become tolerant sooner.[30] Well before there was a major interest in probiotics, Vanderhoof energetically pursued research into their potential efficacy in child development, making him a true pioneer in the field.

Landscape for Probiotic Development: Challenges and Opportunities

Fast-forward to today, and the probiotics industry has ballooned into a multibillion-dollar operation. Along with great expectations, it also has raised major concerns, mainly linked to the lack of a clear legislative path or plan to regulate the use of probiotics for therapeutic interventions.

There is much confusion about informing health care professionals on when and why to use probiotics, and which indications and which probiotics demonstrate therapeutic efficacy.

This confusion is partly fueled by a lack of agreement on how to categorize probiotics; producers want to keep them as food additives promoting health. With this categorization, they are not required to go through the scrutiny of the U.S. Food and Drug Administration (FDA) exam to prove safety and efficacy. However, patient advocates and legislators safeguarding the interests of patients are demanding the same scrutiny for probiotic products that is required for any intervention aimed at treating, preventing, or ameliorating disease.

According to Vanderhoof, this is a tricky situation, since the FDA is used to dealing with drugs in which pharmacokinetics and lethal dose are clear parameters to monitor. But to date, there has been no clear guidance on how to regulate the use of live organisms that do not follow a typical pharmacological dynamic. Furthermore, the current path for drug development and commercialization involves several preclinical and clinical phases for commercial approval, with an average of fifteen years and more than $1 billion of investment. While these costs can be recovered by commercializing drugs that are sold at a much higher price than their cost of production, this approach will not work with probiotics, where the revenue margin is much smaller.

Another challenging factor is that industries that produce probiotics have a difficult time intellectually protecting their products with patents, since there is great misunderstanding about the importance of strain specificity. You can use the same species of probiotic—let's say LGG—but depending on the specific strain formulated, the mechanism of action and, therefore, the therapeutic impact on the intended disease can be radically different. Nevertheless, if a study shows efficacy of LGG in treating a condition, there is nothing to prevent another company from using another *Lactobacillus* strain and commercializing it by claiming that a similar efficacy has been reported in the literature, without any proof that this is the case.

Based on these considerations, Vanderhoof believes that a mechanism is needed to ensure that when probiotics are sold, there are data proving that they actually work. He also emphasizes the importance of strain specificity and some type of consumer protection. It is imperative that the strain used in the clinical trial is the same strain that is commercialized, with comparable characteristics. These include that the strain is alive and at a reasonable dose compared to what was used in the clinical study.[31]

The situation becomes even more confused when the intent is not to treat but to prevent. Most probiotic consumers take them because "they are good for my health." If you do not have a specific preventive target, depending on the choice of the probiotic species and strain to be used, you can obtain detrimental rather than beneficial effects. There is a clear legislative need for the federal agencies (see below) to ensure that a specific probiotic marketed for a specific indication has been proven in its efficacy; that it has some value to the consumer; and that the strain's (or strains') composition, strength (number of colony-forming units or CFUs), and viability claimed on the product label are factual.

There have been several reports in the literature showing that the probiotics products you buy at the supermarket or health food store may not contain enough CFUs of probiotics or may not contain probiotics at all. Even worse, there can be pathogens in the products. A related concern is that because of the market opportunity and lack of regulations, some companies commercialize products without the proper safety and efficacy data. This unethical business practice can vilify the use of probiotics as a potentially superb tool to ameliorate inflammation in a variety of conditions.

Another fear with the current use of probiotics is that we are making the same mistake we made with penicillin when it was discovered almost a century ago. This is particularly true in the pediatric community—it is now rare to see children in clinic who have not been exposed to probiotics. After its discovery, penicillin was promiscuously used for a variety of infections. We have since learned that some of these infections would never respond to penicillin because they are caused by Gram-negative bacteria that are not sensitive to penicillin, and the bacteria that initially responded to penicillin have since developed resistance to it.

Similarly, the wide use of probiotics, without clear indications and formulations, may jeopardize this potentially effective tool to modify the microbiome's composition and function to support immune health and gut barrier integrity or to treat specific inflammatory processes caused by dysbiosis. The final point to be made is that because humans are not made genetically and biologically equal, we cannot generalize the use of any given probiotic formulation as beneficial to everyone's health. Vanderhoof's ultimate message is that we must understand that in order to exploit the beneficial potential of probiotics, we must expand our scientific knowledge about specific probiotic strains and test their efficacy on specific, quantifiable targets, such as gut permeability and immune response, with the ultimate goal of ameliorating inflammation.[32]

Current Regulations on the Use of Probiotics

To achieve what Vanderhoof has laid out, there is an urgent need for new rules regulating probiotic therapeutic use in humans by federal agencies. Along with health claims, quality control in manufacturing, safety, and efficacy must be considered. Probiotic formulations that infer prevention, therapy, treatment, relief, or disease diagnosis are classified as medical or pharmaceutical products with concomitant regulation. As live, complex organisms, these products are regulated under biologics.

A comprehensive review of this unregulated market by Claudio de Simone sheds some light on the matter. "The great majority do not make disease-specific claims and are therefore classified as food supplements or dietary supplements. In some cases, when the clinical data are convincing for a certain probiotic formulation, the product can be classified as a medical food intended for the dietary management of a specific disease (ie, pouchitis). Both of these categories are regulated much less stringently than pharmaceutical products."[33]

As de Simone notes, there is no international agreement among nations to regulate these products. Probiotics and food supplements are regulated under the Food Products Directive and Regulation of the European Union (EU). The European Food Safety Authority (EFSA) is the agency for the authorization of health claims for probiotics. A list of microbial cultures with a "Qualified Presumption of Safety" is issued from EFSA, but to date, as de Simone notes, EFSA "has rejected all submitted health claims for probiotics."[34]

Similarly, the FDA has not approved any probiotic as a live biotherapeutic product, defined as a biological product other than a vaccine, which contains live organisms used to prevent or treat a disease or condition in humans. However, most probiotic products in the United States are classified as FDA-regulated foods and include dietary supplements containing probiotics that are legally available. Although dietary supplements must comply with "good manufacturing guidelines," they do not include quality or efficacy.

As in Europe, dietary supplements cannot be marketed legally to cure, mitigate, treat, or prevent any diseases. However, in the United States, they can make structural or functional claims such as "supports healthy digestion," accompanied by an FDA-mandated disclaimer.

According to de Simone:

> Claims must be truthful, not misleading, and substantiated by scientific evidence. There is also a category of probiotics which are formulated to be consumed or administered enterally under the supervision of a physician and which are intended for the dietary management of a specific disease or condition for which nutritional requirements, based on recognized scientific principles, are established by medical evaluation. These formulations fall in the category of medical foods in the United States. Medical foods are not drugs, and, therefore, are not subject to any regulatory requirements that specifically apply to drugs. However, a medical food that bears a false or misleading claim would be considered misbranded under section 403(a)(1) of the Federal Food, Drug, and Cosmetic Act.[35]

Since probiotics are usually viewed as drugs by U.S. regulators, for research intentions, human studies for probiotics, foods, and dietary supplements not intended to be marketed as drugs fall under the FDA's Investigational New Drug (IND) program framework. According to de Simone, safety studies must be completed before efficacy studies proceed, even for probiotics in wide use that have a "Generally Recognized as Safe" designation.[36]

As de Simone states, in the EU and the United States, we are left with a "regulatory void which does not take into account the complex nature of probiotic products—the fact that they are living organisms and therefore dynamic and not static; the fact that their characteristics vary significantly among both species and strains; and the additional complexities that arise in multi-species or multistrain products where the individual components may interact with one another. It is increasingly recognized that the current approach to regulation is inadequate and can lead to problems of quality, safety, and claim validity in commercial probiotic products that are used in a medical context, including those used in vulnerable populations."[37]

In an attempt to address these shortfalls, the FDA issued a guidance document in 2016 that explained how researchers studying probiotics as drugs can meet the manufacturing requirements necessary for early clinical trials.[38] In a subsequent document, the FDA stated that "more work and continued partnership between the FDA and various stakeholders is needed to advance the clinical science necessary to appropriately understand the safety and effectiveness of these products."[39] In 2018, the FDA and the NIH outlined a roadmap to fill this legislative void to protect consumers and maximize the therapeutic potential of probiotics.

Commercialized Probiotics and Clinical Indications

The use of probiotics to implement health has grown exponentially during the past two decades. The 2012 National Health Interview Survey (NHIS) showed that about four million (1.6 percent) U.S. adults had used probiotics or prebiotics in the previous thirty days.[40]

Among adults, probiotics or prebiotics were the third most commonly used dietary supplement other than vitamins and minerals. The use of probiotics by adults quadrupled between 2007 and 2012. The 2012 NHIS also showed that three hundred thousand children aged four to seventeen years (0.5 percent) had taken probiotics or prebiotics in the thirty days before the survey.[41]

The popularity of probiotic use in clinical practice is confirmed by the evidence that commercial probiotics have gained widespread popularity and are now estimated to be a $37 billion market worldwide. These data and the fact that microbiome research is still in its infancy suggest that the clinical use of probiotics is outpacing the science that should justify their therapeutic use.

Oral Probiotics for the Treatment of Intestinal Diseases

Among all the possible indications for the use of probiotics in clinical practice, treatment of infectious diarrheas and antibiotic-associated diarrhea (AAD) seem to be the most supported by data, although evidence from clinical trials is mixed and often of low quality. One of the first reports on a large number of patients was a European multicenter study performed in 2000 and headed by Stefano Guandalini, one of the pioneers of the use of probiotics in pediatrics. He is also coauthor Alessio Fasano's mentor and helped guide Fasano's scientific focus toward pediatric gastroenterology and enteric pathogens.

Guandalini and his investigative team used the probiotic LGG in a multicenter trial to evaluate its efficacy in an oral rehydration solution administered to patients with acute-onset diarrhea of all causes. Children from one month to three years of age with acute-onset diarrhea were enrolled in a double-blind, placebo-controlled investigation and randomly allocated to receive oral rehydration solution plus placebo or the same preparation with a live preparation of LGG (at least 10^{10} CFU/250 ml). After rehydration in the first four to six hours, patients were offered their usual meals plus free access to the same solution until their diarrhea was arrested. Duration of the diarrhea after enrollment was reduced by almost thirteen hours in the LGG group compared to the placebo group.[42]

In rotavirus-positive children, the investigators showed that the diarrhea lasted 76.6 +/- 41.6 hours in the placebo group versus 56.2 +/- 16.9 hours in the LGG group. They also reported that the diarrhea lasted longer than seven days in 10.7 percent of children enrolled in the placebo group versus 2.7 percent of LGG-treated patients, who were hospitalized for a shorter period of time compared to those who did not receive the probiotic.[43] Taken together, the authors' results suggested that administration of LGG to children with acute diarrhea was safe and resulted in a shorter duration of diarrhea, less chance of a protracted course, and a quicker discharge from the hospital.

To assess the effects of probiotics in treating infectious diarrhea, a Cochrane analysis reviewed randomized controlled trials comparing a specified probiotic agent with placebo or no probiotic in people with acute diarrhea proven or presumed to be caused by an infectious agent.[44] Twenty-three studies met the inclusion criteria, with a total of 1,917 participants, mainly in countries with low overall mortality rates. Trials varied in relation to the probiotic(s) tested, dosage, methodological quality, and the diarrheal definitions and outcomes.

Based on the analysis, probiotics reduced the risk of diarrhea at three days and the mean duration of diarrhea by 30.48 hours. Subgroup analysis by probiotic(s) tested, rotavirus diarrhea, national mortality rates, and the age of participants did not fully account for the heterogeneity. Based on these data, the authors concluded that probiotics appear to be a useful adjunct to rehydration therapy in treating acute, infectious diarrhea in both adults and children.[45]

While justifying the use of probiotics in infectious diarrheas, these analyses extrapolate data from a variety of highly heterogenous studies that do not necessarily hold true when challenged by well-designed prospective studies. Indeed, two large-scale clinical trials reported in the *New England Journal of Medicine* suggest that the situation in infectious diarrhea might also be more complex than previously believed. Stephen Freedman and colleagues did a randomized controlled trial of a probiotic containing *L. rhamnosus* and *L. helveticus* in children presenting at the emergency department with gastroenteritis.[46] Contrary to expectations, they found that the probiotic did not prevent development of moderate-to-severe gastroenteritis within fourteen days after enrollment.

Similarly disappointing results were reported by David Schnadower and colleagues using LGG alone. In trials that used probiotics available over the counter in the United States, they showed no significant difference from placebo in the duration of diarrhea and vomiting, the number of unscheduled health care visits, or the length of absence from day care.[47] While these results cannot be generalized to other probiotic strains or preparations, they do show that we have far to go in elucidating which probiotics might provide benefits in which patients and clinical settings.

Antibiotic-Associated Diarrhea

Several studies performed in the early 2000s on AAD suggested the efficacy of probiotics in treating and sometimes preventing this condition. AAD is reported in up to 39 percent of hospitalized patients treated with antibiotics and varies from uncomplicated diarrhea to colitis and *C. difficile*–associated pseudomembranous colitis. The pathogenic factors include any one or all of the following: use of antibiotics causing dysbiosis, antibiotic-resistant pathogenic bacteria taking over the gut ecosystem and causing inflammation and diarrhea, change in bacterial flora causing changes in carbohydrate metabolism, and changes in SCFA metabolism and absorption.

The well-known syndrome of pseudomembranous colitis characterized by high levels of toxin-forming *C. difficile* may take a fulminate course with a high recurrence and mortality rate. To evaluate the evidence for probiotic use in the prevention and treatment of AAD, a systematic review and meta-analysis was performed on a total of sixty-three randomized clinical trials that satisfied inclusion criteria.[48] The majority of the reviewed trials used *Lactobacillus*-based interventions alone or in combination with other genera. However, the specific strains used were poorly documented.

The pooled relative risk of the studies analyzed, which included 11,811 participants, indicated a statistically significant efficacy of probiotic administration in reducing AAD in trials reporting on the number of patients with AAD. This result was relatively insensitive to numerous subgroup analyses. However, the authors reported significant heterogeneity in pooled results, and the results were insufficient to determine whether this association varied systematically by population, antibiotic nature, or probiotic preparation.

Based on these results, the use of probiotics after antibiotic treatment or preemptively to prevent AAD is an increasingly common clinical practice. However, two studies reported in *Cell* question whether taking highly concentrated supplements of probiotics is effective rather than detrimental in favoring the recovery of normal gut flora. Jotham Suez and colleagues, who investigated the recovery of the gut microbiota after antibiotic treatment in a murine model and in a human preclinical model, reported that probiotics might perturb rather than aid this process.[49]

The authors showed that probiotics rapidly colonized the gut mucosa but prevented the normal microbiota from repopulating for up to five months. Also questioning the concept of using one probiotic to fix everything, Niv Zmora and colleagues showed that colonization occurs in highly individualized patterns. Results from their study showed that the gastrointestinal tracts of some individuals rejected the colonization of probiotics while others allowed colonization.[50] Studies like these, which we need many more of, show that many individuals taking probiotic supplements are simply wasting their time and money.

Oral Probiotics for Treatment of Extra-Intestinal Diseases

So far, we have focused on the use of probiotics to treat gastrointestinal diseases. While the previous section confirms that the vast majority of commercially available probiotics are aimed at treating gastrointestinal disorders, there is growing evidence that human microbiota can exert effects at sites far from the microenvironment in which they reside.

We have mentioned that the gut microbiome can affect behavior and metabolic pathways leading to obesity, asthma, and T1D, just to name a few. Therefore, it is implicit in this concept that the beneficial manipulation of gut microbiota through probiotic intervention does not necessarily impact only the gut pathophysiology, but it may also exert beneficial effects at distant sites and organs. For example, it has been shown that the manipulation of gut microbiota by probiotic consumption may mitigate cardiovascular disorders;[51] promote bone health and integrity by elaborating an anti-inflammatory effect;[52] accelerate the wound-healing process; protect against UV-induced photodamage; and alleviate symptoms in skin diseases, such as dermatitis and psoriasis.[53]

One of the most provocative and fascinating opportunities to use probiotics to treat extra-intestinal diseases is their use in resolving dysregulation of the gut-brain axis involved in a variety of mental health problems, neurodegenerative diseases, and neurodevelopmental disorders. This topic will be covered in greater detail in chapter 16.

A Final Word of Caution on the Therapeutic Use of Oral Probiotics

Combined, the studies outlined in this chapter provide some "food for thought" on the indiscriminate use of probiotics for treating or preventing diseases. Not only is the evidence for these benefits open to dispute, but also, because probiotics are sold as supplements, manufacturers in many countries are not required to provide evidence of their safety and efficacy to regulatory bodies. The ubiquity of probiotic products would suggest that, at worst, they are harmless. Nevertheless, some safety concerns have been raised, including the risk of contamination, the possibility of fungaemia or bacteremia (particularly in immunocompromised, elderly, or critically ill individuals), SIBO, and antibiotic resistance.

Adding to these concerns, clinical trials of probiotics have not consistently reported safety outcomes. While the logic behind probiotics might seem sound, it is clear that we have a long way to go before understanding the complexity of the microbiome and the effects—both helpful and detrimental—that probiotics might have on human health. Everyone has a unique gut microbiome, and the effects of different bacteria on different people are likely to be highly variable. As a consequence, probiotic use needs to be personalized for optimal benefits.

Also, commercially available products might not contain the correct strains or quantities of bacteria to provide benefits to all, and most probiotic supplements contain only single strains, vastly oversimplifying the complexity of microbiome modulation. Therefore, although taking supplements for improved health is certainly an attractive prospect, those looking to aid their gut microbiome to treat or prevent diseases might be better served by consuming a healthy, varied diet. In the meantime, we are in dire need of rigorous clinical trials to substantiate potential health benefits and to help develop next-generation probiotics to implement more personalized manipulation of the host microbiome to treat, ameliorate, or prevent diseases.

Next-Generation Probiotics

The limitations on the commercial probiotics currently available that are outlined in this chapter prompt some considerations to optimize their use for microbiome manipulation to improve health. Most probiotics used in clinical practice to specifically target functions linked to chronic inflammation, including regulation of gut barrier function and immune response, have been randomly selected or chosen based on "common sense" associative studies. These studies have outlined their potential beneficial role in a manner similar to what happened for species *Lactobacillus* or *Bifidobacterium*.

While most of them show good safety profiles, and a small subgroup also show some efficacy, the overall general effects and functions in the amelioration of diseases are statistically marginal at best for most probiotics. Conversely, the administration of traditional probiotics does not aim to treat specific diseases but rather to improve health. This concept is based on the original observation that the regular consumption of fermented dairy products containing lactic acid bacteria was associated with enhanced health and longevity in Bulgarian elders.[54]

Based on these premises, there is now a great focus on identifying and characterizing novel and ideally disease-specific next-generation probiotics (NGP).[55] Contrary to ordinary probiotics, NGP should ideally have a clear mechanism of action mechanistically linked to defined targets that are pathogenetically relevant to disease onset. Furthermore, the genetic features and physiological characteristics of bacteria, including growth dynamics, sensitivity pattern to antibiotics, and half-life cycle under specific environmental conditions, should also be well defined.

To achieve this goal, it is necessary to screen and isolate NGP through cutting-edge techniques, including next-generation sequencing and bioinformatics technique platforms, followed by stringent functional validation of the new probiotics. Therefore, compared to the traditional approaches used so far to isolate probiotics, the strategies needed to isolate and validate NGP will require radically different strategies, including the need for longitudinal studies that can mechanistically link specific probiotics to clinical outcomes (see chapter 17). More comprehensive bioinformatics analyses of the microbiota composition, of the metagenomics to gain insights on its functions, and of the host responses assessed by metabolomics analysis are also needed.

Only by using these tools along with in-depth multiomic analysis and mathematical modeling can probiotic candidates be identified and selected as potential NGP. Ideally, once identified, these bacterial strain candidates should be characterized by functional validation, including assays of in vitro cell lines, ex vivo and in vivo animal models, and human gut organoids to characterize specific mechanisms of action that, if confirmed, will then move them into human clinical trials. Of course, standardized processing steps for sample harvesting and their isolation location (stools versus mucosa), optimal storage conditions, and detailed sequencing and bioinformatics analyses to be used need to be optimized, standardized, and strictly followed.

Finally, the many big data results obtained from large-scale analyses need to be integrated with metadata to consider specific traits of the host, such as genomic profiling, nutrition, and drug treatment, to have a complete and integrated picture of the mechanisms involved in the interaction between the microbiota and the host dictating clinical outcomes. Table 15.1 lists some of the NGP that have been identified and their possible beneficial effects in specific pathological conditions.

Synbiotics

Synbiotics are defined as "synergistic mixtures of probiotics and prebiotics that beneficially affect the host by improving the survival and colonization of live beneficial microorganisms in the gastrointestinal tract of the host."[56] However, synbiotics are used not only for the improved survival of beneficial microorganisms added to food or animal feed but also for the stimulation of the proliferation of specific native bacterial strains present in the gastrointestinal tract, thus modulating the gut microbiota composition and microbial metabolite production. It should be mentioned that the beneficial effect of synbiotics is probably associated with the individual combination of a probiotic with a specific prebiotic, raising the attractive possibility of customizing specific synbiotics for specific hosts to obtain specific therapeutic results.

Therefore, the selection criteria to find the proper prebiotic-probiotic combination to obtain the desired health outcome should consider the goal of having a prebiotic that selectively

Table 15.1
Next-generation probiotics, their main therapeutic effects, and their main mechanism of action

NGP	Main therapeutic effect	Mechanism of action
B. breve *B. longum* *B. adolescentis* *B. lactis* (Bifidobacterium)	Anticancer therapy	Reduce tumor growth, increase the effect of checkpoint inhibitors, function as the vehicle for transporting anticancer genes to a target tumor
Prevotella copri (Bacteroidetes)	Amelioration of prediabetic syndromes	Improve aberrant glucose tolerance syndromes and enhance hepatic glycogen storage in animals via the production of succinate responsible for glucose homeostasis through modulating intestinal gluconeogenesis
B. fragilis (Bacteroidetes)	Amelioration of inflammation (including neuroinflammation and cancer)	Produce *B. fragilis* polysaccharide A (PSA), an archetypical bacterial capsular polysaccharide that possesses zwitterionic motifs that enrich a specific subset of anti-inflammatory memory CD4+FoxP3 T cells after direct interactions with antigen-presenting cells such as plasmacytoid dendritic cells
A. muciniphila (Verrucomicrobiae)	Improvement of metabolic disorders and obesity	Control glucose and energy metabolism through an immunomodulatory protein "Amuc_1100" located in the bacterial outer membrane. Other studies also suggested modulation of the endocannabinoid (eCB) system
C. minuta (Christensenellaceae)	Reduction of obesity and related syndromes	Alter obesity-associated microbiome, so decreasing body weight

Table 15.1 (continued)

F. prausnitzii (Ruminococcaceae)	Promotion of gut health Improvement in the efficacy of anticancer treatments	Produce butyrate Increase the efficacy of immune checkpoint blockade therapy. Strong positive correlation between the abundance of Faecalibacterium spp. in the gut and CD8+ T cell infiltration within the tumor environment (TME), in addition to the frequency of effector CD4+ and CD8+ T cells in the periphery.
P. goldsteinii (Parabacteroides)	Reduction of obesity	Unknown

Adapted from C.-J. Chang, T.-L. Lin, Y.-L. Tsai, T.-R. Wu, W.-F. Lai, C.-C. Lu, and H.-C. Lai, "Next Generation Probiotics in Disease Amelioration," *Journal of Food and Drug Analysis* 27, no. 3 (July 2019): 615–622, https://doi.org/10.1016/j.jfda.2018.12.011.

stimulates the growth of the associated probiotic and other beneficial microorganisms present in the host's microbiota, without favoring the growth of species that can be detrimental to the host's health. A typical example is the fortification of infant formula with synbiotics that have been shown to support normal growth in infants with cow's milk allergy, modulate intestinal microbiota, and prevent asthma-like symptoms in infants with atopic dermatitis.[57]

In adults, several meta-analyses suggest the positive effects of synbiotics on lowering high fasting blood glucose levels, alleviating constipation, and reducing the risk of developing postoperative sepsis after gastrointestinal surgery.[58] While a clear mechanism of action explaining their beneficial effects has not yet been identified, it has been hypothesized that synbiotics may affect the metabolic activity in the intestine, characterized by increased levels of SCFAs, ketones, carbon disulfides, and methyl acetates. This activity could contribute to the maintenance of the intestinal barrier and biostructure, the development of beneficial microbiota, and the inhibition of potential pathogens present in the gastrointestinal tract.

Furthermore, synbiotics seem to reduce concentrations of undesirable metabolites and contribute to the inactivation of nitrosamines and cancerogenic substances. As a consequence of their beneficial impact on the composition and function of the host microbiome, synbiotics may be capable of preventing osteoporosis, reducing blood fat and sugar levels, regulating the immune system, and treating brain disorders associated with abnormal hepatic function.[59]

Postbiotics

Postbiotics are functional, fermentation-soluble factors (products or metabolic byproducts), including enzymes, SCFAs, peptides, teichoic acids, peptidoglycan-derived muropeptides, polysaccharides, cell surface proteins, and organic acids, actively secreted by live bacteria or released after their lysis (cell death). Therefore, their production and abundance are highly dependent on the gut microbiome composition and its metabolic and functional phenotype, implying that the extent to which components are microbially metabolized may differ between individuals.

Recent insights have contributed to a renewed appreciation of food fermentation and have given rise to the postbiotics concept. Postbiotics are functional fermentation compounds, like the ones previously mentioned, that can be used in combination with nutritional components to promote health. Because of their clear chemical structure; safety dose parameters; lack of live organisms that pose potential safety concerns; long shelf life; and the content of various signaling molecules that may have anti-inflammatory, immunomodulatory, anti-obesogenic, antihypertensive, hypocholesterolemic, antiproliferative, and antioxidant activities, postbiotics have recently been the focus of great interest in scientific and clinical communities.

Other desirable features of postbiotics include eliminating the need to resolve the technical challenge of colonization efficiency and keeping the microorganisms viable and stable in the product at the high dose found in probiotics. Other positive features include the possibility of delivering the active ingredient at the desired location in the intestine, as well as simplified packaging and transport, making postbiotics suitable to be used in situations in which it is harder to control and maintain production and storage conditions, such as in developing countries.

Finally, the use of postbiotics could be an attractive alternative to probiotics in critically ill or high-risk patients, such as premature neonates for whom the risk of bacteremia following oral probiotics administration remains a concern. These properties suggest that postbiotics may contribute to the improvement of host health by targeting specific pathways involved in chronic inflammatory processes, even though their exact mechanisms of action have not been entirely elucidated.[60]

16 Microbiome Research in Gut-Brain Axis
Diseases: Psychobiotics

Role of Psychobiotics

Based on the gut-brain axis concept addressed in chapter 10, one of the most intriguing and provocative aspects of microbiome-related research is the potential for the extra-intestinal impact of probiotics on diseases of the nervous system, including mental health, neurodegenerative diseases, and neurodevelopmental disorders. Given their peculiar mechanism of action and preference for the CNS as a target, the term "psychobiotics" was proposed by Timothy Dinan and colleagues in 2013. It defines a novel class of probiotics that, upon ingestion in adequate amounts, yields a positive influence on mental health, suggesting potential applications in treating psychiatric diseases.[1]

The vast majority of psychobiotic research conducted during the past five years has been performed using animal models to evaluate motivation, anxiety, and depression. These studies have shown that psychobiotics exert their therapeutic effect by regulating the production and release of neurotransmitters and proteins, including gamma-aminobutyric acid (GABA), serotonin, glutamate, and brain-derived neurotrophic factor. These play important roles in controlling the neural excitatory-inhibitory balance, mood, cognitive functions, and learning and memory processes.[2] Specifically, some strains of *Lactobacillus* spp. and *Bifidobacterium* ssp. have been shown to produce GABA, acetylcholine, and serotonin.

The effect of psychobiotics is not limited to the regulation of the neuroimmune axes and diseases that involve the nervous system but is also related to cognition, memory, learning, and behavior. For example, studies in animal models have shown that consumption of psychobiotics can lead to a decreased rate of stress- and anxiety-related symptoms.[3] Additionally, in murine models some psychobiotics have been shown to suppress proinflammatory and oxidative damage responses in the brain and thus minimize the symptoms of neurodegenerative and demyelinating diseases.[4]

Effect of Psychobiotics on Human Mental Health

Based on the promising results obtained in animal studies, pilot human trials have reported beneficial effects of psychobiotics on mental health in humans. Healthy volunteers who were administered *Bifidobacterium longum* strain 1714 for four weeks exhibited reduced stress and improved memory.[5] A randomized, double-blind, placebo-controlled trial investigated the effects of potential psychobiotics in yogurt (*L. acidophilus* LA5 and *B. lactis* BB12) or capsules (*L. casei, L. acidophilus, L. rhamnosus, L. bulgaricus, B. breve, B. longum,* and *Streptococcus thermophiles*) on petrochemical workers.[6]

Recipients using both yogurt and capsules exhibited improved mental health parameters on the depression, anxiety, and stress scale general health questionnaire. A probiotic combination of *L. helveticus* R0052 plus *B. longum* R0175 has been shown to reduce anxiety and depression in healthy subjects compared with controls.[7] Ongoing clinical studies are investigating the effects of probiotic supplements (such as *L. plantarum* PS128, *L. plantarum* 299v, *L. rhamnosus* GG, Bifihappy, Vivomixx®, and Probio'Stick) on depression and anxiety by focusing their analysis on the state of inflammation, stress, and mood of participating patients.[8]

Effect of Psychobiotics on Neurodegenerative Disorders

The gut-brain axis operates with a bidirectional communication in which the gut can influence key neuroinflammatory processes that characterize several neurodegenerative disorders. With this understanding, there has been a surge of interest in psychobiotics that can ameliorate inflammation in a variety of these diseases. Among all neurodegenerative disorders, probably Alzheimer's disease and Parkinson's disease are the two conditions in which the evidence of dysbiosis has a more solid basis, providing the rationale for the potential use of psychobiotics.

Alzheimer's Disease

Alzheimer's disease, the most common form of dementia, is a progressive neurodegenerative disorder characterized by gradual memory loss and subsequent impairment in mental and behavioral functions. Although the primary risk factor for Alzheimer's disease is advancing age, other factors, including T2D, hyperlipidemia, obesity, and vascular factors, seem to play a role in its pathogenesis. The most prevalent hypothesis for Alzheimer's disease is the amyloid hypothesis, suggesting that changes in the proteolytic processing of the amyloid precursor protein lead to the accumulation of the amyloid beta (Aβ) peptide, causing an immune response that drives neuroinflammation and neurodegeneration in Alzheimer's disease.

However, recent evidence generated in germ-free animal models and in animals exposed to pathogenic microbial infections, antibiotics, probiotics, or humanized models suggests a role for the gut microbiota in its pathogenesis.[9] The increased permeability of both the gut and BBBs induced by dysbiosis seems to play a key role in the pathogenesis of Alzheimer's disease and other neurodegenerative disorders, especially those associated with aging, a process that seems to cause deterioration in the functioning of these barriers (see chapter 18). Additionally, research evidence suggests that some gut microbiota species can secrete large amounts of amyloids and LPS, which might contribute to the modulation of signaling pathways and the production of proinflammatory cytokines associated with the pathogenesis of Alzheimer's disease.

Moreover, microbiota seem to be associated with the pathogenesis of obesity and T2D. These clinical conditions, as mentioned earlier, are linked to the pathogenesis of Alzheimer's disease. The rationale for using psychobiotics to efficiently treat symptoms of Alzheimer's disease is based on a series of animal studies suggesting their use can ameliorate neuroinflammation and its clinical consequences, including memory and learning deficits.[10] Specifically, the reduction of oxidative stress; the improvement of learning and memory deficits; reductions in the number of amyloid plaques, inflammation, and oxidative stress; the decrease of insulin levels and insulin resistance; and increases in learning, memory, behavior, antioxidant levels, cognitive behaviors, and gross behavioral activities have been described.

In human trials, the use of a single strain in severe Alzheimer's disease showed lack of efficacy in mitigating cognitive impairment.[11] One explorative intervention study used multiple strains—*L. casei* W56, *Lactococcus lactis* W19, *L. acidophilus* W22,

B. lactis W52, *L. paracasei* W20, *L. plantarum* W62, *B. lactis* W51, *B. bifidum* W23, and *L. salivarius* W24—on subjects with Alzheimer's disease.[12] The results showed a decrease in fecal zonulin concentrations (suggesting an improvement of gut barrier function) and an increase in *F. prausnitzii* (a beneficial microbial component described as decreasing inflammation) compared to baseline, which was associated with changes in tryptophan metabolism in serum.

One randomized, double-blind, controlled clinical trial found that after consuming probiotic-treated milk (*L. acidophilus*, *L. casei*, *B. bifidum*, and *L. fermentum*) for twelve weeks, subjects showed a significant improvement in their Mini–Mental State Examination score and favorable changes in plasma malondialdehyde, serum high-sensitivity C-reactive protein, the homeostasis model of assessment-estimated insulin resistance, beta cell function, serum triglycerides, and the quantitative insulin sensitivity check index compared to the control group.[13] Overall, this study demonstrated that probiotic consumption for twelve weeks positively affects cognitive function and some metabolic statuses in some Alzheimer's disease patients.

Parkinson's Disease

Parkinson's disease is a debilitating neuromotor disorder involving the nigrostriatal pathway in the midbrain. It affects approximately 2 percent of the general population, with an estimated three million patients worldwide. Parkinson's disease is characterized by a multitude of motor symptoms ranging from tremors to rigidity, bradykinesia (often akinesia), and postural abnormalities (characterized by a shuffling gait). These motor problems are often associated with neuropsychiatric symptoms, including anxiety, depression, apathy, cognitive decline, dementia, and psychosis.

In addition, patients with Parkinson's disease often suffer from gastrointestinal symptoms, such as constipation and bloating. Histopathologically, the disease is characterized by accumulation of a misfolded protein, α-synuclein, in the basal ganglia neurons. Neuroinflammation and inflammation of the enteric nervous system have both been described in Parkinson's disease, which points toward the involvement of the gut-brain axis in its pathogenesis. Several factors, such as the presence of glial cell markers in the myenteric and Auerbach's plexus in the intestinal mucosa and increased glial cell dysfunction and oxidative stress in the substantia nigra of patients, provide possible links between the development of inflammation in the nervous system and Parkinson's disease.[14]

Increasing interest in the role of non-neurological factors in the pathogenesis of Parkinson's disease stems from the fact that only 10 percent of cases have a strong genetic predisposition, pointing to the crucial role of environmental and intrinsic host factors in its pathogenesis. Changes in gut microbiome composition and function, along with intestinal permeability and the onset of gastrointestinal symptoms (constipation, dysphagia, hypersalivation, and swallowing disorders) have been shown to precede the onset of neurodegenerative symptoms in Parkinson's disease by five to ten years.[15]

It has been hypothesized that intestinal barrier impairment initiates the spread of bacteria across the tight junctions into the mesenteric lymphoid tissue. This could lead to the access of bacterial components to the lymphoid tissue, thereby activating mucosal immune cells to release inflammatory cytokines and activate the vagal nervous system, with the subsequent release of neuroactive peptides that modulate the central nervous and enteric nervous systems.[16] Since many patients suffer from gastrointestinal symptoms long before the onset of the hallmark motor symptoms of the condition, it has been hypothesized that gut inflammation and deposition of aberrant α-synuclein fibrils in the enteric nervous system initiate the process that leads to a retrograde spread via the vagal nerve trunk to the neuronal tissue in the central nervous system.[17]

Based on these premises, several studies exploring the efficacy of psychobiotics in Parkinson's disease have been performed. In a randomized, double-blind, placebo-controlled clinical trial, subjects with Parkinson's disease were administered multiple probiotics (*L. acidophilus*, *B. bifidum*, *L. reuteri*, and *L. fermentum*) for twelve weeks, and they reported decreased scores on the Unified Parkinson's Disease Rating Scale. Other results showed a reduction in hs-CRP and MDA levels, an increase in glutathione levels, and improvement in insulin function compared to the placebo group.[18]

One randomized controlled study that focused on inflammation, insulin, and lipid-related genes in peripheral blood mononuclear cells from subjects with Parkinson's disease showed that, after twelve weeks, subjects with Parkinson's disease who received the probiotic supplement displayed a significantly downregulated expression of IL-1, IL-8, and TNF-α and an upregulated expression of transforming growth factor beta (TGF-β) and peroxisome proliferator-activated receptor gamma compared with the placebo control.[19] Three other studies found that Parkinson's disease patients using probiotics exhibited improved gastrointestinal functions, including a reduction in constipation,[20] decreased abdominal pain and bloating,[21] and improved stool consistency and bowel habits.[22]

While the majority of clinical studies of probiotics in patients with Parkinson's disease have been focused on gastrointestinal function, only one of them reported that probiotics improved the motor control of patients with Parkinson's disease.[23] Given the supposed pathogenetic role of misfolded α-synuclein in entero-endocrine cells and its propagation to the nervous system in Parkinson's disease, evaluation of the effect of probiotic supplements on abnormal α-synuclein in enteroendocrine cells may represent a useful primary outcome to consider in future investigations into this disease.

Effect of Psychobiotics on Human Neurodevelopmental Disorders

Introduced in chapter 10, ASD is a mixture of neurodevelopmental diseases characterized by deficits in social communication and social interactions across multiple contexts. ASD is accompanied by restricted and repetitive patterns of behaviors, interests, and activities. Children affected by ASD frequently experience gastrointestinal symptoms, including one or more of diarrhea, constipation, and gastroesophageal reflux. For this reason, the use of probiotics to ameliorate these gastrointestinal symptoms has become increasingly popular among ASD patients.

A recent search on the official site for registration of human trials yielded a total of nine trials evaluating psychobiotics for ASD treatment, of which three have already been completed.[24] Most of these studies have used a combination of probiotics or prebiotics and probiotics. Visbiome® extra strength, a formulation containing eight strains of probiotics, showed some efficacy in controlling gastrointestinal symptoms in children with ASD. Another multiple probiotic strains formulation, Vivomixx®, containing 450 billion lyophilized bacterial cells per sachet made up of the same eight probiotic strains present in Visbiome®, is being evaluated in two ongoing trials of ASD patients.

The primary outcome of a trial conducted in Italy is a change in the severity of ASD as measured by the Autism Diagnostic Observational Schedule®-2 assessment test. Subjects were divided into two groups, namely those with or without gastrointestinal symptoms, to take two packets per day of Vivomixx® for one month and one packet for an additional five months. Another trial conducted in the United Kingdom is a crossover trial in which ASD patients are assigned to either the Vivomixx® or placebo group for four weeks, followed by four weeks of washout. After washout, the subjects will cross over to the other group for four weeks. This study is using the Autism Treatment Evaluation Checklist (ATEC) assessment test to measure changes in ASD symptoms.

An open-label trial using three strains of probiotics, *L. acidophilus*, *L. rhamnosus*, and *B. longum*, showed that the severity of autistic and gastrointestinal symptoms, as assessed using ATEC and the Gastrointestinal Severity Index questionnaire, respectively, improved after a three-month intervention with probiotics.[25] The effects of a probiotic product containing *B. lactis* BB-12 (BB-12) and *L. rhamnosus* GG (LGG) (BB-12 + LGG) on ASD symptoms also were investigated. And there are several ongoing placebo-controlled trials assessing the safety and efficacy of several combinations of multistrain probiotics.

Studies using single probiotic strains in ASD also have been performed. However, a lack of information on both the dose used and consideration of the primary outcome, added to the recruitment of only boys, are examples of potential biases or shortcomings of these studies that limit the interpretation of the results obtained. A systematic review of the role of probiotics in ASD showed that six trials involving the use of probiotic supplementation in children with the condition have been published to date. Most of these were prospective, open-label studies, and the results showed limited evidence to support the role of probiotics in alleviating gastrointestinal or behavioral symptoms in children with ASD. The two available double-blind, randomized, placebo-controlled trials found no significant difference in gastrointestinal symptoms and behavior.[26]

Despite promising preclinical findings, probiotics have demonstrated an overall limited efficacy in the management of gastrointestinal or behavioral symptoms in children with ASD. These findings may suggest that the gut microbiome does not play a pivotal role in the pathogenesis of ASD, or that probiotics are not efficient in treating dysbiosis and its effect on the gut-brain axis. However, the lack of personalized interventions (impossible in cross-sectional studies) to specifically correct dysbiosis mechanistically linked with neuroinflammation, the lack of a standardized probiotics regimen, the multiple different strains and concentrations of probiotics used, and the variable duration of treatments are all factors that can explain the inconsistency of results and poor outcomes obtained to date.

Nevertheless, psychobiotics may have opened a very broad and unexpected scenario that is drastically changing the current paradigm of interaction between bacteria and humans, from a merely symbiotic relationship to one that seems to be a more mutually influencing commensalism. The clinical applicability of psychobiotics will strongly depend on more targeted interventions based on prospective longitudinal studies as described in chapters 13 and 14 to personalize their use and maximize their potential beneficial effects.

17 Artificial Intelligence, Synthetic Biology, and the Microbiome

Artificial Intelligence and New Targets for Treatment and Prevention

Thirty years after the launch of the Human Genome Project, we continue to struggle with the complex nature of human biology as we seek therapeutic targets to address a variety of chronic inflammatory diseases. Disappointingly, the map of the human genome led to the identification of clinically relevant mutations involving only 2 percent of possible therapeutic targets. However, the lesson learned from these results was that the remaining 98 percent of human diseases are driven by changes in gene expression rather than their mutations, suggesting a key function of what was initially thought to be noncoding "junk DNA" as targets for epigenetic regulation of coding DNA.

Understanding how the microbiome modulates this "regulatory genome" represents a transformational advance toward identifying novel therapeutic and preventive targets for many human diseases. Technology and costs are no longer limiting factors to achieve these goals. Rather, the limitation with the explosion of data we can generate by genomic, transcriptomic, metabolomic, and proteomic studies of the human cell lines or of the human microbiome is our inability to deal with a staggering amount of high-dimensional data. Analysis of these data often requires the development of new computational tools and algorithms.

Researchers are turning to artificial intelligence (AI) and machine learning to help turn these data into knowledge and to learn from this knowledge in order to make smarter decisions to identify targets for therapeutic interventions and to improve patient outcomes through precision, personalized medicine. If machine learning algorithms are trained on sample data that accumulate over time as more data are collected, these algorithms can grow smarter and smarter in performing these tasks in a manner that surpasses human ability. In a model based on human learning, the computer learns from observational data, figuring out its own solution to the problem at hand.

Deep Learning Approaches to Artificial Intelligence

AI and its subset of machine learning are key tools to make sense of the staggering amount of data that is generated every day. These tools can dramatically improve diagnostic accuracy and benefit prognostic applications aimed at establishing how disease may progress.

The term "AI," which is broadly defined as the science of making intelligent machines to imitate intelligent human behavior, has been credited to a group of mathematicians who met in the summer of 1956 at Dartmouth College in Hanover, New Hampshire. Dartmouth mathematics professor John McCarthy organized the "Dartmouth Summer Research Project on Artificial Intelligence" to "proceed on the basis of the conjecture that every aspect of learning or any other feature of intelligence can in principle be so precisely described that a machine can be made to simulate it."[1]

For insights into AI, machine learning, and biomedical research, we talked to Ali Zomorrodi, our colleague and an engineering expert in computational and systems biology at the MIBRC at MGH. The field of machine learning has witnessed a revolution in the past few years as a consequence of the development of "deep learning" approaches, which can transform applications of AI in medicine and health care. Most deep learning approaches are based on deep neural networks (DNNs), which are the extension of traditional artificial neural networks (ANNs) that have existed for a long time, said Zomorrodi.[2]

ANNs are machine learning models developed with the aim of imitating the learning and decision-making processes of the human brain, which consists of a complex network of interconnected neurons. Traditional ANNs are the simplest form of ANNs, which are composed of three layers of neurons: the input layer, which takes the input data; the hidden layer, which processes the data; and the output layer, which makes the decision based on the processed data. Of course, as Zomorrodi noted, learning and decision making by the human brain is a much more complex process.[3]

The human brain consists of several layers of neurons structured hierarchically, in which each layer "processes" the data and successively passes the results to the next layer for further "processing" until a decision is made in the output layer. DNNs aim at simulating this complex human process through the addition of more hidden layers to traditional ANNs. DNNs can handle "noisy" and sparse data as well as high-dimensional heterogenous data types with highly nonlinear relationships, all of which are characteristic of high-throughput biological data. Nevertheless, according to Zomorrodi, a limitation of deep learning approaches is that they often require very large training datasets, which may not be readily available all the time, and a lengthy and computationally demanding training phase, which may require special hardware such as graphical processing units.[4]

Deep learning approaches have demonstrated outstanding performance on many important applications such as computer vision and speech recognition. Companies including Google, Amazon, Microsoft, and Facebook have started to deploy these approaches on a large scale. Amazon, Google, and Facebook also have started to explore the use of AI technologies to improve health care; one example is by finding protein structures, an early task in the development of new drugs.[5] But the full potential of deep learning in biomedicine and microbiome research has not yet been fully explored and exploited.

Statistical Analysis or Machine Learning

Why do we need these very sophisticated analytic methodologies instead of the typical statistical approaches that we have been using for so long in science? Some people may believe that machine learning and statistics are aimed at the same goals, with machine learning being "the next generation" of a more powerful statistical method. However, they are not synonymous.

In the words of an AI blogger, "The major difference between machine learning and statistics is their purpose. Machine learning models are designed to make the most accurate predictions possible. Statistical models are designed for inference about the relationships between variables."[6] For common mortals like us, this technical explanation can still be obscure in its real meaning.

To be more explicit, statistics is the mathematical study of data, and therefore, you cannot create statistics unless you have data. A statistical model is a model for the data that is used either to infer something about the relationships within the data or to create a model that can predict future values. To make this even more explicit, there are many statistical models that can make predictions, but predictive accuracy is not their strength. Likewise, machine learning models provide varying degrees of interpretability, but they generally sacrifice interpretability for predictive power.

According to Zomorrodi, typical statistical approaches follow rigid rules dictating which method of analysis is to be applied based on data distribution. Some approaches are used if the data are normally distributed, and other approaches are employed if the data are not normally distributed.[7] Conversely, in machine learning, there is more flexibility, because the machines can "learn" from data without having to impose any rigid rules or assumptions about them. Another advantage is that machine learning approaches can simultaneously analyze and learn from multiple, disparate data types and an enormous number of potential predictors, which is not feasible with statistical approaches.

Deep Machine Learning and Multiomic Analysis

In the case of multiomic analysis, there are some interesting questions: Are the host genome and RNA sequencing informative enough to infer specific pathways involved in the pathogenesis of the diseases studied? Which feature or features of metagenomic data (microbes or encoded functional pathways) can be linked to a specific disease? Is metatranscriptomic information useful in understanding the possible epigenetic role of the microbiome? Can metabolites produced by the microbiome or the host potentially be mechanistically linked to the onset of the disease?

According to Zomorrodi, over time, machine learning approaches may achieve even more ambitious goals. For example, once we have established links between the composition and function of the microbiome and specific diseases, we might be able to identify the disease of a specific patient simply by analyzing their microbiome. Even more exciting, if the machine learning approach is trained on longitudinal data, we may be able to predict the risk of developing a specific disease or the severity of the disease in an individual at a specific point in time.[8]

This is the "Holy Grail" of first, disease interception and second, primary prevention by modifying the microbiome composition and function through targeted personalized interventions. These interventions could include changes in diet and the use of specific prebiotics (nutrients that promote the growth of beneficial microbes in the gut), probiotics (beneficial gut microbes, which can be ingested as a pill), or a combination of the two (i.e., synbiotics). (See chapter 15.)

This might seem to be a distant target, but this principle has already been implemented in other applications. One example is the development of advanced driver assistance systems (ADAS) that can monitor driver attention and send alerts to improve road safety and avoid unsafe driving. Real-time estimation of driver gaze can be coupled with an alerting system to enhance the effectiveness of the ADAS. However, these real-time systems like ADAS have been faced with many challenges in obtaining reliable eye off-the-road estimation and classification of the gaze zones.[9] To overcome these limitations, DNNs have been applied to predict, on the basis of erratic and variable data acquisition, drivers' distractions and prompt the car to activate ADAS in time to prevent accidents.[10]

Similar approaches can be applied to microbiome analysis, which in the future will be able to provide us with extensive information about the person from whom the sample has been obtained, including age, geographical region, and diet, to name only a few. But the goal that represents a transformational change in the field of microbiome studies is the use of deep learning to develop predictive models for personalized therapeutic interventions or disease interception, two concepts that pervade this book. According to Zomorrodi, the major limitation in reaching these goals is the availability of large-scale, robust data that account for all variables noted to train the system to develop predictive machine learning models.[11]

This process is comparable to a situation in which medical students attend medical school courses to acquire robust data on the foundations of medicine. After they graduate and begin practicing, they use these data for ongoing learning during their practice, but they face situations they did not learn about in school. If medicine is practiced in rural or urban areas, in poor or affluent communities, in northern or southern regions, to name only a few variables, the new doctors can face extremely different situations. Once the physician's learning process is consolidated from the new patient data, she develops the predictive power to diagnose and treat diseases encountered in previous learning experiences in a standardized and efficient manner.[12]

A vast amount of robust microbiome data related to specific diseases is needed in order to employ deep learning to develop reliable predictive models for prediction outcomes. One major challenge is that most of these data, even if obtained from many subjects, are typically cross-sectional, meaning that we are comparing microbiome information between patients affected by a specific disease and healthy subjects matched by age, sex, and possibly lifestyle. These studies assume that, if we control other variables, the difference of moving from the state of health to the state of disease is all related to differences in microbiome composition and function. Unfortunately, this generally is not the case, which limits the strength and reliability of these machine learning models.

As mentioned in chapter 14, for this reason, the most powerful data come from longitudinal prospective studies with at-risk individuals followed over time; some of these people develop the disease later and some do not. In this case, the use of data before, during, and after the onset of a given disease for training machine learning algorithms can result in more accurate predictive models.

These models would be able to link specific microbiome components, ideally at the strain level, to specific time points in order to determine disease susceptibility or protection. These findings can then be validated in germ-free murine models to confirm that a specific microbiome strain affects specific metabolic pathways linked to disease pathogenesis or protection. This would be the ideal path to identify the next-generation probiotics we discussed in chapter 15.

However, even next-generation probiotics may generate results that are too rudimental or inadequate to mechanistically and specifically target pathways to influence an individual's state of health. This is because the potential therapeutic effect of next-generation probiotics can be mitigated by interaction with other microbiome components. So, the next question is, could synthetic biology be a more effective way to exploit these therapeutic opportunities?

Synthetic Biology Targets for Treatment and Prevention

To answer this question, we reached out to Timothy "Tim" Lu of MIT in Cambridge, Massachusetts, a friend, colleague, long-standing collaborator, and one of the best experts worldwide in synthetic biology. He agrees that there certainly have been many efforts to try to modulate the microbiome in different ways, using prebiotics, probiotics, postbiotics, FMT, and other methods. These are all legitimate approaches, some leading to remarkable results. One example is the treatment of antibiotic-resistant *C. difficile* colitis with FMT, which has led to some dramatic results in improving the quality of life in patients, proving that it is possible to manipulate the microbiome for therapeutic purposes.[13]

However, according to Lu, currently available strategies to manipulate the gut microbiome may be somewhat limited, because they mainly rely on the natural features of the prebiotics, or they are based on the unproved notion that the probiotics that we are administering to patients are solely and directly responsible for the therapeutic effect. He believes that this kind of approach is somewhat akin to the early days of cell therapy.[14]

When investigators first started applying cell therapy to patients, they took nonengineered cells, processed them in certain ways, and put them back into the patient. What we saw from those results was that, yes, you can see good effects in some anecdotal cases. But to achieve truly transformative effects, like those we have seen in immunotherapy for certain types of cancer, we had to achieve a higher level of engineering to exploit the potency of these products in greater detail and define their mechanisms.

Therefore, Lu's general feeling is that we will have a similar evolution when it comes to the manipulation of the microbiome, starting with a "give-it-a-try approach" with current tools.[15] But once we have gained more insight into key pathways and key mechanisms that modulate specific biological events in humans, we can then move to strategies that allow more reproducible, highly potent ways to try to impact those pathways. And one of the most straightforward and rational methods of this approach is to genetically engineer these mechanisms into specific therapeutic products.

This goal can be achieved through a stepwise approach, as researchers continue their efforts first to decipher, second to develop, and third to explore nonengineered versions of next-generation probiotics to manipulate microbiome composition and function for specific treatment strategies. However, once we have clearer insights into which pathways are affected by dysbiosis, thanks to deep machine learning and mathematical modeling, we can envision making products that will be able to modulate those pathways in a more targeted fashion through synthetic biology approaches.

Replicating an Amazing Machine

The definition of synthetic biology is applying engineering techniques to develop genetic circuitry. This circuitry can be implanted into a microorganism to perform a specific metabolic function and record events, which we can use as information to achieve the goal of targeted therapeutic interventions.

Lu describes synthetic biology as "a sort of an outgrowth of what we know as genetic engineering."[16] A classic example of old-fashioned, traditional genetic engineering is the cloning and production of insulin, which is focused on engineering bacteria to produce abundant amounts of a single protein. This technology has been developed and has been robustly used to manufacture drugs. Synthetic biology engineers want to advance this technology to address more complex questions. For example: How do you put together multiple genes, not just a single one, and how do those genes interact with each other in a way that is basically turning the cell into a miniature computer?

One analogy Lu likes to use is to compare traditional genetic engineering to pasting words from the *Wall Street Journal* and trying to write a book with the result.[17] But if you want to put together a more complex sort of program or story, that's more difficult. During the last decade, with continuously cheaper DNA synthesis and sequencing technology, we have figured out how to write and type arbitrary letters into the genome, and that gives us the ability to program cells in a much more complex way. Now we can start thinking about cells as computing elements.

The development of a human being from the union of two single cells to a functioning, complex, and multicellular organism is similar to the construction of an amazing machine. Humans have all the genetic instructions in our DNA that allow us to grow autonomously into this multifaceted being with distinct patterns and coordinated functions and movements. Ultimately, you can view this intricate dance as a computational problem, one that even the world's best computers today are not necessarily able to mimic.

What Lu and his colleagues are trying to do with synthetic biology is to replicate this "amazing machine," but to do it with some sort of control of its dynamics and its "decision-making" abilities, and ultimately, the combination of these two factors. The initial application of genetic engineering to microorganisms was aimed at transforming them into a type of factory with a unidimensional function. The bioengineer would insert genes, perhaps an insulin gene, in the microorganism's genome. The only thing the microbe was required to do, the only "decision" it had to make, was to replicate itself, and by doing that it would make a large amount of insulin. That's classical genetic engineering.

Now, we have the ability to go to the next level. Instead of asking microorganisms to make a single product, we ask the microbe to make its own decisions based on contingent environmental situations. These conditions dictate if and when to enact a series of actions and gene expressions based on specific programming to obtain a specific outcome. But if the environmental conditions change, then the microorganism's genetic circuitry will activate a different decision-making process leading to a different outcome.

This almost sounds like science fiction, because this concept unlocks an unbelievable number of opportunities. But we can now capitalize on these novel technologies to achieve the necessary next step in the world of microbiome research, namely, to move from a scientific proof of concept or a purely theoretical possibility to its applicability in human therapies.

Synthetic Biology Strategies

Lu represents the quintessential entrepreneur working to translate all this information into making this technology available for the treatment of human disease, and he is the founder of several start-up companies aimed at this goal. We asked him about the strategies that these companies are implementing to capitalize on synthetic biology to exploit the microbiome for therapeutic purposes.

From his perspective there are a couple of approaches to manipulating the microbiome. One is to try to add things to the microbiome. And when you try adding things to the microbiome, you can add a single probiotic, or you can add a defined mixture of bacteria, or FMT material. Another option is to try to add a genetically engineered mixture that is programmed to have a specific and well-defined function.[18]

With the attention on bioengineering technology training, we have been primarily focused on adding material to the microbiome, such as engineered probiotic strains or, in the near future, an engineered commensal strain that already lives in the gut. An alternative way of manipulating the microbiome is to try to remove material from it. This is somewhat less explored in current clinical settings, but researchers have applied several methods of achieving this goal.

The most advanced approach is the use of bacteriophages, which are the natural predators for bacteria (see chapter 5). There are ongoing efforts to create bacteriophage-based therapeutics to manipulate certain bugs in the microbiome. These therapeutics are in early stages of development, and there are a lot of open questions as we try to develop this sort of manipulation. But Lu is hopeful that through human testing and learning more about how the microbiome works, we will be able to drive these techniques toward successful clinical applications.[19]

Challenges Faced in Synthetic Biology

Based on Lu's vision, and with current technology, let's imagine that in the near future we will be able to manipulate the microbiome to perform a specific metabolic function. In other words, we will be able to exploit synthetic biology to deploy a microorganism or a group of microorganisms to produce and deliver molecules that could, in one example, counterbalance the metabolic consequences of dysbiosis and therefore ameliorate inflammation. Lu believes that this is a tangible possibility, as this approach has been already used in animal models.[20]

The key questions in terms of human translation are: What do we want these bacteria to actually make and deliver? What are the key disease biomarkers they should respond to? And what should they actually do if they sense the presence of these molecules? For example, should they target IL-22 or TNF-α, or certain biological pathways?

This is more of a fundamental biology question. The limiting factor is no longer whether synthetic biology manipulation is feasible, but rather the lack of sufficiently full and robust knowledge of the biological problem we want to target. We need to acquire this robust knowledge before we can actually design specific therapies.

Only when we arrive at this stage can we tackle the second challenge, namely, how to design a therapeutic that can actually achieve that particular target product profile or perform that function. This is more of an engineering challenge, and Lu believes that this is where synthetic biology comes into full force.

The key question for major therapies will be: How potent can we make those therapies, given that the gut is a complex place and we are engrafting bacteria to manipulate a fairly complex genotype? Lu agrees that it is essential to have a comprehensive understanding of the degree of therapeutic impact that each bacterium needs to deploy for the desired biological outcome that translates into beneficial biological human experiences.[21]

This challenge is reminiscent of the strategies used to develop drugs in the past and how the first biologics were developed. At that time, we had the same question about dosing and about how to deliver and achieve the desired functionality in a specific, targeted manner. Most likely, we are going to go through the same learning process with microbiome manipulation, as we think about it as another organ that we will try to manipulate in vivo.

This might be easy to say, but it's much harder to implement. For this reason, we asked Lu to share his hypothetical ideal plan to move therapeutic translation from the scientific approach to synthetic biology, assuming that money and time would not be limiting factors. According to Lu, there are a few key things we need to do.

Broadening the Bacterial "Library"

The first step is to develop a broader tool kit of engineered bacteria. Currently, the synthetic biology approach can only be used on a fairly limited number of well-characterized bacterial strains, for example, specific *E. coli* or *Bacteroides* strains. But there are thousands of different types of bacteria in the microbiome that could be targeted for continuing production and manipulation. Improving our ability to culture a much larger array of bacterial strains and developing genetic tools capable of genetically manipulating them at will, to do whatever we want to their genome, is a major need in the field in order to fully exploit synthetic biology for therapeutic purposes, according to Lu.[22]

The second step Lu outlined is to develop better translational models that can be adequately tested.[23] Strictly speaking, this is not a synthetic biology problem but rather a general biology problem.

To engineer and optimize as fast as we can, we will need a predictive and reliable model, since we cannot design therapeutic strategies against a system that we do not fully understand. Without these tools, deploying synthetic biology techniques to manipulate the microbiome to ameliorate human diseases will take too long. Therefore, it's imperative to dedicate more time and effort to developing those translational assays, such as human gut organoids (discussed earlier in this book) or innovative animal models.

The third and final step is to figure out how to accelerate the time from research to clinical application. Obviously, the ultimate goal of all of these efforts is to advance these technologies to human trials, but the current approaches are extremely expensive and time consuming. As an engineer, Lu is often frustrated with how long it takes to test synthetic biology products. Basic scientists can perform an experiment and get an answer in real time. To understand if an approach works in a clinical setting, the steps dictated by the FDA to bring any therapeutic strategy to human use (assuming it works) require an average of fifteen years and more than $1 billion.

While Lu has no magic solutions, he expressed his wish for more investment in policy studies to make clinical trial strategies more flexible. He would like to see the implementation of innovative methods to pilot products more quickly and the use of exploratory approaches with continual updating and adjustments of the approach "on the go." Lu views this as a promising vision for more proactive and responsive clinical trials.[24]

Based on Lu's and Zomorrodi's considerations, we are left wondering whether AI and deep machine learning can accelerate the process further by creating theoretical models such as in silico, germ-free models. By developing these models, researchers can work faster and become more efficient in translating the data obtained in clinical trials to more generalized information that will better guide genetic engineers to more efficient approaches.

Lu agrees that machine learning has powerful potential but, as we have stressed in this chapter, he contends that it is only useful when applied to a good dataset. If we do not have datasets that tell us what is real and what is not, we will train a model that will not be useful. To overcome this problem, ideally, the scientific community should impose a rule that all data generated in the field of microbiome research must be shared in an integrated format, so that all researchers can access all of the data in a central location. This truly collaborative approach would greatly accelerate microbiome research and innovative techniques, including synthetic biology. Given the enormity of the challenge of data integration and collaboration on such a huge scale, this is an unlikely development. However, it is important to recognize that whatever approach we take to tackle these challenges could either help or hinder millions of people around the world.

Safety Concerns

Even if we could overcome all of the challenges discussed, so that synthetic biology can be applied successfully to manipulate our microbiome at will, a key question remains: Is it possible to unintentionally unleash a technology that can go out of control? The biological targets are microorganisms that reproduce every twenty minutes and are capable of changing their genetic fingerprints very rapidly. This biological reality raises concerns around the complex topics of both safety and ethics for human clinical applications.

Lu agrees that, as with any living therapy, there is always a different safety profile to consider than with the conventional small-molecule drugs or biologics, which are not "alive."[25] As we develop living remedies for a variety of clinical applications, we are likely to encounter challenges that we haven't anticipated.[26] Therefore, we must have robust preclinical studies trying to troubleshoot as many novel problems as possible to minimize any unintended effects of using these innovative therapeutic strategies.[27]

When it comes to manipulating bacteria and their genetic plasticity, Lu noted, this is an essential thing to remember, for the manufacturing of the product as well as safety considerations. We must have procedures in place to make sure that we have extremely rigid, well-controlled quality, so that we can monitor the manufacture of synthetic biology products, ensuring that only the intended function is achieved. Once the engineered strain is deployed in patients, then we should try to maintain those genes to last long enough to perform their function and ensure that they cannot mutate.[28]

Lu tells us that researchers are trying to mitigate these risks by using bacterial strains that are more accommodating of immune function. Some bacteria that he calls "super-optimized" do not tolerate genetic manipulation very well. According to Lu, the microplasma developed by synthetic biology engineers is optimized to be a "minimal organism—if you shovel additional pieces into it, it doesn't like it." He noted that other organisms, like yeast, are much more tolerant of genetic manipulation.[29] Additionally, researchers are trying to make far more energy-efficient versions of these gene circuits, which are much less susceptible to mutational issues.

A large group of synthetic biology researchers are aware of these challenges and are working to develop safeguards to deal with this. Lu has been involved in projects to address the risk of genetic mutations, and others are slowly developing tools they hope to use to build kill switches or ways of limiting horizontal gene transfer. This would ensure that what we engineer in one bacterium cannot easily escape and create unintended consequences.

Nevertheless, as typically happens in biology, we can aim at minimizing risks, but we can't bring them to an absolute zero. There is no 100 percent guarantee that the use of synthetic biology in microbiome manipulation will be safe. But as with any other intervention developed so far to treat human diseases, time will tell.

18 Maintaining a Resilient Microbiome through Old Age

The Biology of Aging

Aging is a syndrome of changes that are deleterious, progressive, universal, and—to date—irreversible. Between 1997 and 2019, PubMed published approximately 369,000 articles on senescence and aging. Most of them were published in the last five years, and more than two thousand articles were focused on the role of the microbiome in aging. Based on this body of work, we now have a much better understanding of the biological bases of the aging process.

Aging damage occurs to molecules (DNA, proteins, and lipids), to cells, and to organs with the accumulation of changes in a person over time. Aging in humans refers to a multidimensional process of physical, psychological, and social change. The complex process of biological aging is the result of genetic and, to a greater extent, environmental factors and time itself, occurring heterogeneously across multiple cells and tissues. As the rate of aging is not the same in all humans, biological age does not always correspond to chronological age.

Normal aging implies sensory changes like visual acuity, hearing loss, and dizziness; muscle weakening; reduced agility and mobility; and changes in fat. At the same time, there is a functional reduction of several systems, including the renal, respiratory, and gastrointestinal systems, while the body increasingly succumbs to certain diseases, including hypertension, cardiovascular diseases, diabetes, osteoarthritis, osteoporosis, cancer, and several neurological disorders.

In the background of all the changes that occur during aging are three key factors: inflammation, immune aging, and senescence. With this premise, two basic questions on aging remain: (1) Why do we age? and (2) How do we age? To answer these questions, we need a better understanding of the molecular basis of the process of aging. Making the story even more intriguing is recent research showing that even late in life, the potential exists for physical, mental, and social growth and development.

Recent scientific successes in rejuvenation and extending the life span in animal models offer the promise of achieving negligible senescence and reversing aging, or at least significantly delaying it. These results are possibly due to a better understanding of the pathogenesis of aging, now based on the following nonmutually exclusive theories:

1. **The free radical (oxidative stress) theory:** It is known that a senescent phenotype may be stimulated or induced by various types of stressors. These include changes induced by reactive oxygen species (ROS), a natural byproduct of normal oxygen metabolism. ROS is thought to regulate several physiological functions, including signal transduction and gene expression and proliferation. The major cellular sources of ROS are mitochondria, cell membranes, and endoplasmic reticulum. While lengthening of the organismal life span is associated with low ROS concentration, senescent phenotype maintenance is endangered with high ROS concentrations. The oxidant/

antioxidant imbalance causes structural damage to macromolecules (DNA, proteins, and lipids). Age-related accumulation of damaged macromolecules is one of the mechanisms that contribute to the aging processes. The balance between oxidant generation and antioxidant processes in healthy tissues is maintained with a predominance of various antioxidants.

2. **The cellular senescence and apoptosis theory:** The relationship between cellular aging and the aging of the whole organism is complex. Cellular "immortality" is essential for stem cells, but an "immortal" somatic cell is cancerous. Apoptosis is programmed cell suicide—a genetically controlled cell death that causes cells to shrink and be eliminated without the tissue trauma associated with inflammation that accompanies necrosis, which is uncontrolled cell death.

3. **The immune system theory:** According to this theory, many aging effects are due to the declining ability of the immune system to differentiate "foreign" from "self" proteins. There is evidence that histocompatibility genes, genes affecting DNA repair, and genes for superoxide dismutase production, all of which affect longevity, are located close together on human chromosome 6.

4. **The inflammation theory:** With aging, the body contains increasing quantities of proinflammatory cytokines. Aging is associated with increasing activity of the proinflammatory transcription factor NF-kB.

5. **The intestinal permeability theory:** Several reports in animal models and humans link gut permeability to noninfectious, chronic inflammation leading to senescence. In fruit flies, the increase in intestinal permeability is the best predictor of imminent death, even more than the actual age of the insect.

The Proinflammatory State of Aging: "Inflammaging"

No matter which theory is considered, they all recognize that aging phenotypes result from an imbalance between stressors and stress-buffering mechanisms and a resultant loss of compensatory reserve leading to the accumulation of unrepaired damage. Consequently, this results in increased disease susceptibility, reduced functional reserve, reduced healing capacity and stress resistance, unstable health, and finally failure to thrive. The resultant physical and cognitive decline that culminates with the frailty syndrome is the tipping point of an individual's health span and implies a high risk of system decompensation and death.

Preserving physical and cognitive function is the main focus of geriatric and gerontological research, but it is important to recognize that accomplishing this goal requires a profound understanding of the molecular, cellular, and physiological mechanisms that ultimately determine functional changes. In this context, the proinflammatory state of aging, or "inflammaging," plays a major role. Longitudinal studies have shown that with aging, most individuals tend to develop a chronic, low-grade proinflammatory state. This state is viewed as a strong risk factor for multimorbidity, increased susceptibility to infections such as COVID-19, physical and cognitive disability, frailty, and death.

But how and why does this chronic, low-grade inflammation occur? Once again, the gut dysbiosis–permeability–immune activation triangulation, a theme that permeates this book, seems to be in play. An emerging theory suggests that inflammaging is associated with alterations of the intestinal microbiome that substantially differ between young and elderly human hosts (more than seventy years) and even between centenarians and frail elderly people with a history of cancer.[1]

Aging-associated gut dysbiosis, which is characterized by a shift toward proinflammatory commensals and reduction of beneficial microbes, causes impairment and leakiness of the intestinal barrier.[2] Subsequent leakage of microbial LPS and other microbial products upregulates interferons, TNF-α, IL-6, and IL-1 in circulation, promoting a mild proinflammatory state that is highly prevalent in elderly people, predicting the accelerated decline of fitness and health.[3]

While this seems plausible, do we have strong evidence supporting this theory and the sequential events outlined earlier? It's not surprising to learn that this is not a recent proposition. As already mentioned in chapter 15, in 1907 Elie Metchnikoff proposed that tissue destruction and senescence were consequences of chronic systemic inflammation, which occurred as a result of increased permeability in the colon and the escape of bacteria and their products.[4] He believed that these bacterial products activated phagocytes, and that the resulting inflammatory response caused the deterioration of surrounding tissues.

The more recent evidence that aging is characterized by a state of chronic, low-grade systemic inflammation supports Metchnikoff's hypothesis.[5] Although age-associated inflammation influences the aging process, it is unclear why levels of cytokines in the tissues and circulation increase with age—no matter which mechanisms are at play.

Intestinal Permeability and Aging

There is now fairly solid evidence that increased gut permeability triggered by dysbiosis and the subsequent passage of circulating endotoxins are key factors leading to inflammaging. Studies in *Drosophila* demonstrated that intestinal barrier dysfunction was caused by age-related changes in the microbiota composition.[6] This finding is correlated with life span across a range of *Drosophila* genotypes and environmental conditions, including mitochondrial dysfunction and dietary restriction. Regardless of chronological age, intestinal barrier dysfunction predicted impending death in individual fruit flies. This was more accurate than chronological age in identifying individual fruit flies with systemic metabolic defects and inflammatory profiles previously linked to aging.[7]

Similar findings have been reported in human studies. Yan-Fei Qi and colleagues studied two healthy cohorts of young (18–30 years) and older (70+ years) adults. The researchers measured serum concentrations of TNF-α and IL-6 (indicators of systemic inflammation), zonulin used as a biomarker of gut permeability, and high-mobility group box protein (HMGB1, a nuclear protein triggering inflammation). Correlations of serum levels of zonulin and HMGB1 with strength of plantar flexor muscles and number of steps taken per day were analyzed.

Serum concentration of zonulin and HMGB1 were 22 percent and 16 percent higher in the older versus young adults. Serum zonulin was positively associated with concentrations of TNF-α and IL-6. Intriguingly, IL-6 has been reported as a promoter of the zonulin gene, raising the possibility of a self-feeding, vicious loop between increased gut permeability and systemic inflammation.[8] Finally, both zonulin and HMGB1 were negatively correlated with skeletal muscle strength and habitual physical activity.[9]

Based on these results, the authors concluded that gut permeability assessed by serum zonulin levels was associated with both systemic inflammation and two key indices of physical frailty. This finding suggests that loss of intestinal barrier function may play a critical role in the development of age-related inflammation and frailty.[10]

Similar results were obtained by Pedro Carrera-Bastos and colleagues, who monitored serum zonulin and serum LPS as an indicator of endotoxemia in disease-free centenarians, young healthy controls, and patients with precocious acute myocardial infarction. Disease-free centenarians showed significantly lower levels of serum zonulin and LPS than young patients with acute myocardial infarction, and the centenarians had significantly lower concentrations of serum LPS than young healthy controls.[11]

Additionally, the two variables correlate with each other in centenarians and patients affected by acute myocardial infarction, suggesting that intestinal hyperpermeability may cause endotoxemia, which in turn could lead to inflammation. Based on these results, the authors concluded that increased intestinal permeability and endotoxemia may play a pathogenetic role not only in coronary heart disease but also in life span modulation, by accelerating aging if altered and prolonging life span if within normal limits.

Inflammation and Aging

Unlike acute, transient inflammation in which damaged tissue can heal once the causative agents are removed, chronic inflammation persists for a long time. During chronic inflammation, affected tissues are infiltrated by specific immune cells, mainly macrophages and lymphocytes, and fibrosis and necrosis of the affected tissue may occur. Chronic inflammation is associated with many age-related physiological or pathophysiological processes and diseases. In subjects of normal, healthy aging, researchers found serum concentrations of proinflammatory cytokines, including IL-1, IL-2, IL-6, IL-8, IL-12, IL-15, IL-17, IL-18, IL-22, IL-23, TNF-α, and interferon-gamma.[12]

At the same time, in elderly people the concentration of anti-inflammatory cytokines interleukin1 receptor antagonist (IL-1Ra), IL-4, IL-10, IL-37, and transforming growth factor beta 1 (TGF-β1) are higher than in young persons. The role of anti-inflammatory cytokines is to neutralize proinflammatory cytokine activity, reduce chronic inflammation, and thus act protectively on tissues. These results suggest that a large part of the aging phenotype, including immunosenescence, is the consequence of an imbalance between inflammatory networks. This disparity is fueled by loss of gut barrier function and subsequent systemic endotoxin exposure, along with anti-inflammatory networks trying to counterbalance the established low-grade, chronic proinflammatory status that leads to inflammaging.

Within this perspective, healthy aging and longevity are likely the result not only of a lower propensity to mount inflammatory responses but also of efficient anti-inflammatory networks. In normal aging, these networks fail to fully neutralize the inflammaging process.[13] Such a global imbalance can be a major driving force for frailty and common age-related pathologies and should be addressed and studied within an evolutionary-based systems biology perspective. The bottom line is that it is likely that "aging gracefully" is characterized by a proper balance between the action of proinflammatory and anti-inflammatory mediators, while their imbalance leads to accelerated aging and to the development of various age-related pathological conditions.

Dysbiosis and Aging

While the results presented in this chapter are supportive of Metchnikoff's original theory, it remains unclear whether dysbiosis, by causing loss of barrier function and subsequent systemic passage of endotoxins, triggers a proinflammatory and anti-inflammatory network imbalance as a driver of age-associated inflammation, or this is merely association without causation. If the former is true, it would indicate that these age-related changes in microbiome composition and function are a form of microbial dysbiosis.

More direct proof has come from Netusha Thevaranjan and colleagues, who used a murine model to show that age-associated microbial dysbiosis causes loss of gut barrier function, systemic inflammation, and macrophage dysfunction.[14] Interestingly, when maintained under germ-free conditions, mice did not display an age-related increase in circulating proinflammatory cytokine levels, and on average they lived longer than their conventional counterparts. These data suggest that aging-associated microbiota promote inflammation, and that reversing these age-related microbiota changes represents a potential strategy for reducing age-associated inflammation and the accompanying morbidity.

Moving to human studies, in the early 2000s Elena Biagi and colleagues compared gut microbiomes across the life spans of adults. They showed that the microbial composition and diversity of the gut ecosystem of young adults and seventy-year-old adults is highly similar but differs significantly from that of centenarians.[15] After one hundred years of symbiotic association with the human host, the microbiota radically shifts by a rearrangement in the *Firmicutes* population and an enrichment in facultative anaerobes, notably pathobionts.

The presence of such an imbalance in microbiota in the centenarians resulted in associations with inflammaging, as determined by a range of peripheral blood inflammatory markers. This may be explained by a remodeling of the centenarians' microbiota, with a marked decrease in *F. prausnitzii* and its relatives, which are symbiotic species with reported anti-inflammatory properties. *Eubacterium limosum* and related bacteria seem to be the signature bacteria of a long life, since they were found in the centenarians in a more than tenfold increase compared to the young adults. These data suggest that the aging process deeply affects the structure of the human gut microbiota, as well as its interplay with the host's immune system.[16]

Fifteen years later, Biagi's group reported in more detail what should be considered the ideal symbiotic relationship between the gut microbiota and a long-living host. By analyzing the phylogenetic microbiota of adults, elderly people, and "semi-supercentenarians," including individuals aged 105 to 109 years, the authors reconstructed the longest available human microbiota trajectory along aging generated so far, which is seen in figure 18.1. They found the presence of a core microbiota of highly occurring, symbiotic bacterial taxa that primarily belonged to the dominant *Ruminococcaceae*, *Lachnospiraceae*, and *Bacteroidaceae* families, with their cumulative abundance gradually decreasing with age progression.[17]

Aging seems to be characterized by an increasing abundance of subdominant species, as well as a rearrangement in their co-occurrence network. These features are maintained in longevity and extreme longevity, but interestingly enough, some peculiarities emerged, especially in semi-supercentenarians, describing adaptations that might possibly support health maintenance during aging through an enrichment or higher prevalence of

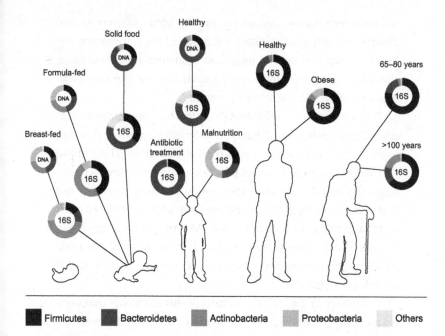

Firmicutes   Bacteroidetes   Actinobacteria   Proteobacteria   Others

Figure 18.1

How the microbiome changes with age and environmental factors. Adapted from M. Levy, A. Kolodziejczyk, C. A. Thaiss, and E. Elinav, "Dysbiosis and the Immune System," *Nature Reviews Immunology* 17, no. 4 (March 2017): 219–232, https://doi.org/10.1038/nri.2017.7.

health-associated groups, or both, including *Akkermansia, Bifidobacterium,* and *Christensenellaceae.*[18]

Diet and the Aging Microbiome

Focusing on environmental factors affecting the aging process, it should not come as a surprise that diet seems to have a tremendous impact. Marcus Claesson and colleagues have shown that the fecal microbiota composition of 178 elderly subjects formed groups correlating with residence location in the community, day hospital, rehabilitation, or long-term residential care.[19] However, clustering of subjects by diet separated them by the same residence location and microbiota groupings.

The separation of microbiota composition significantly correlated with measures of frailty, comorbidity, nutritional status, markers of inflammation, and fecal metabolites. They found that individual microbiota of people in long-term care were significantly less diverse than those of community dwellers, and that loss of community-associated microbiota correlated with increased frailty. Collectively, their data support a relationship between diet and microbiota and health status and indicate a role for diet-driven microbiota alterations in varying rates of health decline upon aging.

A corollary of the studies outlined in this chapter is that once we more deeply understand the natural evolution of the microbiome's composition and function with aging, we will be able to develop strategies to favor a resilient microbiome capable of preserving the host-microbe homeostasis to counteract inflammaging and increased intestinal permeability, and their clinical consequences related to cellular, organ, functional, and mental senescence. Although many factors undoubtedly contribute to health decline and are difficult to completely adjust for in retrospective studies like the ones cited here, the composition and function of the microbiome does seem to be at play.

Among other factors affecting gut microbiome composition and function, diet also seems to be the variable with the most impact in older people. Diet-determined differences in microbiota composition may have subtle impacts in young adults in developed countries. These would be difficult to correlate with health parameters, but these disparities become far more evident in the elderly. This is supported by the stronger microbiota-health associations evident in the long-term residential cohort, and there is now a reasonable case for microbiome-related acceleration of aging-related health deterioration.

An aging population is now a general feature of Western countries and an emerging phenomenon among developing countries. The association of the intestinal microbiome of older people with inflammaging, and the clear association between diet and microbiota, argue in favor of an approach of modulating the microbiota with dietary interventions designed to promote healthier aging. Dietary supplements with defined prebiotics promoting specific components of the microbiome may prove useful for maintaining health in older people. Therefore, microbiome profiling, potentially coupled with metabolomics, offers the potential for biomarker-based identification of individuals at risk for or already undergoing less-healthy aging in specific community-based settings.

Epilogue: Why Studying Our Microbiome Is Important for Our Future

All truth passes through three stages:
First: it is ridiculed;
Second: it is violently opposed;
Third: it is accepted as being self-evident.
—A. Schopenhauer, 1840

Why did we decide to embark on this major undertaking of a book on the human microbiome? Do we really need another book on this topic? The answer is affirmative. We are at a crucial crossroads in the field of microbiome science. If we are smart enough to work together to benefit from our current knowledge and invest the time, talent, and resources to develop a roadmap for translating current and future scientific information into implementable clinical interventions, the outcome of this collaborative effort could change our collective destiny—for the better.

Understanding that microbiome composition and function can influence antigen trafficking, the immune system, and metabolism led to the realization of the fundamental role the microbiome may play in human health. The epidemiological evidence of "epidemics" of noninfectious, chronic inflammatory diseases recorded in industrialized countries, initially explained by the hygiene hypothesis, is in line with this paradigm. However, despite persistent suboptimal hygiene conditions, now we are witnessing similar epidemics in developing countries. This has prompted some people to question the real role of the microbiome in human health. If we revisit the lifestyle trajectory that brought us to our current situation, we will have even stronger evidence of the role of our microbiome in health and disease.

Our Evolutionary Journey

For most of our evolutionary journey, we lived as hunters and gatherers. We traveled in small groups, practicing a nomadic lifestyle with few chances to encounter other hominids. Then, three major lifestyle changes—agriculture, urbanization, and globalization—completely revolutionized our evolutionary plan. These changes caused a radical departure from a carefully crafted and ideal symbiotic relationship in which specific lineages of microbes coevolved with humans over millions of years, passing through hundreds of thousands of generations, shaping our biology throughout evolution until the first disruptor, agriculture, arrived.

Agriculture

The domestication of livestock and the cultivation of crops made food procurement much more predictable and less time consuming. No longer tied to animal migrations and crop cycles, we became settlers, increasing the density of human communities and making interpersonal microbial exchanges more frequent. Living in close contact with animals led to another unplanned consequence, namely the risk of zoonosis (the passage of micro-organisms from animal to human host). Combined with a higher consumption of animal protein, these changes caused a major deviation from the planned evolution of the human microbiome's composition and function.

Urbanization

The second disruptor, urbanization, marked another major milestone in human history. It caused an even greater concentration and interconnection of people, which increased the speed at which exchanges of microorganisms occurred. When this exchange involved pathogens, it led to the spread of new infections. Fast-forward to the twentieth century, when these infectious diseases were tackled by the advent and extensive use of antibiotics. The implementation of a highly sanitized environment also had a major impact on the "urban microbiota," which became less diverse compared to the "rural microbiota" that more closely resembled our original microbiota.

Another consequence of urbanization was far-reaching changes to the global habitat, with the expansion of large cities and highly dense populations, thus limiting areas for extensive agricultural production. This posed additional challenges to human evolution in terms of food procurement and sustainability and created major environmental and social shifts, including concentration of resources—power, knowledge, wealth, and human density—that contrasted with scattered resources in rural areas.

This power differential was found between rural and urban environments. Within urban areas, the same power differential was characterized by extreme inequality between rich and poor populations living in close proximity. This dynamic caused the marginalization of part of the population due to exclusion from the production system, in which mechanization gradually replaced human labor. The segregation between highly populated cities and food supplies coming from scarcely populated rural areas created economic inequities with the multiplication of intermediaries between agricultural producers and consumers.

Globalization

The challenge of maintaining food sustainability for a disproportionately urban consumer community, supplied by a shrinking farming community, was met through globalization, the third disruptor. Now we are in a global village of communication, with the instant exchange of ideas and goods and the constant mobility of people. We can move from one end of the world to another in a matter of hours. However, globalization arrived with a high price tag.

The closer integration of the world economy has facilitated a much faster and unplanned exchange of microorganisms, including the global spread of pathogens through trade and travel. But the globalization of the food supply has had an even greater impact on shifts in microorganisms. The dominant role of the globalized, corporate food system in our modern societies implies that processed foods and, more specifically, mass-produced, empty-calorie nonfoods, like snacks, sweetened beverages, prepared frozen meals, and fast-food items, occupy an exponentially increasing part of the diet of typical consumers in these societies.

To save cost and maintain demand, processed fats, sugar, and salt are used as low-cost ingredients in these foods. The prevalence of these dietary choices means that consumers eat a large proportion of "empty calories" without fiber, high-quality fats, sufficient vitamins, and minerals. Even more worrisome is the fact that what was once an occasional choice, the consumption of unhealthy food, is now the norm as the backbone of the typical Western diet. This is especially true as consumers become more urbanized, undertaking sedentary lifestyles without time to cook from scratch using healthy ingredients.

With the appreciation that diet is the most influential factor shaping our gut microbiome, and that dysbiosis can be associated with a variety of chronic inflammatory diseases, more affluent people are now moving away from junk food and making healthier food choices. The impact of globalization on human health has changed the landscape to the point that the old paradigm of describing noninfectious, chronic inflammatory diseases as "diseases of affluence" typical of Western societies has become misleading. In fact, it is low-income people in industrialized countries as well as in the developing world who currently face the greatest impact from these diseases.

Empty calories are often very cheap calories for people who live in poorer sectors around the world. Consumption of processed or dominantly carbohydrate diets with insufficient whole grains, fruits, and vegetables is more common among the economically disadvantaged, and these dietary traits have a negative impact on microbiome composition and function. Accordingly, the "hygiene hypothesis" is now being challenged by the "microbiome hypothesis." This postulates that by having an influence on the evolutionary, symbiotic relationship between humans and our microbiota, lifestyle changes and, most important, dietary changes are the driving force fueling the epidemics of noninfectious, chronic inflammatory diseases worldwide.

How we can capitalize on this information to slow down if not completely reverse these "epidemics"? Trying to understand what features define a healthy gut community, and how these are host-dependent, will be key to identifying new strategies for disease treatment and prevention. This knowledge may also define how lifestyle-driven changes affecting the microbiota can impact health across populations. For this, we need novel and strategic tools employed by members of an integrated scientific and clinical community working in partnership to effectively study our microbiota. But before we move forward, let us revisit key milestones in microbiome research to appreciate how and where we began this journey, where we are now, and where we go from here.

Key Milestones in Human Microbiome Research

> I then most always saw, with great wonder, that in the said matter there were many very little living animalcules, very prettily a-moving.
> —A. van Leeuwenhoek, 1683

Now that we have a better understanding of what we did wrong, we may have a path to correct our mistakes and bring the relationship with our microbiome back to symbiotic terms. For a summary of key milestones in microbiome science, coauthor Alessio Fasano has capitalized on an outstanding overview created for *Nature*'s website by a group of very talented colleagues.[1] Below are his thoughts on their timeline as it relates to the contents of this book, which are summarized in figure E.1.

Milestone 1: "When we began this book project," Fasano recalls, "I was convinced that I had experienced in person, both as a spectator and for a minor part as an actor, most of the history of the field of research related to the human microbiome. However, this was a major oversight of scientific history dating back to the 1680s."

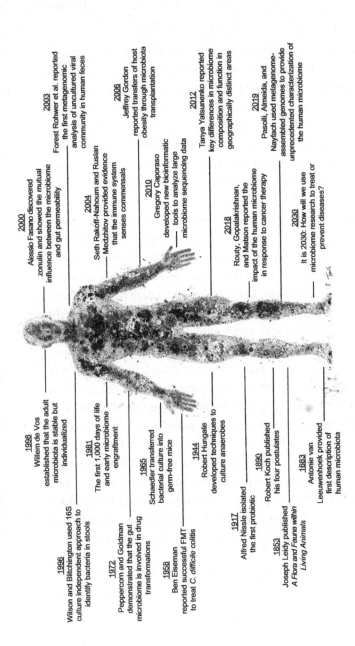

Figure E.1
Key milestones in microbiome science from its inception in 1683 to future projections of its potential exploitation for personalized medicine and disease intervention. (Adapted from N. Pariente, "Milestones in Human Microbiota Research," *Milestones*, June 18, 2019, https://www.nature.com/immersive/d42859-019-00041-z/index.html, accessed February 2, 2020.)

Antonie van Leeuwenhoek, making use of his newly developed microscopes, described and illustrated in a letter written in 1683 to the Royal Society of London, discovered five different kinds of "animalcules" (the term he used to describe bacteria) present in his own mouth. He subsequently compared his own oral and fecal microbiota, determining that there are differences between body sites as well as between health and disease. This was among the earliest reports suggesting the existence of a human microbiota.

Milestone 2: Almost two centuries later in 1853, Joseph Leidy published the book *A Flora and Fauna within Living Animals*, which most likely represents the official document considered by many to be the origin of microbiota research.[2] Then the work of Pasteur, Metchnikoff, Koch, Theodor Escherich, Arthur Kendall, and many others laid the foundations of modern microbiology and the modern understanding of infectious diseases by providing key information on host-microorganism interactions. Besides postulating the germ theory of disease, Pasteur also was convinced that nonpathogenic microorganisms might have an important role in normal human physiology. Metchnikoff believed that microbiome composition and interactions with the host were both essential for healthy aging. And Escherich was convinced that understanding the endogenous flora was essential for understanding the physiology and pathology of key gastrointestinal functions. These postulations implied that besides a belligerent relationship with pathogens, the human host also was engaged in a symbiotic interaction with commensals.

Milestone 3: By publishing his famous four postulates in 1890, Robert Koch provided the fundamentals establishing the causative relationship between the presence of a microorganism and a specific infectious disease. His approach was limited, because in that era bacteria could only be cultivated in the presence of oxygen. This limitation meant that the vast majority of nonpathogenic human commensals, which are typically anaerobes, were overlooked.

Milestone 4: During World War I, German physician Alfred Nissle noticed that one particular soldier did not succumb to dysentery. He wondered if the cause was a protective microorganism in the soldier's gut. In 1917, Nissle isolated the *E. coli* Nissle 1917 strain, which remains a commonly used probiotic. He later showed that it antagonized pathogens, so establishing the concept of colonization resistance, whereby human-associated microorganisms prevent the establishment of pathogens in the same niche.

Milestone 5: Milestones 1–4 provided the foundations for the research field of human microbiota that accelerated in the 1940s, when Robert Hungate described in detail the methods, still used nowadays, to grow microorganisms in the absence of oxygen—this is milestone 5. Thanks to these culture techniques, we began to appreciate the complexity of the human microbiome well beyond the boundaries of what was then known. By using anaerobic culture approaches, we could classify different microorganisms occupying many of the human host niches and appreciate their impact on many human physiological functions.

Milestone 6: The consequence of an unbalanced microbiome, when pathogens take over specific human host niches, was further appreciated with the use of FMT as a method to "push the reset button" of an ecosystem that has become detrimental to the host. The use of FMT to treat a variety of human diseases, mainly gastrointestinal problems, dates back to fourth-century China, where "yellow soup" was used in cases of severe food poisoning and diarrhea. By the sixteenth century, the Chinese had developed a variety of feces-derived products for gastrointestinal complaints as well as systemic symptoms such as fever and pain. Anecdotal reports suggest that Bedouin groups consumed the stools of their camels as a remedy for bacterial dysentery. Italian anatomist and surgeon Fabricius Acquapendente (1537–1619) further extended this to a concept he called "transfaunation," the transfer of gastrointestinal contents from a healthy to a sick animal, which has since been applied extensively in the field of veterinary medicine. Interestingly, many animal species are found to naturally practice coprophagy, a sort of self-administered FMT, leading to a greater diversity of microorganisms in their intestines. Slowly, these ideas began to spark interest in eighteenth-century European physicians, but with no major success until Ben Eiseman and colleagues. With the start of the "microbiome revolution," in 1958 they published results from the successful treatment of four people suffering from pseudomembranous colitis, before *C. difficile* was determined to be the cause.

Milestone 7: In 1965, Russell Schaedler and colleagues added another major cornerstone to microbiome research by reporting the transfer of bacterial cultures to germ-free mice to study the effects of the gut microbiome on the host physiopathology. They found that feeding bacterial cultures isolated from the gut of albino mice free of ordinary mouse pathogens, as well as intestinal *E. coli* and *Proteus* spp., to germ-free mice led to the engraftment of the microbiome in a way comparable to the donor mice.

They also showed that the gut microbiota of these mice remained stable for several months, and that specific metabolic activities reported from some bacterial strains were not detected unless a complex and diversified microbiota was present, confirming the importance of a balanced ecosystem for an ideal symbiotic relationship between microorganisms and their host.

Milestone 8: In 1972, Mark Peppercorn and Peter Goldman demonstrated that an anti-inflammatory drug could be degraded in conventional rats when cultured with human gut bacteria, but not in germ-free rats, indicating a role for the gut microbiome in drug transformations. From this initial observation, several studies have confirmed the role of the microbiome in drug metabolism as not limited only to the gut, highlighting implications for drug inactivation, efficacy, and toxicity.

Milestone 9: In early 1980, the symbiotic relationship established between the engrafting microbiome and its human host during the first one thousand days of life—and how this relationship will dictate our health trajectory for the years to come—was first recognized. And while the succession of events leading to the establishment of a stable microbiome has been studied for decades, three pivotal studies published in 1981 quantitatively characterized the early acquisition of gut commensals and the study of how feeding shapes our initial microbiome.

Milestone 10: Until the early 1990s, studies of the human microbiota were based on culture-dependent methods, which undermined understanding of the great biodiversity of the human-associated microbial communities. Thanks to techniques developed during the Human Genome Project, Kenneth Wilson and Rhonda Blitchington compared the diversity of cultivated and noncultivated bacteria within a human fecal sample in 1996. Thanks to their pioneering work, 16S ribosomal (r) RNA sequencing has become a powerful tool for assessing microbial diversity in the human microbiome.

Milestone 11: The search for the "normal human microbiome" to identify departures from its composition linked to diseases has been elusive and frustrating. In 1998, a study by Willem de Vos and colleagues used polymerase chain reaction amplification of regions of the 16S rRNA gene coupled with temperature gradient gel electrophoresis (TGGE) to visualize the diversity of the amplified gene. Comparisons of the banding profiles generated by TGGE from sixteen adult fecal samples showed unequivocally that everyone has a unique microbial community. Furthermore, by monitoring two individuals over time, the researchers showed that the TGGE profiles were stable over a period of at least six months, suggesting that, once an ideal and highly personalized symbiotic relationship is established between the microbiome and its host, there is a strong effort to maintain the status quo as the ideal equilibrium.

Milestone 12: Until the early 1990s, little was known about whether, how, and why gut permeability was modulated. With the growing awareness of the complexity of the intercellular space controlled by tight junction competency and the discovery of zonulin as a physiological modulator of intestinal permeability, several studies have been published linking this molecule to a variety of chronic inflammatory diseases in which dysbiosis has been hypothesized as a pathogenetic component. The key interplay between intestinal permeability, including zonulin-mediated changes, and gut dysbiosis contributed to mechanistically linking changes in microbiome composition and function to altered antigen trafficking involved in disease pathogenesis.

Milestone 13: While bacteria have been the focus of almost the entirety of the microbiome-related literature, it is well appreciated that viruses, fungi, and archaea are also important members of the human ecosystem, with potential effects on human health. In 2001, marine microbial ecologist Forest Rohwer's research group published a randomized, shotgun library-sequencing method to analyze genomic DNA from a single bacteriophage. This was a crucial step toward the much more complex task of analyzing the human virome, a task achieved by Rohwer's research group in 2003, which provided the first quantitative description of the composition of the uncultured virome in human feces collected from a single healthy adult.

Milestone 14: The interplay between the host immune system and microorganisms typically has been interpreted as a war in which host defense is principally aimed at eliminating pathogens. The observation that in germ-free animals the immune system matures inappropriately opened a new interpretation of this inter-action, suggesting that the previously reductive, belligerent view should be revisited in a much more complex programming of immune system maturation and function by the engrafted micro-biome. A key element to distinguish pathogens from commensals involves the recognition by the host of colonizing microorgan-isms via pattern recognition receptors (PRR), which sense con-served microbial molecules. In 2004, Seth Rakoff-Nahoum and Ruslan Medzhitov provided evidence that the immune system senses commensals through PRRs under normal conditions, and that this sensing is crucial for tissue repair. This finding opened a new perspective on immune response to microorganisms, not as simply a host defense, but also as a symbiotic physiological process in a mutually triangulating effect among the gut barrier (see milestone 12), the immune system, and the microbiome.

Milestone 15: The rising prevalence of chronic inflammatory diseases recorded in industrialized countries during the past few decades has been associated with a Westernized diet that highly influences microbiome composition and function. Early studies using germ-free mice showed that body fat content and insulin resistance are transferable from obese to lean mice through exposure to fecal material. In a 2006 pioneering paper, Jeff Gordon and his collaborators reported that the microbiota of obese mice are more efficient at extracting energy from the host diet compared to the microbiota of lean ones.[3] This phenotype was transferable by transplanting the microbiota from the cecum of obese mice into lean, germ-free animals. The same group of researchers highlighted the crucial impact that diet can have on gut microbiota and host metabolism, opening up the development of nutrition-based interventions to manipulate the host microbiome affecting human health.

Milestone 16: The staggering quantity of data generated with microbiome sequencing required innovative bioinformatics tools to facilitate their analysis. In 2010, Gregory Caporaso and coworkers described the software pipeline QIIME, which stands for "quantitative insights into microbial ecology," as a tool that enables the analysis and interpretation of the increasingly large datasets generated by microbiome sequencing.

Milestone 17: Human adaptation to different geographic areas has always been considered a premise of genetic variability. However, with the appreciation that the host microbiome may play a crucial epigenetic role, studying differences in human microbiomes related to different geographic regions became an important focus of research to link lifestyle, environment, and clinical outcome. In 2012, Tanya Yatsunenko and colleagues characterized bacterial species in fecal samples from cohorts living in different regions, including the Amazonas of Venezuela, rural Malawi, and metropolitan areas in the United States. Yatsunenko and colleagues found pronounced differences in the composition and functions of the gut microbiomes between these geographically distinct cohorts and age groups, inferring that there is a strong need to consider the microbiome when evaluating human development, nutritional needs, physiological variations, and the impact of a "Westernized" lifestyle.

Milestone 18: In 2018, three independent reports showed that the human microbiome can affect a person's response to cancer therapy. Following earlier studies in mouse models, these investigators reported that gut microbiota composition may affect the response of melanoma patients, as well as patients suffering from advanced lung or kidney cancer, to immune checkpoint therapy and tumor control.

Milestone 19: Advances in computational methods have enabled the reconstruction of bacterial genomes from metagenomic datasets. This approach was used in 2019 by three research groups to identify thousands of new, uncultured, candidate bacterial species from the gut and other body sites of global populations from rural and urban settings. This substantially expanded the known phylogenetic diversity and improved classification of understudied, non-Western populations.

Milestone 20: This story is set in the year 2030. It's a story summarizing coauthor Alessio Fasano's vision of how microbiome research will radically change the future of medicine. And it's about the future of a little girl who just happens to be fictional, but who is, in fact, a lot like millions of very real children all around the world. She is an example of someone whose life may be transformed by the kind of research-driven clinical care that will be developed and provided thanks to the amazing work of many individuals, some mentioned in this book and many, many others we have been unable to recognize for their superlative contributions to the microbiome research domain. Without them, this 2030 story would not be conceivable.

Gemma is finally asleep. Melanie stands by the window in Dr. Fasano's office, in the warmth of the late afternoon sun. She is gently swaying, baby in her arms, as she watches her husband in the park across the street with their three-year-old son, Bobby. Bobby and his father are looking for airplanes. Planes flying overhead. Contrails. Any evidence of flight. Like many children with ASD, Bobby has an all-consuming obsession. His obsession is airplanes. Melanie closes her eyes and breathes in rhythm with the sleeping child resting on her shoulder. She thinks about how easily she could fall asleep, right here on her feet. It has been a long week. Gemma's ear infection has eased, thanks to a three-day course of a targeted oral antibiotic. But now the baby is constipated and her tummy hurts. There have been stool collections and blood draws and anxious moments. Melanie snaps awake as Gemma's doctor opens the door. Alternating waves of hyper-alert attention and out-of-body panic wash over her as he begins to speak.

There is good news. And bad news. And more good news. The good news, Dr. Fasano explains, is that Gemma's whole genome was sequenced at birth—enabling him to use that data, along with gut permeability tests, immune profiling based on a blood sample, and microbiome, metatranscriptomic, and metabolomic analyses performed on a stool sample, to look for both the underlying causes of Gemma's acute illness and biomarkers known to be predictors of ASD. The tests have revealed that Gemma's zonula (a marker for gut permeability) appears elevated; her gut microbiome appears unbalanced, with low amounts of *F. prausnitzii*; her Enterobacteria count is a bit high; and the genes controlling lactate production by Lactobacilli have been downregulated. A metabolomic analysis confirms a reduction of lactate in Gemma's stools. The whole genome sequencing and epigenetic changes reveal that genes controlling Gemma's immune response have been activated. For this reason, Dr. Fasano requested a positron emission tomography (PET) scan of Gemma's brain, which shows neuroinflammation. Armed with these test results, which Dr. Fasano explains to Melanie, he then turns to his computer and performs a risk analysis revealing that the combination of Gemma's positive biomarkers, immune profile, specific gene variants, gut microbiome, and metabolome composition carry a fifty-five-fold increased risk that she will develop ASD within nine months.

Melanie catches her breath. She flashes back to the moment, nearly two years ago, when she first heard the word *autism* used in connection with her son. But this is now. Another time. Another child. A child who is apparently in danger, despite having hit

every growth or developmental milestone in her first twelve
months of life. Melanie wills herself back into this room, which
suddenly seems strangely devoid of oxygen, and into this conversation. Dr. Fasano is saying that there is, in fact, good news. He
is prescribing a change in diet specifically tailored to Gemma's
profile to favor the growth of protective microorganisms and a
three-month course of a genetically engineered probiotic capable
of sensing changes in the gut micromilieu and reestablishing
proper microbiome composition and metabolic profiles—thereby
preventing the onset of ASD. "Can I allow myself to believe this?"
Melanie thinks, remembering that when Bobby was diagnosed
ASD wasn't even treatable—and certainly not preventable. "Could
this be true?" she asks out loud. "*It is possible*," Dr. Fasano affirms,
"*because of the amazing work of thousands of researchers from all
over the globe.*" For the last 350 years, these inspired and persistent individuals have generated an enormous body of work leading to the exploitation of microbiome manipulation for disease
interception.

The future looks bright for Melanie's children. In three months,
Gemma will be back in this room for a checkup. Her biomarkers will be back to normal. Her PET scan will be normal. Her
childhood will be healthy and happy, and her life will be full
of promise. And Bobby will be enrolled in a new treatment
protocol, building on the same research that yielded Gemma's
therapies. Bobby's doctors hope that by reducing the feedback
mechanism between the body's immune response mechanisms
and some specific microbiome-derived biomarkers, they can ease
his symptoms and drive toward long-term improvement.

Milestone 20 is more than just a wish; it is the reason we get up
in the morning extremely excited to start another day of work.

Acknowledgments

Just as it takes many different components—bacterial, viral, fungal, and more—to make up the human microbiome, it takes many people to write a book about the human microbiome. Since we began this project several years ago, the field of microbiome research has witnessed an explosion of research studies and initiatives across the globe. These new studies and avenues of research have only been made possible by building on decades of research in microbiology, immunology, environmental biology, biostatistics, and other scientific disciplines.

To help us address the complex issue of the human microbiome in health and disease, we interviewed fifteen researchers in the microbiome sciences. Some have been traveling this road for decades, and others are new to the field. We are deeply grateful to the following people, listed in the order of their contributions, for providing insights and material to this book: Rita Colwell, Claire Fraser, Josef Neu, Jacques Ravel, W. Allan Walker, Forest Rohwer, Lee Kaplan, Sarkis Mazmanian, Laurence Zitzvogel, Lauren Fiechtner, Maureen Leonard, Victoria Martin, Jon Vanderhoof, Ali Zomorrodi, and Timothy Lu.

We would also like to thank Massachusetts General Hospital (MGH) and MGH colleagues for their support of this project: Ronald Kleinman, Lisa Milone, Joanne O'Brien, Kristin Hasselschwert, and members of the Division of Pediatric Gastroenterology and Nutrition, the Center for Celiac Research and Treatment, and the MIBRC at MGH, including Christina Faherty for providing material and reviewing the manuscript. We thank Robert Prior, Anne-Marie Bono, and Kathleen Caruso of the MIT Press for their remarkable patience and flexibility. And we thank Jennifer Ottman and Christian Zang for editing and proofreading support.

Finally, along with the dedicated scientists and clinicians who will forge new paths in disease treatment and prevention, we thank the generous patients who participate in research studies to advance findings that will lead to new therapies for age-old diseases. We know that by the time this book is in print, the field of microbiome science will have leapt forward with new and unexpected discoveries. We have made our best attempt to include breakthrough research and provocative theories in these chapters, with the full realization that the rapid pace of change in microbiome science may make some of the material redundant. Lastly, any errors and omissions are our own.

Notes

1 Evolutionary Biology Explains Bacterial Adaptability

1. T. Djokic, M. J. Van Kranendonk, K. A. Campbell, M. R. Walter, and C. R. Ward, "Earliest Signs of Life on Land Preserved in ca. 3.5 Ga Hot Spring Deposits," *Nature Communications* 8 (May 9, 2017): 15263, https://doi.org/10.1038/ncomms15263.

2. J. Fox-Skelly, "What Is the Hottest Temperature Life Can Survive?," *BBC Earth*, February 10, 2016, http://www.bbc.com/earth/story/20160209-this-is-how-to-survive-if-you-spend-your-life-in-boilin-water, accessed November 7, 2019.

3. N. Touchette, "World's Hottest Microbe: Loving Life in Hell," *Genome News Network*, August 22, 2003, http://www.genomenewsnetwork.org/articles/08_03/hottest.shtml, accessed November 7, 2019.

4. J. Scott, "CU-Boulder-Led Team Finds Microbes in Extreme Environment in South American Volcanoes," *Colorado Arts and Sciences Magazine Archive*, June 8, 2012, https://www.colorado.edu/asmagazine-archive/node/982, accessed June 19, 2020.

5. P. Sommers, R. S. Fontenele, T. Kringen, S. Kraberger, D. L. Porazinska, J. L. Darcy, S. K. Schmidt, and A. Varsani, "Single-Stranded DNA Viruses in Antarctic Cryoconite Holes," *Viruses* 11, no. 11 (November 2019): 1022, https://doi.org/10.3390/v11111022.

6. B. J. MacFadden, "Fossil Horses—Evidence for Evolution," *Science* 307, no. 5716 (March 18, 2005): 1728–1730, https://doi.org/10.1126/science.1105458.

7. H. Imachi, M. K. Nobu, N. Nakahara, Y. Morono, M. Ogawara, Y. Takaki, Y. Takano, et al., "Isolation of an Archaeon at the Prokaryote-Eukaryote Interface," *Nature* 577 (January 23, 2020): 519–525, https://doi.org/10.1038/s41586-019-1916-6.

8. M. Blaser, *Missing Microbes: How the Overuse of Antibiotics Is Fueling Our Modern Plagues* (New York: Henry Holt, 2014).

9. E. R. Davenport, J. G. Sanders, S. J. Song, K. R. Amato, A. G. Clark, and R. Knight, "The Human Microbiome in Evolution," *BMC Biology* 15 (December 27, 2017): 127, https://doi.org/10.1186/s12915-017-0454-7.

10. J. Snow, "The Cholera near Golden-Square, and at Deptford," *Medical Times and Gazette*, September 23, 1854, 321–322.

11. A. C. Ghose, "Lessons from Cholera and *Vibrio cholerae*," *Indian Journal of Medical Research* 133, no. 2 (February 2011): 164–170, http://www.ijmr.org.in/text.asp?2011/133/2/164/78117.

12. Ghose, "Lessons from Cholera."

13. D. Lippi, E. Gotuzzo, and S. Caini, "Cholera," *Microbiology Spectrum* 4, no. 4 (August 2016): PoH-0012-2015, https://doi.org/10.1128/microbiolspec.poh-0012-2015.

14. M. Trucksis, J. Michalski, Y. K. Deng, and J. B. Kaper, "The *Vibrio cholerae* Genome Contains Two Unique Circular Chromosomes," *Proceedings of the National Academy of Sciences of the United States of America* 95, no. 24 (November 24, 1998): 14464–14469, https://doi.org/10.1073/pnas.95.24.14464.

15. National Institutes of Health, National Human Genome Research Institute, "Human Genome Project FAQ," https://www.genome.gov/human-genome-project/Completion-FAQ, accessed November 7, 2019.

16. Human Genome Project Information Archive 1990–2003, https://web.ornl.gov/sci/techresources/Human_Genome/project/press4_2003.shtml, accessed November 7, 2019.

17. C. Fraser, oral interview with authors, November 1, 2017.

18. M. J. Behbehani, H. V. Jordan, and D. L. Santoro, "Simple and Convenient Method for Culturing Anaerobic Bacteria," *Applied and Environmental Microbiology* 43, no. 1 (January 1982): 255–256, https://aem.asm.org/content/43/1/255.

2 The Ancestral Microbiome

1. D. P. Strachan, "Hay Fever, Hygiene, and Household Size," *British Medical Journal* 299, no. 6710 (November 18, 1989): 1259–1260, https://doi.org/10.1136/bmj.299.6710.1259.

2. S. F. Bloomfield, R. Stanwell-Smith, R. W. R. Crewel, and J. Pickup, "Too Clean, or Not Too Clean: The Hygiene Hypothesis and Home Hygiene," *Clinical and Experimental Allergy* 36, no. 4 (April 2006): 402–425, https://doi.org/10.1111/j.1365-2222.2006.02463.x.

3. R. Pahwa, A. Goyal, P. Bansal, and I. Jialal, *Chronic Inflammation* (Treasure Island, FL: StatPearls, 2018), https://www.ncbi.nlm.nih.gov/books/nbk493173/.

4. L. S. Weyrich, S. Duchene, J. Soubrier, L. Arriola, B. Llamas, J. Breen, A. G. Morris, et al., "Neanderthal Behaviour, Diet, and Disease Inferred from Ancient DNA in Dental Calculus," *Nature* 544, no. 7650 (April 20, 2017): 357–361, https://doi.org/10.1038/nature21674.

5. Weyrich et al., "Neanderthal Behaviour."

6. A. Humar, I. McGilvray, M. J. Phillips, and G. A. Levy, "Severe Acute Respiratory Syndrome and the Liver," *Hepatology* 39, no. 2 (February 2004): 291–294, https://doi.org/10.1002/hep.20069; World Health Organization, "SARS (Severe Acute Respiratory Syndrome)," https://www.who.int/ith/diseases/sars/en/, accessed May 14, 2020.

7. Humar et al., "Severe Acute Respiratory Syndrome."

8. L.-F. Wang, Z. Shi, S. Zhang, H. E. Field, P. Daszak, and B. T. Eaton, "Review of Bats and SARS," *Emerging Infectious Diseases* 12, no. 12 (December 2006): 1834–1840, https://doi.org/10.3201/eid1212.060401.

9. D. Cyranoski, "Mystery Deepens over Animal Source of Coronavirus," *Nature* 579 (March 5, 2020): 18–19, https://doi.org/10.1038/d41586-020 -00548-w.

10. J. Howard, "Plague Explained," *National Geographic*, August 20, 2019, https://www.nationalgeographic.com/science/health-and-human-body /human-diseases/the-plague/, accessed November 9, 2019.

11. Naval History and Heritage Command, Navy Department Library, "Influenza of 1918 (Spanish Flu) and the U.S. Navy," April 6, 2015, https://www.history.navy.mil/content/history/nhhc/research/library /online-reading-room/title-list-alphabetically/i/influenza/influenza-of -1918-spanish-flu-and-the-us-navy.html, accessed November 9, 2019.

12. New England Historical Society, "Exactly How New England's Indian Population Was Decimated," updated in 2018, http://www.newengland-historicalsociety.com/exactly-new-englands-indian-population-decimated/, accessed November 9, 2019.

13. National Museum Australia, "Defining Moments: Smallpox Epidemic," https://www.nma.gov.au/defining-moments/resources/smallpox-epi demic, accessed November 9, 2019.

14. HIV.gov, "A Timeline of HIV and AIDS," https://www.youtube.com /watch?v=EyaryYcXjho, accessed November 9, 2019.

15. Centers for Disease Control and Prevention, "Coronavirus Disease 2019 (COVID-19): World Map," https://www.cdc.gov/coronavirus/2019 -ncov/global-covid-19/world-map.html, accessed May 15, 2020.

16. Fraser, oral interview.

17. T. Bosch, interview with M. Dominguez-Bello, "How Western Civiliza-tion Affects the Microbiome," at Kiel University, October 27, 2017, https:// www.youtube.com/watch?v=EyaryYcXjho, accessed November 9, 2019.

18. Bosch, interview with Dominguez-Bello.

19. J. C. Clemente, E. C. Pehrsson, M. J. Blaser, K. Sandhu, Z. Gao, B. Wang, M. Magris, et al., "The Microbiome of Uncontacted Amerindians," *Science Advances* 1, no. 3 (April 2015): e1500183, https://doi.org/10.1126/sciadv.1500183.

20. Clemente et al., "The Microbiome of Uncontacted Amerindians."

21. Clemente et al., "The Microbiome of Uncontacted Amerindians."

22. Clemente et al., "The Microbiome of Uncontacted Amerindians."

23. A. J. Obregon-Tito, R. Y. Tito, J. Metcalf, K. Sankaranarayanan, J. C. Clemente, L. K. Ursell, Z. Z. Xu, et al., "Subsistence Strategies in Traditional Societies Distinguish Gut Microbiomes," *Nature Communications* 6 (March 25, 2015): 6505, https://doi.org/10.1038/ncomms7505.

24. Obregon-Tito et al., "Subsistence Strategies."

25. Blaser, *Missing Microbes*.

26. A. Keys, ed., "Coronary Heart Disease in Seven Countries," supplement, *Circulation* 41, no. S1 (April 1970).

27. C. A. Thaiss, D. Zeevi, M. Levy, G. Zilberman-Schapira, J. Suez, A. C. Tengeler, L. Abramson, et al., "Transkingdom Control of Microbiota Diurnal Oscillations Promotes Metabolic Homeostasis," *Cell* 159, no. 3 (October 23, 2014): 514–529, https://doi.org/10.1016/j.cell.2014.09.048.

28. S. A. Smits, J. Leach, E. D. Sonnenburg, C. G. Gonzalez, J. S. Lichtman, G. Reid, R. Knight, et al., "Seasonal Cycling in the Gut Microbiome of the Hadza Hunter-Gatherers of Tanzania," *Science* 357, no. 6353 (August 25, 2017): 802–806, https://doi.org/10.1126/science.aan4834.

29. G. Dubois, C. Girard, F.-J. Lapointe, and B. J. Shapiro, "The Inuit Gut Microbiome Is Dynamic over Time and Shaped by Traditional Foods," *Microbiome* 5 (November 16, 2017): 151, https://doi.org/10.1186/s40168-017-0370-7.

30. Dubois et al., "The Inuit Gut Microbiome."

31. Dubois et al., "The Inuit Gut Microbiome."

32. C. Girard, N. Tromas, M. Amyot, and B. J. Shapiro, "Gut Microbiome of the Canadian Arctic Inuit," *mSphere* 2, no. 1 (January–February 2017): e00297-16, https://doi.org/10.1128/msphere.00297-16.

3   Early Factors Influencing the Human Microbiome

1. T. A. Manuck, "Racial and Ethnic Differences in Preterm Birth: A Complex, Multifactorial Problem," *Seminars in Perinatology* 41, no. 8 (December 2017): 511–518, https://doi.org/10.1053/j.semperi.2017.08.010.

2. A. C. Skinner, S. N. Ravanbakht, J. A. Skelton, E. M. Perrin, and S. C. Armstrong, "Prevalence of Obesity and Severe Obesity in US Children, 1999–2016," *Pediatrics* 141, no. 3 (March 2018): e20173459, https://doi.org/10.1542/peds.2017-3459.

3. E. Jašarević, C. L. Howerton, C. D. Howard, and T. L. Bale, "Alterations in the Vaginal Microbiome by Maternal Stress Are Associated with Metabolic Reprogramming of the Offspring Gut and Brain," *Endocrinology* 156, no. 9 (September 1, 2015): 3265–3276, https://doi.org/10.1210/en.2015-1177.

4. T. M. Nelson, J. C. Borgogna, R. D. Michalek, D. W. Roberts, J. M. Rath, E. D. Glover, J. Ravel, M. D. Shardell, C. J. Yeoman, and R. M. Brotman, "Cigarette Smoking Is Associated with an Altered Vaginal Tract Metabolomic Profile," *Scientific Reports* 8 (January 16, 2018): 852, https://doi.org/10.1038/s41598-017-14943-3.

5. J. Ravel, personal communication, January 14, 2020.

6. M. E. Perez-Muñoz, M.-C. Arrieta, A. E. Ramer-Tait, and J. Walter, "A Critical Assessment of the 'Sterile Womb' and 'In Utero Colonization' Hypotheses: Implications for Research on the Pioneer Infant Microbiome," *Microbiome* 5 (April 28, 2017): 48, https://doi.org/10.1186/s40168-017-0268-4.

7. J. Ravel, personal communication, January 14, 2020.

8. M. Mshvildadze, J. Neu, J. Shuster, D. Theriaque, N. Li, and V. Mai, "Intestinal Microbial Ecology in Premature Infants Assessed with

Non-Culture-Based Techniques," *Journal of Pediatrics* 156, no. 1 (January 2010): 20–25, https://doi.org/10.1016/j.jpeds.2009.06.063.

9. C. Willyard, "Could Baby's First Bacteria Take Root Before Birth?," *Nature* 553 (January 18, 2018): 264–266, https://doi.org/10.1038/d41586 -018-00664-8.

10. J. Neu, oral interview with authors, May 23, 2018.

11. L. J. Funkhouser and S. R. Bordenstein, "Mom Knows Best: The Universality of Maternal Microbial Transmission," *PLoS Biology* 11, no. 8 (August 20, 2013): e1001631, https://doi.org/10.1371/journal.pbio.1001631.

12. Mshvildadze et al., "Intestinal Microbial Ecology."

13. T. Tapiainen, N. Paalanne, M. V. Tejesvi, P. Koivusaari, K. Korpela, T. Pokka, J. Salo, et al., "Maternal Influence on the Fetal Microbiome in a Population-Based Study of the First-Pass Meconium," *Pediatric Research* 84, no. 3 (September 2018): 371–379, https://doi.org/10.1038 /pr.2018.29.

14. J. Ravel, oral interview with authors, September 21, 2017.

15. Ravel, oral interview.

16. A. P. Lauder, A. M. Roche, S. Sherrill-Mix, A. Bailey, A. L. Laughlin, K. Bittinger, R. Leite, M. A. Elovitz, S. Parry, and F. D. Bushman, "Comparison of Placenta Samples with Contamination Controls Does Not Provide Evidence for a Distinct Placenta Microbiota," *Microbiome* 4 (June 23, 2016): 29, https://doi.org/10.1186/s40168-016-0172-3.

17. Ravel, oral interview.

18. Perez-Muñoz et al., "A Critical Assessment."

19. M. C. de Goffau, S. Lager, U. Sovio, F. Gacciolli, E. Cook, S. J. Peacock, J. Parkhill, D. S. Charnock-Jones, and G. C. S. Smith, "Human Placenta Has No Microbiome but Can Contain Potential Pathogens," *Nature* 572, no. 7769 (August 15, 2019): 329–334, https://doi.org/10.1038/s41586-019 -1451-5.

20. Funkhouser and Bordenstein, "Mom Knows Best."

21. M. A. Zarate, M. D. Rodriguez, E. I. Chang, J. T. Russell, T. J. Arndt, E. M. Richards, B. A. Ocasio, et al., "Post-Hypoxia Invasion of the Fetal Brain by Multidrug Resistant *Staphylococcus*," *Scientific Reports* 7 (July 25, 2017): 6458, https://doi.org/10.1038/s41598-017-06789-6.

22. M. D. Seferovic, R. M. Pace, M. Carroll, B. Belfort, A. M. Major, D. M. Chu, D. A. Racusin, et al., "Visualization of Microbes by 16S In Situ Hybridization in Term and Preterm Placentas without Intraamniotic Infection," *American Journal of Obstetrics and Gynecology* 221, no. 2 (August 1, 2019): 146e1–146e23, https://doi.org/10.1016/j.ajog.2019.04.036.

23. N. Younge, J. R. McCann, J. Ballard, C. Plunkett, S. Akhtar, F. Araújo-Pérez, A. Murtha, D. Brandon, and P. C. Seed, "Fetal Exposure to the Maternal Microbiota in Humans and Mice," *JCI Insight* 4, no. 19 (October 3, 2019): e127806, https://doi.org/10.1172/jci.insight.127806.

24. Ravel, oral interview.

25. L. R. McKinnon, S. L. Achilles, C. S. Bradshaw, A. Burgener, T. Crucitti, D. N. Fredricks, H. B. Jaspan, et al., "The Evolving Facets of Bacterial Vaginosis: Implications for HIV Transmission," *AIDS Research and Human Retroviruses* 35, no. 3 (March 2019): 219–228, https://doi.org/10.1089/aid.2018.0304.

26. J. Leizer, D. Nasioudis, L. J. Forney, G. M. Schneider, K. Gliniewicz, A. Boester, and S. S. Witkin, "Properties of Epithelial Cells and Vaginal Secretions in Pregnant Women When *Lactobacillus crispatus* or *Lactobacillus iners* Dominate the Vaginal Microbiome," *Reproductive Sciences* 25, no. 6 (June 1, 2018): 854–860, https://doi.org/10.1177/1933719117698583; S. S. Witkin, D. Nasioudis, J. Leizer, E. Minis, A. Boester, and L. J. Forney, "Epigenetics and the Vaginal Microbiome: Influence of the Microbiota on the Histone Deacetylase Level in Vaginal Epithelial Cells from Pregnant Women," *Minerva Ginecologica* 71, no. 2 (April 2019): 171–175, https://doi.org/10.23736/s0026-4784.18.04322-8; V. L. Edwards, S. B. Smith, E. J. McComb, J. Tamarelle, B. Ma, M. S. Humphrys, P. Gajer, et al., "The Cervicovaginal Microbiota-Host Interaction Modulates *Chlamydia trachomatis* Infection," *mBio* 10, no. 4 (August 2019): e01548-19, https://doi.org/10.1128/mbio.01548-19.

27. Ravel, oral interview.

28. B. Ma, L. J. Forney, and J. Ravel, "Vaginal Microbiome: Rethinking Health and Disease," *Annual Review of Microbiology* 66 (2012): 371–389, https://doi.org/10.1146/annurev-micro-092611-150157.

29. Ravel, oral interview.

30. Ravel, oral interview.

31. H. M. Dunsworth, "There Is No 'Obstetrical Dilemma': Towards a Braver Medicine with Fewer Childbirth Interventions," *Perspectives in Biology and Medicine* 61, no. 2 (Spring 2018): 249–263, https://doi.org /10.1353/pbm.1018.0040; H. M. Dunsworth, "Thank Your Intelligent Mother for Your Big Brain," *Proceedings of the National Academy of Sciences of the United States of America* 113, no. 25 (June 21, 2016): 6816–6818, https://doi.org/10.1073/pnas.1606596113.

32. Ravel, oral interview.

33. M. A. Elovitz, P. Gjer, V. Riis, A. G. Brown, M. S. Humphrys, J. B. Holm, and J. Ravel, "Cervicovaginal Microbiota and Local Immune Response Modulate the Risk of Spontaneous Preterm Delivery," *Nature Communications* 10 (March 21, 2019): 1305, https://doi.org/10.1038 /s41467-019-09285-9.

34. D. B. DiGiulio, B. J. Callahan, P. J. McMurdie, E. K. Costello, D. J. Lyell, A. Robaczewska, C. L. Sun, et al., "Temporal and Spatial Variation of the Human Microbiota during Pregnancy," *Proceedings of the National Academy of Sciences of the United States of America* 112, no. 35 (September 1, 2015): 11060–11065, https://doi.org/10.1073/pnas.1502875112.

35. DiGiulio et al., "Temporal and Spatial Variation."

36. M. J. Stout, Y. Zhou, K. M. Wylie, P. I. Tarr, G. A. Macones, and M. G. Tuuli, "Early Pregnancy Vaginal Microbiome Trends and Preterm Birth," *American Journal of Obstetrics and Gynecology* 217, no. 3 (September 1, 2017): 356e1–356e18, https://doi.org/10.1016/j.ajog.2017.05.030.

37. Stout et al., "Early Pregnancy Vaginal Microbiome Trends."

38. Stout et al., "Early Pregnancy Vaginal Microbiome Trends."

39. B. J. Callahan, D. B. DiGiulio, D. S. Aliaga Goltsman, C. L. Sun, E. K. Costello, P. Jeganathan, J. R. Biggio, et al., "Replication and Refinement of a Vaginal Microbial Signature of Preterm Birth in Two Racially Distinct Cohorts of US Women," *Proceedings of the National Academy of Sciences of the United States of America* 114, no. 37 (September 12, 2017): 9966–9971, https://doi.org/10.1073/pnas.1705899114.

40. Callahan et al., "Replication and Refinement."

41. Elovitz et al., "Cervicovaginal Microbiota."

42. Ravel, oral interview.

43. C. E. Cho and M. Norman, "Cesarean Section and Development of the Immune System in the Offspring," *American Journal of Obstetrics and Gynecology* 208, no. 4 (April 1, 2013): 249–254, https://doi.org/10.1016/j.ajog.2012.08.009; R. Romero and S. J. Korzeniewski, "Are Infants Born by Elective Cesarean Delivery without Labor at Risk for Developing Immune Disorders Later in Life?," *American Journal of Obstetrics and Gynecology* 208, no. 4 (April 1, 2013): 243–246, https://doi.org/10.1016/j.ajog.2012.12.026.

44. A. Sevelsted, J. Stokholm, K. Bønnelykke, and H. Bisgaard, "Cesarean Section and Chronic Immune Disorders," *Pediatrics* 135, no. 1 (January 2015): e92–e98, https://doi.org/10.1542/peds.2014-0596.

45. N. N. Schommer and R. L. Gallo, "Structure and Function of the Human Skin Microbiome," *Trends in Microbiology* 21, no. 12 (December 1, 2013): 660–668, https://doi.org/10.1016/j.tim.2013.10.001.

46. Neu, oral interview.

47. American College of Obstetricians and Gynecologists, Committee on Obstetric Practice, "Vaginal Seeding," ACOG Committee Opinion 725, November 2017, https://www.acog.org/clinical/clinical-guidance/committee-opinion/articles/2017/11/vaginal-seeding.

48. L. F. Stinson, M. S. Payne, and J. A. Keelan, "A Critical Review of the Bacterial Baptism Hypothesis and the Impact of Cesarean Delivery on the Infant Microbiome," *Frontiers in Medicine* (Lausanne) 5 (May 2018): 135, https://doi.org/10.3389/fmed.2018.00135.

49. Y.-C. Shi, H. Guo, J. Chen, G. Sun, R.-R. Ren, M.-Z. Guo, L.-H. Peng, and Y.-S. Yang, "Initial Meconium Microbiome in Chinese Neonates Delivered Naturally or by Cesarean Section," *Scientific Reports* 8 (February 19, 2018): 3255, https://doi.org/10.1038/s41598-018-21657-7.

50. E. Rutayisire, K. Huang, Y. Liu, and F. Tao, "The Mode of Delivery Affects the Diversity and Colonization Pattern of the Gut Microbiota during the First Year of Infants' Life: A Systematic Review," *BMC Gastroenterology* 16 (July 30, 2016): 86, https://doi.org/10.1186/s12876-016-0498-0.

51. P. S. La Rosa, B. B. Warner, Y. Zhou, G. M. Weinstock, E. Sodergren, C. M. Hall-Moore, H. J. Stevens, et al., "Patterned Progression of Bacterial Populations in the Premature Infant Gut," *Proceedings of the National Academy of Sciences of the United States of America* 111, no. 34 (August 26, 2014): 12522–12527, https://doi.org/10.1073/pnas.1409497111.

52. I. I. Carvalho-Ramos, R. T. D. Duarte, K. G. Brandt, M. B. Martinez, and C. R. Taddei, "Breastfeeding Increases Microbial Community Resilience," *Jornal de Pediatria* (Porto Alegre) 94, no. 3 (May–June 2018): 258–267, https://doi.org/10.1016/j.jped.2017.05.013.

53. K. E. Gregory and W. A. Walker, "Immunologic Factors in Human Milk and Disease Prevention in the Preterm Infant," *Current Pediatrics Reports* 1, no. 4 (December 2013): 222–228, https://doi.org/10.1007/s40124-013-0028-2.

54. N. Colliou, Y. Ge, B. Sahay, M. Gong, M. Zadeh, J. L. Owen, J. Neu, et al., "Commensal *Propionibacterium* Strain UF1 Mitigates Intestinal Inflammation via Th17 Cell Regulation," *Journal of Clinical Investigation* 127, no. 11 (November 1, 2017): 3970–3986, https://doi.org/10.1172/jci95376.

55. J. Shulhan, B. Dicken, L. Hartling, and B. M. K. Larsen, "Current Knowledge of Necrotizing Enterocolitis in Preterm Infants and the Impact of Different Types of Enteral Nutrition Products," *Advances in Nutrition* 8, no. 1 (January 2017): 80–91, https://doi.org/10.3945/an.116.013193.

56. Neu, oral interview.

57. S. Senger, L. Ingano, R. Freire, A. Anselmo, W. Zhu, R. Sadreyev, W. A. Walker, and A. Fasano, "Human Fetal-Derived Enterospheres Provide Insights on Intestinal Development and a Novel Model to Study Necrotizing Enterocolitis (NEC)," *Cellular and Molecular Gastroenterology and Hepatology* 5, no. 4 (February 4, 2018): 549–568, https://doi.org/10.1016/j.jcmgh.2018.01.014.

58. R. G. Brown, J. R. Marchesi, Y. S. Lee, A. Smith, B. Lehne, L. M. Kindinger, V. Terzidou, et al., "Vaginal Dysbiosis Increases Risk of Preterm Fetal Membrane Rupture, Neonatal Sepsis and Is Exacerbated by Erythromycin," *BMC Medicine* 16 (January 24, 2018): 9, https://doi.org/10.1186/s12916-017-0999-x.

4 Cracking the Codes

1. R. Colwell, oral interview with authors, January 19, 2018.

2. Colwell, oral interview.

3. Colwell, oral interview.

4. Colwell, oral interview.

5. R. R. Colwell, "Polyphasic Taxonomy of the Genus *Vibrio*: Numerical Taxonomy of *Vibrio cholerae*, *Vibrio parahaemolyticus*, and Related *Vibrio* Species," *Journal of Bacteriology* 104, no. 1 (October 1970): 410–433, https://jb.asm.org/content/104/1/410.

6. A. H. Sturtevant and G. W. Beadle, *An Introduction to Genetics* (Philadelphia: W. B. Saunders, 1940).

7. J. G. Mendel, "Versuche über Pflanzenhybriden," *Verhandlungen des naturforschenden Vereines in Brünn* 4 (1865): 3–47.

8. Nobel Prize, "The Nobel Prize in Physiology or Medicine 1933," https://www.nobelprize.org/prizes/medicine/1933/summary/, accessed May 24, 2020.

9. Nobel Prize, "The Nobel Prize in Physiology or Medicine 1962," https://www.nobelprize.org/prizes/medicine/1962/summary/, accessed November 9, 2019.

10. National Institutes of Health, U.S. National Library of Medicine, "Rosalind Franklin: The Rosalind Franklin Papers," *Profiles in Science,* https://profiles.nlm.nih.gov/spotlight/kr/feature/biographical-overview, accessed November 9, 2019.

11. National Institutes of Health, U.S. National Library of Medicine, "Rosalind Franklin."

12. B. Maddox, "The Double Helix and the 'Wronged Heroine,'" *Nature* 421 (January 23, 2003): 407–408, https://doi.org/10.1038/nature01399.

13. J. D. Watson and F. H. C. Crick, "Molecular Structure of Nucleic Acids: A Structure for Deoxyribose Nucleic Acid," *Nature* 171, no. 4356 (April 25, 1953): 737–738, https://doi.org/10.1038/171737a0.

14. M. H. F. Wilkins, A. R. Stokes, and H. R. Wilson, "Molecular Structure of Nucleic Acids: Molecular Structure of Deoxypentose Nucleic Acids," *Nature* 171, no. 4356 (April 25, 1953): 738–740, https://doi.org /10.1038/171738a0; R. E. Franklin and R. G. Gosling, "Molecular Configuration in Sodium Thymonucleate," *Nature* 171, no. 4356 (April 25, 1953): 740–741, https://doi.org/10.1038/171740a0.

15. Watson and Crick, "Molecular Structure of Nucleic Acids."

16. L. Osman Elkin, "Rosalind Franklin and the Double Helix," *Physics Today* 56, no. 3 (March 2003): 42–48, https://doi.org/10.1063/1 .1570771.

17. Nobel Prize, "Frederick Sanger: Facts," https://www.nobelprize.org /prizes/chemistry/1958/sanger/facts/, accessed November 10, 2019.

18. Nobel Prize, "The Nobel Prize in Physiology or Medicine 1978," https://www.nobelprize.org/prizes/medicine/1978/summary/, accessed November 10, 2019.

19. D. W. Hood, M. E. Deadman, M. P. Jennings, M. Bisercic, R. D. Fleischmann, J. C. Venter, and E. R. Moxon, "DNA Repeats Identify Novel Virulence Genes in *Haemophilus influenzae,*" *Proceedings of the National Academy of Sciences of the United States of America* 93, no. 20 (October 1, 1996): 11121–11125, https://doi.org/10.1073/pnas.93.20.11121.

20. National Institutes of Health, National Human Genome Research Institute, "NHGRI History and Timeline of Events," https://www.genome .gov/about-nhgri/Brief-History-Timeline, accessed June 17, 2020.

21. National Institutes of Health, National Human Genome Research Institute, "Human Genome Project FAQ."

22. Fraser, oral interview.

23. Fraser, oral interview.

24. Fraser, oral interview.

25. Hood et al., "DNA Repeats Identify Novel Virulence Genes."

26. C. M. Fraser, J. D. Gocayne, O. White, M. D. Adams, R. A. Clayton, R. D. Fleischmann, C. J. Bult, et al., "The Minimal Gene Complement of *Mycoplasma genitalium*," *Science* 270, no. 5235 (October 20, 1995): 397–404, https://doi.org/10.1126/science.270.5235.397.

27. J. Gallagher, "More Than Half Your Body Is Not Human," BBC News, https://www.bbc.com/news/health-43674270, accessed November 10, 2019.

28. National Institutes of Health, Human Microbiome Project, "HMP1," https://www.hmpdacc.org/overview/, accessed November 10, 2019.

29. National Institutes of Health, Human Microbiome Project, "HMP1."

30. National Institutes of Health, Human Microbiome Project, "HMP1."

31. National Institutes of Health, Human Microbiome Project, "HMP1."

32. A. Gibbons, "Hadza on the Brink," *Science* 360, no. 6390 (May 18, 2018): 700–704, https://doi.org/10.1126/science.360.6390.700.

33. Gibbons, "Hadza on the Brink."

34. Fraser, oral interview.

35. Colwell, oral interview.

36. Fraser, oral interview.

37. Fraser, oral interview.

38. MetaHIT Metagenomics of the Human Intestinal Tract, "Catalog of Genes," http://www.metahit.eu/index.php?id=360, accessed November 9, 2019.

39. O. Koren, D. Knights, A. Gonzalez, L. Waldron, N. Segata, R. Knight, C. Huttenhower, and R. E. Ley, "A Guide to Enterotypes across the Human Body: Meta-Analysis of Microbial Community Structures in Human Microbiome Datasets," *PLoS Computational Biology* 9, no. 1 (January 2013): e1002863, https://doi.org/10.1371/journal.pcbi.1002863.

40. P. I. Costea, F. Hildebrand, M. Arumugam, F. Bäckhed, M. J. Blaser, F. D. Bushman, W. M. de Vos, et al., "Enterotypes in the Landscape of Gut Microbial Community Composition," *Nature Microbiology* 3, no. 1 (January 2018): 8–16, https://doi.org/10.1038/s41564-017-0072-8.

41. Colwell, oral interview.

42. A. Huq, M. Yunus, S. S. Sohel, A. Bhuiya, M. Emch, S. P. Luby, E. Russek-Cohen, G. B. Nair, R. B. Sack, and R. R. Colwell, "Simple Sari Cloth Filtration of Water Is Sustainable and Continues to Protect Villagers from Cholera in Matlab, Bangladesh," *mBio* 1, no. 1 (May 2010): e00034-10, https://doi.org/10.1128/mbio.00034-10.

43. Colwell, oral interview.

5   Beyond Bacteria

1. B. Techatraisak and W. M. Gesler, "Traditional Medicine in Bangkok, Thailand," *Geographical Review* 79, no. 2 (April 1989): 172–182, https://doi.org/10.2307/215524.

2. World Health Organization, "Background of WHO Congress on Traditional Medicine," https://www.who.int/medicines/areas/traditional /congress/congress_background_info/en/, accessed November 11, 2019.

3. American College of Traditional Chinese Medicine, "Chinese Medicine," https://www.actcm.edu/chinese-medicine/, accessed November 11, 2019.

4. American College of Traditional Chinese Medicine, "Chinese Medicine."

5. K. Xu, H. Cai, Y. Shen, Q. Ni, Y. Chen, S. Hu, J. Li, et al., "[Management of Corona Virus Disease-19 (COVID-19): The Zhejiang Experience]," *Zhejiang Da Xue Xue Bao Yi Xue Ban* 49, no. 1 (February 21, 2020), https://pubmed.ncbi.nlm.nih.gov/32096367/.

6. E. de Divitiis, P. Cappabianca, and O. de Divitiis, "The 'Schola Medica Salernitana': The Forerunner of the Modern University Medical Schools," *Neurosurgery* 55, no. 4 (October 2004): 722–745, https://doi.org/10.1227/01.neu.0000139458.36781.31.

7. C. Yapijakis, "Hippocrates of Kos, the Father of Clinical Medicine, and Asclepiades of Bithynia, the Father of Molecular Medicine," *In Vivo* 23, no. 4 (July–August 2009): 507–514, http://iv.iiarjournals.org/content/23/4/507.short.

8. R. Virchow, *Cellular Pathology: As Based upon Physiological and Pathological Histology: Twenty Lectures Delivered in the Pathological Institute of Berlin during the Months of February, March, and April, 1858*, trans. F. Chance (London: John Churchill, 1860), https://archive.org/details/b20418310, accessed November 11, 2019.

9. Colwell, oral interview.

10. E. Scarpellini, G. Ianiro, F. Attili, C. Bassanelli, A. De Santis, and A. Gasbarrini, "The Human Gut Microbiota and Virome: Potential Therapeutic Implications," *Digestive and Liver Disease* 47, no. 12 (December 2015): 1007–1012, https://doi.org/10.1016/j.dld.2015.07.008.

11. "Microbiology by Numbers," editorial, *Nature Reviews Microbiology* 9 (September 2011): 628, https://doi.org/10.1038/nrmicro2644.

12. S. R. Carding, N. Davis, and L. Hoyles, "Review Article: The Human Intestinal Virome in Health and Disease," *Alimentary Pharmacology and Therapeutics* 46, no. 9 (November 2017): 800–815, https://doi.org/10.1111/apt.14280.

13. S. L. Smits, C. M. E. Schapendonk, J. van Beek, H. Vennema, A. C. Schürch, D. Schipper, R. Bodewes, B. L. Haagmans, A. D. M. E. Osterhaus, and M. P. Koopmans, "New Viruses in Idiopathic Human Diarrhea

Cases, the Netherlands," *Emerging Infectious Diseases* 20, no. 7 (July 2014): 1218–1222, https://doi.org/10.3201/eid2007.140190.

14. S. Sternberg, "Tracking a Mysterious Killer Virus in the Southwest," *Washington Post*, June 14, 1994, https://www.washingtonpost.com/archive/lifestyle/wellness/1994/06/14/tracking-a-mysterious-killer-virus-in-the-southwest/5e074ccd-7d88-41c0-9dc4-c0edcc1cd16e/.

15. Centers for Disease Control and Prevention, "Tracking a Mystery Disease: The Detailed Story of Hantavirus Pulmonary Syndrome (HPS)," https://www.cdc.gov/hantavirus/outbreaks/history.html, accessed November 11, 2019.

16. Sternberg, "Tracking a Mysterious Killer Virus."

17. Sternberg, "Tracking a Mysterious Killer Virus."

18. C. K. Johnson, P. L. Hitchens, T. S. Evans, T. Goldstein, K. Thomas, A. Clements, D. O. Joly, J. K. Mazet, et al., "Spillover and Pandemic Properties of Zoonotic Viruses with High Host Plasticity," *Scientific Reports* 5 (October 7, 2015): 14830, https://doi.org/10.1038/srep14830.

19. R. J. Jose and A. Manuel, "COVID-19 Cytokine Storm: The Interplay between Inflammation and Coagulation," *The Lancet Respiratory Medicine* 8 (June 2020): e46–e47, https://doi.org/10.1016/s2213-2600(20)30216-2.

20. F. Rohwer, oral interview with S. M. Flaherty, November 1, 2018.

21. R. Young and S. Raphelson, "How Bacteria Could Affect Outcomes of COVID-19 Patients," WBUR, April 16, 2020, https://www.wbur.org/hereandnow/2020/04/16/bacteria-covid-19-outcomes, accessed May 11, 2020.

22. Rohwer, oral interview.

23. Carding, Davis, and Hoyles, "The Human Intestinal Virome."

24. Rohwer, oral interview.

25. C. B. Silveira and F. L. Rohwer, "Piggyback-the-Winner in Host-Associated Microbial Communities," *NPJ Biofilms and Microbiomes* 2 (July 6, 2016): 16010, https://doi.org/10.1038/npjbiofilms.2016.10.

26. Rohwer, oral interview.

27. N. Van Stralen, "New Immune System Discovered," SDSU News-Center, May 20, 2013, http://newscenter.sdsu.edu/sdsu_newscenter /news_story.aspx?sid=74269, accessed November 11, 2019.

28. Van Stralen, "New Immune System Discovered"; J. J. Barr, R. Auro, M. Furlan, K. L. Whiteson, M. L. Erb, J. Pogliano, A. Stotland, et al., "Bacteriophage Adhering to Mucus Provide a Non-Host-Derived Immunity," *Proceedings of the National Academy of Sciences of the United States of America* 110, no. 26 (June 25, 2013): 10771–10776, https://doi.org/10 .1073/pnas.1305923110.

29. Van Stralen, "New Immune System Discovered."

30. Barr et al., "Bacteriophage Adhering to Mucus."

31. Silveira and Rohwer, "Piggyback-the-Winner."

32. Rohwer, oral interview.

33. S. Nguyen, K. Baker, B. S. Padman, R. Patwa, R. A. Dunstan, T. A. Weston, K. Schlosser, et al., "Bacteriophage Transcytosis Provides a Mechanism to Cross Epithelial Cell Layers," *mBio* 8, no. 6 (November 2017): e01874-17, https://doi.org/10.1128/mbio.01874-17.

34. Nguyen et al., "Bacteriophage Transcytosis."

35. Rohwer, oral interview.

36. Rohwer, oral interview.

37. Rohwer, oral interview.

38. R. T. Schooley, B. Biswas, J. J. Gill, A. Hernandez-Morales, J. Lancaster, L. Lessor, J. J. Barr, et al., "Development and Use of Personalized Bacteriophage-Based Therapeutic Cocktails to Treat a Patient with a Disseminated Resistant *Acinetobacter baumannii* Infection," *Antimicrobial Agents and Chemotherapy* 61, no. 10 (September 2017): e00954-17, https://doi.org/10.1128/aac.00954-17.

39. S. Strathdee and T. Patterson, *The Perfect Predator: A Scientist's Race to Save Her Husband from a Deadly Superbug* (New York: Hachette Book Group, 2019).

40. Rohwer, oral interview.

41. A. Llanos-Chea, R. J. Citorik, K. P. Nickerson, L. Ingano, G. Serena, S. Senger, T. K. Lu, A. Fasano, and C. S. Faherty, "Bacteriophage Therapy Testing against *Shigella flexneri* in a Novel Human Intestinal Organoid–Derived Infection Model," *Journal of Pediatric Gastroenterology and Nutrition* 68, no. 4 (April 2019): 509–516, https://doi.org/10.1097/mpg.0000000000002203.

42. Llanos-Chea et al., "Bacteriophage Therapy Testing."

43. J. B. Harley, X. Chen, M. Pujato, D. Miller, A. Maddox, C. Forney, A. F. Magnusen, et al., "Transcription Factors Operate across Disease Loci, with EBNA2 Implicated in Autoimmunity," *Nature Genetics* 50 (May 2018): 699–707, https://doi.org/10.1038/s41588-018-0102-3.

44. G. Zhao, T. Vatanen, L. Droit, A. Park, A. D. Kostic, T. W. Poon, H. Vlamakis, et al., "Intestinal Virome Changes Precede Autoimmunity in Type 1 Diabetes–Susceptible Children," *Proceedings of the National Academy of Sciences of the United States of America* 114, no. 30 (July 25, 2017): E6166–E6175, https://doi.org/10.1073/pnas.1706359114.

45. P. Nagappan, "Common Foods Can Help 'Landscape' the Jungle of Our Gut Microbiome," SDSU NewsCenter, January 15, 2020, https://newscenter.sdsu.edu/sdsu_newscenter/news_story.aspx?sid=77862, accessed January 31, 2020; L. Boling, D. A. Cuevas, J. A. Grasis, H. S. Kang, B. Knowles, K. Levi, H. Maughan, et al., "Dietary Prophage Inducers and Antimicrobials: Toward Landscaping the Human Gut Microbiome," *Gut Microbes* 11, no. 4 (January 13, 2020): 721–734, https://doi.org/10.1080/19490976.2019.1701353.

46. J. D. Forbes, C. N. Bernstein, H. Tremlett, G. Van Domselaar, and N. C. Knox, "A Fungal World: Could the Gut Mycobiome Be Involved in Neurological Disease?," *Frontiers in Microbiology* 9 (January 2019): 3249, https://doi.org/10.3389/fmicb.2018.03249.

47. A. K. Nash, T. A. Auchtung, M. C. Wong, D. P. Smith, J. R. Gesell, M. C. Ross, C. J. Stewart, et al., "The Gut Mycobiome of the Human Microbiome Project Healthy Cohort," *Microbiome* 5 (November 25, 2017): 153, https://doi.org/10.1186/s40168-017-0373-4.

48. Forbes et al., "A Fungal World."

49. C. L. Hager and M. A. Ghannoum, "The Mycobiome: Role in Health and Disease, and as a Potential Probiotic Target in Gastrointestinal Disease," *Digestive and Liver Disease* 49, no. 11 (November 1, 2017): 1171–1176, https://doi.org/10.1016/j.dld.2017.08.025.

50. E. C. Dinleyici, M. Eren, M. Ozen, Z. A. Yargic, and Y. Vandenplas, "Effectiveness and Safety of *Saccharomyces boulardii* for Acute Infectious Diarrhea," *Expert Opinion on Biological Therapy* 12, no. 4 (April 2012): 395–410, https://doi.org/10.1517/14712598.2012.664129.

51. K. E. Murfin, A. R. Dillman, J. M. Foster, S. Bulgheresi, B. E. Slatko, P. W. Sternberg, and H. Goodrich-Blair, "Nematode-Bacterium Symbioses—Cooperation and Conflict Revealed in the 'Omics' Age," *The Biological Bulletin* 223, no. 1 (August 2012): 85–102, https://doi.org/10.1086/bblv223n1p85.

52. V. Marzano, L. Mancinelli, G. Bracaglia, F. Del Chierico, P. Vernocchi, F. Di Girolamo, S. Garrone, et al., "'Omic' Investigations of Protozoa and Worms for a Deeper Understanding of the Human Gut 'Parasitome,'" *PLoS Neglected Tropical Diseases* 11, no. 11 (November 2, 2017): e0005916, https://doi.org/10.1371/journal.pntd.0005916.

53. World Health Organization, "Malaria: Key Facts," March 27, 2019, https://www.who.int/en/news-room/fact-sheets/detail/malaria, accessed November 11, 2019.

54. Centers for Disease Control and Prevention, "Malaria: The History of Malaria, an Ancient Disease," https://www.cdc.gov/malaria/about/history/index.html, accessed November 11, 2019.

55. World Health Organization, "Fact Sheet: World Malaria Report 2016," December 13, 2016, https://www.who.int/malaria/media/world-malaria-report-2016/en/, accessed November 11, 2019.

56. World Health Organization, "Malaria: Key Facts."

57. World Health Organization, "Malaria: Key Facts."

58. A. Trevett and D. Lalloo, "A New Look at an Old Drug: Artemesinin and Qinghaosu," *Papua New Guinea Medical Journal* 35, no. 4 (December 1992): 264–269.

59. Y. Dong, F. Manfredini, and G. Dimopoulos, "Implication of the Mosquito Midgut Microbiota in the Defense against Malaria Parasites," *PLoS Pathogens* 5, no. 5 (May 2009): e1000423, https://doi.org/10.1371/journal.ppat.1000423.

60. Dong, Manfredini, and Dimopoulos, "Implication of the Mosquito Midgut Microbiota."

61. Dong, Manfredini, and Dimopoulos, "Implication of the Mosquito Midgut Microbiota."

62. Dong, Manfredini, and Dimopoulos, "Implication of the Mosquito Midgut Microbiota."

6 The Microbiome Hypothesis

1. J. Riedler, C. Braun-Fahrländer, W. Eder, M. Schreuer, M. Waser, S. Maisch, D. Carr, W. Schierl, D. Nowak, and E. von Mutius, "Exposure to Farming in Early Life and Development of Asthma and Allergy: A Cross-Sectional Survey," *The Lancet* 358, no. 9288 (October 6, 2001): 1129–1133, https://doi.org/10.1016/s0140-6736(01)06252-3.

2. M. M. Stein, C. L. Hrusch, J. Gozdz, C. Igartua, V. Pivniouk, S. E. Murray, J. G. Ledford, et al., "Innate Immunity and Asthma Risk in Amish and Hutterite Farm Children," *New England Journal of Medicine* 375, no. 5 (August 4, 2016): 411–421, https://doi.org/10.1056/nejmoa1508749.

3. T. S. Böbel, S. B. Hackl, D. Langgartner, M. N. Jarczok, N. Rohleder, G. A. Rook, C. A. Lowry, H. Gündel, C. Waller, and S. O. Reber, "Less Immune Activation Following Social Stress in Rural vs. Urban Participants Raised with Regular or No Animal Contact, Respectively," *Proceedings of the National Academy of Sciences of the United States of America* 115, no. 20 (May 15, 2018): 5259–5264, https://doi.org/10.1073/pnas.1719866115.

4. S. V. Lynch and H. A. Boushey, "The Microbiome and Development of Allergic Disease," *Current Opinion in Allergy and Clinical Immunology* 16, no. 2 (April 2016): 165–171, https://doi.org/10.1097/aci.0000000000000255.

5. Böbel et al., "Less Immune Activation."

6. Strachan, "Hay Fever, Hygiene, and Household Size."

7. G. A. W. Rook, C. A. Lowry, and C. L. Raison, "Microbial 'Old Friends,' Immunoregulation and Stress Resilience," *Evolution, Medicine, and Public Health* 2013, no. 1 (April 9, 2013): 46–64, https://doi.org/10.1093/emph/eot004.

8. D. Leonhardt, "Life Expectancy Data," *New York Times*, September 27, 2006, https://www.nytimes.com/2006/09/27/business/27leonhardt_sidebar.html, accessed November 12, 2019.

9. J. V. Weinstock, "The Worm Returns," *Nature* 491, no. 7423 (November 8, 2012): 183–185, https://doi.org/10.1038/491183a; C. M. Ferreira, A. T. Vieira, M. A. R. Vinolo, F. A. Oliveira, R. Curi, and F. dos Santos Martins, "The Central Role of the Gut Microbiota in Chronic Inflammatory Diseases," *Journal of Immunology Research* 2014 (September 18, 2014): 689492, https://doi.org/10.1155/2014/689492; J. V. Weinstock and D. E. Elliott, "Helminth Infections Decrease Host Susceptibility to Immune-Mediated Diseases," *Journal of Immunology* 193, no. 7 (October 1, 2014): 3239–3247, https://doi.org/10.4049/jimmunol.1400927; M. J. Blaser, "The Theory of Disappearing Microbiota and the Epidemics of Chronic Diseases," *Nature Reviews Immunology* 17, no. 8 (August 2017): 461–463, https://doi.org/10.1038/nri.2017.77.

10. P. Vangay, A. J. Johnson, T. L. Ward, G. A. Al-Ghalith, R. R. Shields-Cutler, B. M. Hillmann, S. K. Lucas, et al., "US Immigration Westernizes the Human Gut Microbiome," *Cell* 175, no. 4 (November 1, 2018): 962–972, https://doi.org/10.1016/j.cell.2018.10.029.

11. A. Berger, "Th1 and Th2 Responses: What Are They?," *British Medical Journal* 321, no. 7258 (August 12, 2000): 424, https://doi.org/10.1136/bmj.321.7258.424.

12. A. S. Clem, "Fundamentals of Vaccine Immunology," *Journal of Global Infectious Diseases* 3, no. 1 (January–March 2011): 73–78, https://doi.org/10.4103/0974-777x.77299.

13. M. E. Duffey, B. Hainau, S. Ho, and C. J. Bentzel, "Regulation of Epithelial Tight Junction Permeability by Cyclic AMP," *Nature* 294, no. 5840 (December 3, 1981): 451–453, https://doi.org/10.1038/294451a0.

14. M. Furuse, T. Hirase, M. Itoh, A. Nagafuchi, S. Yonemura, Sa. Tsukita, and Sh. Tsukita, "Occludin: A Novel Integral Membrane Protein Localizing at Tight Junctions," *Journal of Cell Biology* 123, no. 6 (December 1993): 1777–1788, https://doi.org/10.1083/jcb.123.6.1777.

15. A. Fasano, "Intestinal Zonulin: Open Sesame!" *Gut* 49, no. 2 (August 2001): 159–162, https://doi.org/10.1136/gut.49.2.159.

16. A. Fasano, "Zonulin and Its Regulation of Intestinal Barrier Function: The Biological Door to Inflammation, Autoimmunity, and Cancer," *Physiological Reviews* 91, no. 1 (January 2011): 151–175, https://doi.org/10.1152/physrev.00003.2008.

17. L. Scheffler, A. Crane, H. Heyne, A. Tönjes, D. Schleinitz, C. H. Ihling, M. Stumvoll, et al., "Widely Used Commercial ELISA Does Not Detect Precursor of Haptoglobin2, but Recognizes Properdin as a Potential Second Member of the Zonulin Family," *Frontiers in Endocrinology* (Lausanne) 9 (February 2018): 22, https://doi.org/10.3389/fendo.2018.00022.

18. D. Rittirsch, M. A. Flierl, B. A. Nadeau, D. E. Day, M. S. Huber-Lang, J. J. Grailer, F. S. Zetoune, A. V. Andjelkovic, A. Fasano, and P. A. Ward, "Zonulin as Prehaptoglobin2 Regulates Lung Permeability and Activates the Complement System," *American Journal of Physiology Lung Cellular and Molecular Physiology* 304, no. 12 (June 2013): L863–L872, https://doi.org/10.1152/ajplung.00196.2012; K. A. Shirey, W. Lai, M. C. Patel, L. M. Pletneva, C. Pang, E. Kurt-Jones, M. Lipsky, et al., "Novel Strategies for Targeting Innate Immune Responses to Influenza," *Mucosal Immunology* 9, no. 5 (September 2016): 1173–1182, https://doi.org/10.1038/mi.2015.141.

19. K. M. Lammers, R. Lu, J. Brownley, B. Lu, C. Gerard, K. Thomas, P. Rallabhandi, et al., "Gliadin Induces an Increase in Intestinal Permeability and Zonulin Release by Binding to the Chemokine Receptor CXCR3," *Gastroenterology* 135, no. 1 (July 2008): 194–204, https://doi.org/10.1053/j.gastro.2008.03.023.

20. Lammers et al., "Gliadin Induces an Increase."

21. C. Sturgeon, J. Lan, and A. Fasano, "Zonulin Transgenic Mice Show Altered Gut Permeability and Increased Morbidity/Mortality in the DSS Colitis Model," *Annals of the New York Academy of Sciences* 1397, no. 1 (June 2017): 130–142, https://doi.org/10.1111/nyas.13343.

22. A. Miranda-Ribera, M. Ennamorati, G. Serena, M. Cetinbas, J. Lan, R. I. Sadreyev, N. Jain, A. Fasano, and M. Fiorentino, "Exploiting the Zonulin Mouse Model to Establish the Role of Primary Impaired Gut Barrier Function on Microbiota Composition and Immune Profiles," *Frontiers in Immunology* 10 (September 2019): 2233, https://doi.org/10.3389/fimmu.2019.02233.

23. Sturgeon, Lan, and Fasano, "Zonulin Transgenic Mice."

24. Miranda-Ribera et al., "Exploiting the Zonulin Mouse Model."

7   The Microbiome and Gut Inflammatory Disorders

1. B. E. Lacy and N. K. Patel, "Rome Criteria and a Diagnostic Approach to Irritable Bowel Syndrome," *Journal of Clinical Medicine* 6, no. 11 (November 2017): 99, https://doi.org/10.3390/jcm6110099.

2. S. A. Leong, V. Barghout, H. G. Birnbaum, C. E. Thibeault, R. Ben-Hamadi, F. Frech, and J. J. Ofman, "The Economic Consequences of Irritable Bowel Syndrome: A US Employer Perspective," *Archives of Internal Medicine* 163, no. 8 (April 28, 2003): 929–935, https://doi.org/10.1001/archinte.163.8.929.

3. C. Canavan, J. West, and T. Card, "The Economic Impact of the Irritable Bowel Syndrome," *Alimentary Pharmacology and Therapeutics* 40, no. 9 (November 2014): 1023–1034, https://doi.org/10.1111/apt.12938.

4. L. Wang, N. Alammar, R. Singh, J. Nanavati, Y. Song, R. Chaudhary, and G. E. Mullin, "Gut Microbial Dysbiosis in the Irritable Bowel Syndrome: A Systematic Review and Meta-Analysis of Case-Control Studies," *Journal of the Academy of Nutrition and Dietetics* 120, no. 4 (April 1, 2020): 565–586, https://doi.org/10.1016/j.jand.2019.05.015..

5. B. B. Warner, E. Deych, Y. Zhou, C. Hall-Moore, G. M. Weinstock, E. Sodergren, N. Shaikh, et al., "Gut Bacteria Dysbiosis and Necrotising Enterocolitis in Very Low Birthweight Infants: A Prospective Case-Control Study," *The Lancet* 387, no. 10031 (May 7, 2016): 1928–1936, https://doi.org/10.1016/s0140-6736(16)00081-7.

6. Senger et al., "Human Fetal-Derived Enterospheres."

7. A. Collison, E. Percival, J. Mattes, and R. Bhatia, "What Are Allergies and Why Are We Getting More of Them?," *The Conversation*, October 7, 2015, https://theconversation.com/what-are-allergies-and-why-are-we-getting-more-of-them-40318, accessed January 2, 2020.

8. G. Du Toit, G. Roberts, P. H. Sayre, H. T. Bahnson, S. Radulovic, A. F. Santos, H. A. Brough, et al., "Randomized Trial of Peanut Consumption in Infants at Risk for Peanut Allergy," *New England Journal of Medicine* 372, no. 9 (February 26, 2015): 803–813, https://doi.org/10.1056/nejmoa1414850.

9. A. T. Stefka, T. Feehley, P. Tripathi, J. Qiu, K. McCoy, S. K. Mazmanian, M. Y. Tjota, et al., "Commensal Bacteria Protect against Food Allergen Sensitization," *Proceedings of the National Academy of Sciences of the United States of America* 111, no. 36 (September 9, 2014): 13145–13150, https://doi.org/10.1073/pnas.1412008111.

10. Z. Ling, Z. Li, X. Liu, Y. Cheng, Y. Luo, X. Tong, L. Yuan, et al., "Altered Fecal Microbiota Composition Associated with Food Allergy in Infants," *Applied and Environmental Microbiology* 80, no. 8 (April 2014): 2546–2554, https://doi.org/10.1128/aem.00003-14.

11. T. Feehley, C. H. Plunkett, R. Bao, S. M. C. Hong, E. Culleen, P. Belda-Ferre, E. Campbell, et al., "Healthy Infants Harbor Intestinal Bacteria that Protect against Food Allergy," *Nature Medicine* 25, no. 3 (March 2019): 448–453, https://doi.org/10.1038/s41591-018-0324-z.

12. M. Sellitto, G. Bai, G. Serena, W. F. Fricke, C. Sturgeon, P. Gajer, J. R. White, et al., "Proof of Concept of Microbiome-Metabolome Analysis and Delayed Gluten Exposure on Celiac Disease Autoimmunity in Genetically At-Risk Infants," *PLoS One* 7, no. 3 (March 2012): e33387, https://doi.org/10.1371/journal.pone.0033387.

13. Sellitto et al., "Proof of Concept of Microbiome-Metabolome Analysis."

14. M. Olivares, A. W. Walker, A. Capilla, A. Benítez-Páez, F. Palau, J. Parkhill, G. Castillejo, and Y. Sanz, "Gut Microbiota Trajectory in Early Life May Predict Development of Celiac Disease," *Microbiome* 6, no. 1 (February 20, 2018): 36, https://doi.org/10.1186/s40168-018-0415-6.

15. B. P. Willing, J. Dicksved, J. Halfvarson, A. F. Andersson, M. Lucio, Z. Zheng, G. Järnerot, C. Tysk, J. K. Jansson, and L. Engstrand, "A Pyrosequencing Study in Twins Shows That Gastrointestinal Microbial Profiles Vary with Inflammatory Bowel Disease Phenotypes," *Gastroenterology* 139, no. 6 (December 2010): 1844–1854, https://doi.org/10.1053/j.gastro.2010.08.049.

16. Willing et al., "A Pyrosequencing Study."

17. S. Michail, M. Durbin, D. Turner, A. M. Griffiths, D. R. Mack, J. Hyams, N. Leleiko, H. Kenche, A. Stolfi, and E. Wine, "Alterations in the Gut Microbiome of Children with Severe Ulcerative Colitis," *Inflammatory Bowel Diseases* 18, no. 10 (October 1, 2012): 1799–1808, https://doi.org/10.1002/ibd.22860.

18. D. Gevers, S. Kugathasan, L. A. Denson, Y. Vázquez-Baeza, W. Van Treuren, B. Ren, E. Schwager, et al., "The Treatment-Naïve Microbiome in New-Onset Crohn's Disease," *Cell Host and Microbe* 15, no. 3 (March 12, 2014): 382–392, https://doi.org/10.1016/j.chom.2014.02.005.

19. Gevers et al., "The Treatment-Naïve Microbiome."

8 The Microbiome and Obesity

1. World Health Organization, "Obesity," https://www.who.int/topics/obesity/en/, accessed January 3, 2020.

2. L. Stoner and J. Cornwall, "Did the American Medical Association Make the Correct Decision Classifying Obesity as a Disease?," *Australasian Medical Journal* 7, no. 11 (November 2014): 462–464, https://doi.org/10.21767/amj.2014.2281.

3. World Health Organization, "Obesity and Overweight," https://www.who.int/en/news-room/fact-sheets/detail/obesity-and-overweight, accessed January 3, 2020.

4. L. Kaplan, oral interview with authors, August 29, 2018.

5. Kaplan, oral interview.

6. Kaplan, oral interview.

7. F. K. Ndiaye, M. Huyvaert, A. Ortalli, M. Canouil, C. Lecoeur, M. Verbanck, S. Lobbens, et al., "The Expression of Genes in Top Obesity-Associated Loci Is Enriched in Insula and Substantia Nigra Brain Regions Involved in Addiction and Reward," *International Journal of Obesity* 44, no. 2 (February 2020): 539–543, https://doi.org/10.1038/s41366-019-0428-7.

8. Kaplan, oral interview.

9. A. E. Locke, B. Kahali, S. I. Berndt, A. E. Justice, T. H. Pers, F. R. Day, C. Powell, et al., "Genetic Studies of Body Mass Index Yield New Insights for Obesity Biology," *Nature* 518, no. 7538 (February 12, 2015): 197–206, https://doi.org/10.1038/nature14177.

10. A. V. Khera, M. Chaffin, K. H. Wade, S. Zahid, J. Brancale, R. Xia, M. Distefano, et al., "Polygenic Prediction of Weight and Obesity Trajectories from Birth to Adulthood," *Cell* 177, no. 3 (April 18, 2019): 587–596, https://doi.org/10.1016/j.cell.2019.03.028.

11. Kaplan, oral interview.

12. Kaplan, oral interview.

13. Kaplan, oral interview.

14. Kaplan, oral interview.

15. J. H. Park, D. I. Park, H. J. Kim, Y. K. Cho, C. I. Sohn, W. K. Jeon, B. I. Kim, K. H. Won, and S. M. Park, "The Relationship between

Small-Intestinal Bacterial Overgrowth and Intestinal Permeability in Patients with Irritable Bowel Syndrome," *Gut and Liver* 3, no. 3 (September 2009): 174–179, https://doi.org/10.5009/gnl.2009.3.3.174.

16. M. Augustyn, I. Grys, and M. Kukla, "Small Intestinal Bacterial Overgrowth and Nonalcoholic Fatty Liver Disease," *Clinical and Experimental Hepatology* 5, no. 1 (March 2019): 1–10, https://doi.org/10.5114/ceh.2019 .83151.

17. R. E. Ley, F. Bäckhed, P. Turnbaugh, C. A. Lozupone, R. D. Knight, and J. I. Gordon, "Obesity Alters Gut Microbial Ecology," *Proceedings of the National Academy of Sciences of the United States of America* 102, no. 31 (August 2, 2005): 11070–11075, https://doi.org/10.1073/pnas .0504978102.

18. P. J. Turnbaugh, R. E. Ley, M. A. Mahowald, V. Magrini, E. R. Mardis, and J. I. Gordon, "An Obesity-Associated Gut Microbiome with Increased Capacity for Energy Harvest," *Nature* 444, no. 7122 (December 21, 2006): 1027–1031, https://doi.org/10.1038/nature051414.

19. R. E. Ley, P. J. Turnbaugh, S. Klein, and J. I. Gordon, "Human Gut Microbes Associated with Obesity," *Nature* 444, no. 7122 (December 21, 2006): 1022–1023, https://doi.org/10.1038/4441022a.

20. V. K. Ridaura, J. J. Faith, F. E. Rey, J. Cheng, A. E. Duncan, A. L. Kau, N. W. Griffin, et al., "Cultured Gut Microbiota from Twins Discordant for Obesity Modulate Metabolism in Mice," *Science* 341, no. 6150 (September 6, 2013): 1241214, https://doi.org/10.1126/science.1241214.

21. Ridaura et al., "Cultured Gut Microbiota from Twins."

22. Kaplan, oral interview.

23. Kaplan, oral interview.

24. Richard Saltus, "Reaping Benefits of Exercise Minus the Sweat," *Harvard Gazette*, January 11, 2012, https://news.harvard.edu/gazette/story /2012/01/reaping-benefits-of-exercise-minus-the-sweat/, accessed June 16, 2020.

25. Kaplan, oral interview.

26. Kaplan, oral interview.

27. Kaplan, oral interview.

28. N. Ouldzeidoune, J. Keating, J. Bertrand, and J. Rice, "A Description of Female Genital Mutilation and Force-Feeding Practices in Mauritania: Implications for the Protection of Child Rights and Health," *PLoS One* 8, no. 4 (April 2013): e60594, https://doi.org/10.1371/journal.pone .0060594.

29. Ouldzeidoune et al., "A Description."

30. Kaplan, oral interview.

31. Kaplan, oral interview.

32. C. R. Martin, V. Osadchiy, A. Kalani, and E. A. Mayer, "The Brain-Gut-Microbiome Axis," *Cellular and Molecular Gastroenterology and Hepatology* 6, no. 2 (April 12, 2018): 133–148, https://doi.org/10.1016/j .jcmgh.2018.04.003.

33. J. I. Gordon, "Nutrition and Microbiota in Health," keynote for "Gut Health, Microbiota and Probiotics throughout the Lifespan 2018: Dietary Influences," Division of Nutrition, Harvard Probiotics Symposium, Harvard Medical School, Boston, October 10, 2018, http://nutrition.med .harvard.edu/1810-edu-gut-health.html, accessed June 16, 2020.

34. Kaplan, oral interview.

35. Kaplan, oral interview.

36. Kaplan, oral interview.

37. Kaplan, oral interview.

38. Kaplan, oral interview.

39. Kaplan, oral interview.

40. Kaplan, oral interview.

41. Kaplan, oral interview.

42. Kaplan, oral interview.

43. Kaplan, oral interview.

44. Kaplan, oral interview.

9 The Microbiome and Autoimmunity

1. T. Watts, I. Berti, A. Sapone, T. Gerarduzzi, T. Not, R. Zielke, and A. Fasano, "Role of the Intestinal Tight Junction Modulator Zonulin in the Pathogenesis of Type 1 Diabetes in BB Diabetic-Prone Rats," *Proceedings of the National Academy of Sciences of the United States of America* 102, no. 8 (February 22, 2005): 2916–2921, https://doi.org/10.1073/pnas.0500178102.

2. J. Visser, J. Rozing, A. Sapone, K. Lammers, and A. Fasano, "Tight Junctions, Intestinal Permeability, and Autoimmunity: Celiac Disease and Type 1 Diabetes Paradigms," *Annals of the New York Academy of Sciences* 1165, no. 1 (May 2009): 195–205, https://doi.org/10.1111/j.1749-6632.2009.04037.x.

3. Watts et al., "Role of the Intestinal Tight Junction Modulator."

4. Watts et al., "Role of the Intestinal Tight Junction Modulator"; J. T. J. Visser, K. Lammers, A. Hoogendijk, M. W. Boer, S. Brugman, S. Beijer-Liefers, A. Zandvoort, et al., "Restoration of Impaired Intestinal Barrier Function by the Hydrolysed Casein Diet Contributes to the Prevention of Type 1 Diabetes in the Diabetes-Prone BioBreeding Rat," *Diabetologia* 53, no. 12 (December 2010): 2621–2628, https://doi.org/10.1007/s00125-010-1903-9.

5. Visser et al., "Restoration of Impaired Intestinal Barrier Function."

6. C. Sorini, I. Cosorich, M. Lo Conte, L. De Giorgi, F. Facciotti, R. Lucianò, M. Rocchi, F. Sanvito, F. Canducci, and M. Falcone, "Loss of Gut Barrier Integrity Triggers Activation of Islet-Reactive T Cells and Autoimmune Diabetes," *Proceedings of the National Academy of Sciences of the United States of America* 116, no. 30 (July 23, 2019): 15140–15149, https://doi.org/10.1073/pnas.1814558116.

7. Sorini et al., "Loss of Gut Barrier Integrity Triggers Activation."

8. D. S. Nielsen, L. Krych, K. Buschard, C. H. F. Hansen, and A. K. Hansen, "Beyond Genetics: Influence of Dietary Factors and Gut Microbiota on Type 1 Diabetes," *Federation of European Biomedical Societies Letters* 588, no. 22 (November 17, 2014): 4234–4243, https://doi.org/10.1016/j.febslet.2014.04.010; Visser et al., "Restoration of Impaired Intestinal Barrier Function."

9. Visser et al., "Restoration of Impaired Intestinal Barrier Function."

10. A. M. Henschel, S. M. Cabrera, M. L. Kaldunski, S. Jia, R. Geoffrey, M. F. Roethle, V. Lam, et al., "Modulation of the Diet and Gastrointestinal Microbiota Normalizes Systemic Inflammation and β-Cell Chemokine Expression Associated with Autoimmune Diabetes Susceptibility," *PLoS One* 13, no. 1 (January 2, 2018): e0190351, https://doi.org/10.1371/journal.pone.0190351.

11. M. C. de Goffau, K. Luopajärvi, M. Knip, J. Ilonen, T. Ruohtula, T. Härkönen, L. Orivuori, et al., "Fecal Microbiota Composition Differs between Children with β-Cell Autoimmunity and Those without," *Diabetes* 62, no. 4 (April 2013): 1238–1244, https://doi.org/10.2337/db12-0526.

12. A. K. Alkanani, N. Hara, P. A. Gottlieb, D. Ir, C. E. Robertson, B. D. Wagner, D. N. Frank, and D. Zipris, "Alterations in Intestinal Microbiota Correlate with Susceptibility to Type 1 Diabetes," *Diabetes* 64, no. 10 (October 2015): 3510–3520, https://doi.org/10.2337/db14-1847; A. D. Kostic, D. Gevers, H. Siljander, T. Vatanen, T. Hyötyläinen, A-M. Hämäläinen, A. Peet, V. Tillman, P. Pöhö, and R. J. Xavier, et al., "The Dynamics of the Human Infant Gut Microbiome in Development and in Progression toward Type 1 Diabetes," *Cell Host Microbe* 17, no. 2 (February 2015): 260–273, https://doi.org/10.1016/j.chom.2015.01.001.

13. Y. Huang, S.-C. Li, J. Hu, H.-B. Ruan, H.-M. Guo, H.-H. Zhang, X. Wang, Y.-F. Pei, Y. Pan, and C. Fang, "Gut Microbiota Profiling in Han Chinese with Type 1 Diabetes," *Diabetes Research and Clinical Practice* 141 (July 1, 2018): 256–263, https://doi.org/10.1016/j.diabres.2018.04.032.

14. P. G. Gavin, J. A. Mullaney, D. Loo, K.-A. Lê Cao, P. A. Gottlieb, M. M. Hill, D. Zipris, and E. E. Hamilton-Williams, "Intestinal

Metaproteomics Reveals Host-Microbiota Interactions in Subjects at Risk for Type 1 Diabetes," *Diabetes Care* 41, no. 10 (October 2018): 2178–2186, https://doi.org/10.2337/dc18-0777.

15. J. Ochoa-Repáraz, D. W. Mielcarz, L. E. Ditrio, A. R. Burroughs, D. M. Foureau, S. Haque-Begum, and L. H. Kasper, "Role of Gut Commensal Microflora in the Development of Experimental Autoimmune Encephalomyelitis," *Journal of Immunology* 183, no. 10 (November 15, 2009): 6041–6050, https://doi.org/10.4049/jimmunol.0900747.

16. Y. K. Lee, J. S. Menezes, Y. Umesaki, and S. K. Mazmanian, "Proinflammatory T-Cell Responses to Gut Microbiota Promote Experimental Autoimmune Encephalomyelitis," *Proceedings of the National Academy of Sciences of the United States of America* 108, supplement 1 (March 15, 2011): 4615–4622, https://doi.org/10.1073/pnas.1000082107.

17. J. Ochoa-Repáraz, D. W. Mielcarz, Y. Wang, S. Begum-Haque, S. Dasgupta, D. L. Kasper, and L. H. Kasper, "A Polysaccharide from the Human Commensal *Bacteroides fragilis* Protects against CNS Demyelinating Disease," *Mucosal Immunology* 3, no. 5 (September 2010): 487–495, https://doi.org/10.1038/mi.2010.29.

18. K. N. Chitrala, H. Guan, N. P. Singh, B. Busbee, A. Gandy, P. Mehrpouya-Bahrami, M. S. Ganewatta, et al., "CD44 Deletion Leading to Attenuation of Experimental Autoimmune Encephalomyelitis Results from Alterations in Gut Microbiome in Mice," *European Journal of Immunology* 47, no. 7 (July 2017): 1188–1199, https://doi.org/10.1002/eji.201646792.

19. S. Miyake, S. Kim, W. Suda, K. Oshima, M. Nakamura, T. Matsuoka, N. Chihara, et al., "Dysbiosis in the Gut Microbiota of Patients with Multiple Sclerosis, with a Striking Depletion of Species Belonging to *Clostridia* XIVa and IV Clusters," *PLoS One* 10, no. 9 (September 14, 2015): e0137429, https://doi.org/10.1371/journal.pone.0137429.

20. S. Jangi, R. Gandhi, L. M. Cox, N. Li, F. von Glehn, R. Yan, B. Patel, "Alterations of the Human Gut Microbiome in Multiple Sclerosis," *Nature Communications* 7 (June 28, 2016): 12015, https://doi.org/10.1038/ncomms12015.

21. E. Cekanaviciute, B. B. Yoo, T. F. Runia, J. W. Debelius, S. Singh, C. A. Nelson, R. Kanner, et al., "Gut Bacteria from Multiple Sclerosis Patients Modulate Human T Cells and Exacerbate Symptoms in Mouse Models," *Proceedings of the National Academy of Sciences of the United States of America* 114, no. 40 (October 3, 2017): 10713–10718, https://doi.org/10.1073/pnas.1711235114.

22. H.-J. Wu, I. I. Ivanov, J. Darce, K. Hattori, T. Shima, Y. Umesaki, D. R. Littman, C. Benoist, and D. Mathis, "Gut-Residing Segmented Filamentous Bacteria Drive Autoimmune Arthritis via T Helper 17 Cells," *Immunity* 32, no. 6 (June 25, 2010): 815–827, https://doi.org/10.1016/j.immuni.2010.06.001.

23. N. Tajik, M. Frech, O. Schulz, F. Schälter, S. Lucas, V. Azizov, K. Dürholz, F. Steffen, et al., "Targeting Zonulin and Intestinal Epithelial Barrier Function to Prevent Onset of Arthritis," *Nature Communications* 11 (April 24, 2020): 1995, https://doi.org/10.1038/s41467-020-15831-7.

24. M. Okada, T. Kobayashi, S. Ito, T. Yokoyama, A. Abe, A. Murasawa, and H. Yoshie, "Periodontal Treatment Decreases Levels of Antibodies to *Porphyromonas gingivalis* and Citrulline in Patients with Rheumatoid Arthritis and Periodontitis," *Journal of Periodontology* 84, no. 12 (December 2013): e74–e84, https://doi.org/10.1902/jop.2013.130079.

25. S. B. Brusca, S. B. Abramson, and J. U. Scher, "Microbiome and Mucosal Inflammation as Extra-Articular Triggers for Rheumatoid Arthritis and Autoimmunity," *Current Opinion in Rheumatology* 26, no. 1 (January 2014): 101–107, https://doi.org/10.1097/bor.0000000000000008.

26. J. D. Corrêa, G. R. Fernandes, D. C. Calderaro, S. M. S. Mendonça, J. M. Silva, M. L. Albiero, F. Q. Cunha, et al., "Oral Microbial Dysbiosis Linked to Worsened Periodontal Condition in Rheumatoid Arthritis Patients," *Scientific Reports* 9 (June 10, 2019): 8379, https://doi.org/10.1038/s41598-019-44674-6.

27. A. Pianta, S. L. Arvikar, K. Strle, E. E. Drouin, Q. Wang, C. E. Costello, and A. C. Steere, "Two Rheumatoid Arthritis–Specific Autoantigens Correlate Microbial Immunity with Autoimmune Responses in Joints," *Journal of Clinical Investigation* 127, no. 8 (August 1, 2017): 2946–2956, https://doi.org/10.1172/jci93450.

28. C. S. Guerreiro, Â. Calado, J. Sousa, and J. E. Fonseca, "Diet, Microbiota, and Gut Permeability—The Unknown Triad in Rheumatoid Arthritis," *Frontiers in Medicine* (Lausanne) 5 (December 2018): 349, https://doi.org/10.3389/fmed.2018.00349.

29. A. Watad, C. Bridgewood, T. Russell, H. Marzo-Ortega, R. Cuthbert, and D. McGonagle, "The Early Phases of Ankylosing Spondylitis: Emerging Insights from Clinical and Basic Science," *Frontiers in Immunology* 9 (November 2018): 2668, https://doi.org/10.3389/fimmu.2018.02688.

30. F. Ciccia, G. Guggino, A. Rizzo, R. Alessandro, M. M. Luchetti, S. Milling, L. Saieva, et al., "Dysbiosis and Zonulin Upregulation Alter Gut Epithelial and Vascular Barriers in Patients with Ankylosing Spondylitis," *Annals of the Rheumatic Diseases* 76, no. 6 (June 2017): 1123–1132, https://doi.org/10.1136/annrheumdis-2016-210000.

31. Ciccia et al., "Dysbiosis and Zonulin Upregulation."

32. Ciccia et al., "Dysbiosis and Zonulin Upregulation."

33. Ciccia et al., "Dysbiosis and Zonulin Upregulation."

34. Shirey et al., "Novel Strategies."

35. Shirey et al., "Novel Strategies"; Rittirsch et al., "Zonulin as Prehaptoglobin2."

36. S. K. Bantz, Z. Zhu, and T. Zheng, "The Atopic March: Progression from Atopic Dermatitis to Allergic Rhinitis and Asthma," *Journal of Clinical and Cellular Immunology* 5, no. 2 (April 7, 2014): 202, https://doi.org/10.4172/2155-9899.1000202.

37. Y. Liang, C. Chang, and Q. Lu, "The Genetics and Epigenetics of Atopic Dermatitis—Filaggrin and Other Polymorphisms," *Clinical Reviews in Allergy and Immunology* 51, no. 3 (December 2016): 315–328, https://doi.org/10.1007/s12016-015-8508-5.

38. B. Shi, N. J. Bangayan, E. Curd, P. A. Taylor, R. L. Gallo, D. Y. M. Leung, and H. Li, "The Skin Microbiome Is Different in Pediatric versus Adult Atopic Dermatitis," *Journal of Allergy and Clinical Immunology* 138,

no. 4 (October 1, 2016): 1233–1236, https://doi.org/10.1016/j.jaci.2016.04.053.

39. A. S. Paller, H. H. Kong, P. Seed, S. Naik, T. C. Scharschmidt, R. L. Gallo, T. Luger, and A. D. Irvine, "The Microbiome in Patients with Atopic Dermatitis," *Journal of Allergy and Clinical Immunology* 143, no. 1 (January 1, 2019): 26–35, https://doi.org/10.1016/j.jaci.2018.11.015.

40. J. Sakabe, M. Yamamoto, S. Hirakawa, A. Motoyama, I. Ohta, K. Tatsuno, T. Ito, K. Kabashima, T. Hibino, and Y. Tokura, "Kallikrein-Related Peptidase 5 Functions in Proteolytic Processing of Profilaggrin in Cultured Human Keratinocytes," *Journal of Biological Chemistry* 288, no. 24 (June 14, 2013): 17179–17189, https://doi.org/10.1074/jbc.m113.476820.

10 The Microbiome and Neurological and Behavioral Disorders

1. E. A. Mayer, R. Knight, S. K. Mazmanian, J. F. Cryan, and K. Tillisch, "Gut Microbes and the Brain: Paradigm Shift in Neuroscience," *Journal of Neuroscience* 34, no. 46 (November 12, 2014): 15490–15496, https://doi.org/10.1523/jneurosci.3299-14.2014.

2. Mayer et al., "Gut Microbes and the Brain."

3. M. C. Cenit, Y. Sanz, and P. Codoñer-Franch, "Influence of Gut Microbiota on Neuropsychiatric Disorders," *World Journal of Gastroenterology* 23, no. 30 (August 14, 2017): 5486–5498, https://doi.org/10.3748/wjg.v23.i30.5486.

4. Mayer et al., "Gut Microbes and the Brain."

5. L. M. Cox and H. L. Weiner, "Microbiota Signaling Pathways That Influence Neurologic Disease," *Neurotherapeutics* 15, no. 1 (January 2018): 135–145, https://doi.org/10.1007/s13311-017-0598-8.

6. G. Sharon, T. R. Sampson, D. H. Gschwind, and S. K. Mazmanian, "The Central Nervous System and the Gut Microbiome," *Cell* 167, no. 4 (November 3, 2016): 915–932, https://doi.org/10.1016/j.cell.2016.10.027.

7. D. Erny, A. L. Hrabě de Angelis, D. Jaitin, P. Wieghofer, O. Staszewski, E. David, H. Keren-Shaul, et al., "Host Microbiota Constantly Control

Maturation and Function of Microglia in the CNS," *Nature Neuroscience* 18, no. 7 (July 2015): 965–977, https://doi.org/10.1038/nn.4030.

8. M. T. Rahman, C. Ghosh, M. Hossain, D. Linfield, F. Rezaee, D. Janigro, N. Marchi, and A. H. H. van Boxel-Dexaire, "IFN-γ, IL-17A, or Zonulin Rapidly Increase the Permeability of the Blood-Brain and Small Intestinal Epithelial Barriers: Relevance for Neuro-Inflammatory Diseases," *Biochemical and Biophysical Research Communications* 507, no. 1–4 (December 9, 2018): 274–279, https://doi.org/10.1016/j.bbrc.2018.11.021.

9. V. Braniste, M. Al-Asmakh, C. Kowal, F. Anuar, A. Abbaspour, M. Tóth, A. Korecka, et al., "The Gut Microbiota Influences Blood-Brain Barrier Permeability in Mice," *Science Translational Medicine* 6, no. 263 (November 19, 2014): 263ra158, https://doi.org/10.1126/scitranslmed.3009759.

10. J. Humann, B. Mann, G. Gao, P. Moresco, J. Ramahi, L. N. Loh, A. Farr, et al., "Bacterial Peptidoglycan Traverses the Placenta to Induce Fetal Neuroproliferation and Aberrant Postnatal Behavior," *Cell Host and Microbe* 19, no. 3 (March 9, 2016): 388–399, https://doi.org/10.1016/j .chom.2016.02.009.

11. J. M. Wong, R. de Souza, C. W. C. Kendall, A. Eman, and D. J. A. Jenkins, "Colonic Health: Fermentation and Short Chain Fatty Acids," *Journal of Clinical Gastroenterology* 40, no. 3 (March 2006): 235–243, https://doi.org/10.1097/00004836-200603000-00015.

12. H. Liu, J. Wang, T. He, S. Becker, G. Zhang, D. Li, and X. Ma, "Butyrate: A Double-Edged Sword for Health?," *Advances in Nutrition* 9, no. 1 (January 2018): 21–29, https://doi.org/10.1092/advances/nmx009.

13. Cox and Weiner, "Microbiota Signaling Pathways."

14. Sharon et al., "The Central Nervous System."

15. C.-Y. Chang, D.-S. Ke, and J.-Y. Chen, "Essential Fatty Acids and Human Brain," *Acta Neurologica Taiwanica* 18, no. 4 (December 2009): 231–241, http://www.ant-tnsjournal.com/mag_files/18-4/18-4p231.pdf.

16. Chang, Ke, and Chen, "Essential Fatty Acids."

17. Sharon et al., "The Central Nervous System."

18. Centers for Disease Control and Prevention, "Data and Statistics on Autism Spectrum Disorder: Prevalence," https://www.cdc.gov/ncbddd/autism/data.html, accessed January 10, 2020.

19. E. Y. Hsiao, S. W. McBride, S. Hsien, G. Sharon, E. R. Hyde, T. McCue, J. A. Codelli, et al., "Microbiota Modulate Behavioral and Physiological Abnormalities Associated with Neurodevelopmental Disorders," *Cell* 155, no. 7 (December 19, 2013): 1451–1463, https://doi.org/10.1016/j.cell.2013.11.024.

20. Mayer et al., "Gut Microbes and the Brain."

21. California Institute of Technology, Division of Biology and Biological Engineering, Sarkis Mazmanian profile, http://www.bbe.caltech.edu/people/sarkis-mazmanian, accessed January 10, 2020.

22. G. Sharon, N. J. Cruz, D.-W. Kang, M. J. Gandal, B. Wang, Y.-M. Kim, E. M. Zink, et al., "Human Gut Microbiota from Autism Spectrum Disorder Promote Behavioral Symptoms in Mice," *Cell* 177, no. 6 (May 30, 2019): 1600–1618, https://doi.org/10.1016/j.cell.2019.05.004.

23. S. K. Mazmanian, oral interview with authors, November 22, 2017.

24. Mazmanian, oral interview.

25. Mazmanian, oral interview.

26. Mazmanian, oral interview.

27. Mazmanian, oral interview.

28. Mazmanian, oral interview.

29. Mazmanian, oral interview.

30. Mazmanian, oral interview.

31. Global Burden of Disease 2015 Disease and Injury Incidence and Prevalence Collaborators, "Global, Regional, and National Incidence, Prevalence, and Years Lived with Disability for 310 Diseases and Injuries, 1990–2015: A Systematic Analysis for the Global Burden of Disease Study 2015," *The Lancet* 388, no. 10053 (October 8, 2016): 1545–1602, https://doi.org/10.1016/s0140-6736(16)31678-6.

32. T. Harach, N. Marungruang, N. Duthilleul, V. Cheatham, K. D. McCoy, G. Frisoni, J. J. Neher, et al., "Reduction of Abeta Amyloid Pathology in APPPS1 Transgenic Mice in the Absence of Gut Microbiota," *Scientific Reports* 7 (February 8, 2017): 41802, https://doi.org/10.1038/srep41802.

33. National Institutes of Health, National Institute of Neurological Disorders and Stroke, "Parkinson's Disease: Hope Through Research," https://www.ninds.nih.gov/disorders/patient-caregiver-education/hope-through-research/parkinsons-disease-hope-through-research, accessed January 10, 2020.

34. A. Mulak and B. Bonaz, "Brain-Gut-Microbiota Axis in Parkinson's Disease," *World Journal of Gastroenterology* 21, no. 37 (October 7, 2015): 10609–10620, https://doi.org/10.3748/wjg.v21.i37.10609.

35. S. K. Dutta, S. Verma, V. Jain, B. K. Surapaneni, R. Vinayek, L. Phillips, and P. P. Nair, "Parkinson's Disease: The Emerging Role of Gut Dysbiosis, Antibiotics, Probiotics, and Fecal Microbiota Transplantation," *Journal of Neurogastroenterology and Motility* 25, no. 3 (July 2019): 363–376, https://doi.org/10.5056/jnm19044.

36. F. MacDonald, "New Evidence Suggests Parkinson's Might Start in the Gut, Not the Brain," *Science Alert: Health*, December 5, 2016, https://www.sciencealert.com/new-evidence-suggests-parkinson-s-might-start-in-the-gut-before-spreading-to-the-brain, accessed January 11, 2020; T. R. Sampson, J. W. Debelius, T. Thron, S. Janssen, G. G. Shastri, Z. E. Ilhan, C. Challis, et al., "Gut Microbiota Regulate Motor Deficits and Neuroinflammation in a Model of Parkinson's Disease," *Cell* 167, no. 6 (December 1, 2016): 1469–1480, https://doi.org/10.1016/j.cell.2016.11.018.

37. R. Svarcbahs, U. H. Julku, and T. T. Myöhänen, "Inhibition of Prolyl Oligopeptidase Restores Spontaneous Motor Behavior in the α-Synuclein Virus Vector–Based Parkinson's Disease Mouse Model by Decreasing α-Synuclein Oligomeric Species in Mouse Brain," *Journal of Neuroscience* 36, no. 49 (December 7, 2016): 12485–12497, https://doi.org/10.1523/jneurosci.2309-16.2016.

38. Sharon et al., "The Central Nervous System."

39. Mazmanian, oral interview.

11 The Microbiome and Environmental Enteropathy

1. J. Louis-Auguste and P. Kelly, "Tropical Enteropathies," *Current Gastroenterology Reports* 19, no. 7 (July 2017): 29, https://doi.org/10.1007/s11894-017-0570-0.

2. R. Lagos, A. Fasano, S. S. Wasserman, V. Prado, O. San Martin, P. Abrego, G. A. Lonsonsky, S. Alegria, and M. M. Levine, "Effect of Small Bowel Bacterial Overgrowth on the Immunogenicity of Single-Dose Live Oral Cholera Vaccine CVD 103-HgR," *Journal of Infectious Diseases* 180, no. 5 (November 1999): 1709–1712, https://doi.org/10.1086/315051.

3. World Health Organization, "Malnutrition," https://www.who.int/news-room/fact-sheets/detail/malnutrition, accessed January 7, 2020.

4. R. L. Guerrant, A. M. Leite, R. Pinkerton, P. H. Q. S. Medeiros, P. A. Cavalcante, M. DeBoer, M. Kosek, et al., "Biomarkers of Environmental Enteropathy, Inflammation, Stunting, and Impaired Growth in Children in Northeast Brazil," *PLoS One* 11, no. 9 (September 30, 2016): e0158772, https://doi.org/10.1371/journal.pone.0158772.

5. I. S. Menzies, M. J. Zuckerman, W. J. Nukajam, S. G. Somasundaram, B. Murphy, A. P. Jenkins, R. S. Crane, and G. G. Gregory, "Geography of Intestinal Permeability and Absorption," *Gut* 44, no. 4 (April 1999): 483–489, https://doi.org/10.1136/gut.44.4.483.

6. J. R. Donowitz, R. Haque, B. D. Kirkpatrick, M. Alam, M. Lu, M. Kabir, S. H. Kakon, et al., "Small Intestine Bacterial Overgrowth and Environmental Enteropathy in Bangladeshi Children," *mBio* 7, no. 1 (January 2016): e02102–15, https://doi.org/10.1128/mbio.02102-15.

7. Guerrant et al., "Biomarkers of Environmental Enteropathy."

8. Guerrant et al., "Biomarkers of Environmental Enteropathy."

9. Guerrant et al., "Biomarkers of Environmental Enteropathy."

10. M. B. de Morais and G. A. Pontes da Silva, "Environmental Enteric Dysfunction and Growth," *Jornal de Pediatra* (Rio de Janeiro) 95, supplement 1 (March–April 2019): 85–94, https://doi.org/10.1016/j.jped.2018.11.004.

11. Morais and Pontes da Silva, "Environmental Enteric Dysfunction."

12. A. Lin, E. M. Bik, E. K. Costello, L. Dethlefsen, R. Haque, D. A. Relman, and U. Singh, "Distinct Distal Gut Microbiome Diversity and Composition in Healthy Children from Bangladesh and the United States," *PloS One* 8, no. 1 (January 2013): e53838, https://doi.org/10.1371/journal.pone.0053838.

13. S. Subramanian, S. Huq, T. Yatsunenko, R. Haque, M. Mahfuz, M. A. Alam, A. Benezra, et al., "Persistent Gut Microbiota Immaturity in Malnourished Bangladeshi Children," *Nature* 510, no. 7505 (June 19, 2014): 417–421, https://doi.org/10.1038/nature13421.

14. Morais and Pontes da Silva, "Environmental Enteric Dysfunction."

15. N. T. Iqbal, S. Syed, K. Sadiq, M. N. Khan, J. Iqbal. J. Z. Ma, F. Umrani, et al., "Study of Environmental Enteropathy and Malnutrition (SEEM) in Pakistan: Protocols for Biopsy Based Biomarker Discovery and Validation," *BMC Pediatrics* 19 (July 22, 2019): 247, https://doi.org/10.1186/s12887-019-1564-x.

16. Iqbal et al., "Study of Environmental Enteropathy."

12   The Microbiome and Cancer

1. R. L. Siegel, K. D. Miller, and A. Jemal, "Cancer Statistics, 2017," *CA: A Cancer Journal for Clinicians* 67, no. 1 (January–February 2017): 7–30, https://doi.org/10.3322/caac.21387.

2. "Targeting Tumour Metabolism," *Nature Reviews Drug Discovery* 9 (July 2010): 503–504, https://doi.org/10.1038/nrd3215.

3. D. Hanahan and R. A. Weinberg, "The Hallmarks of Cancer," *Cell* 100, no. 1 (January 7, 2000): 57–70, https://doi.org/10.1016/s0092-8674(00)81683-9.

4. D. Hanahan and R. A. Weinberg, "Hallmarks of Cancer: The Next Generation," *Cell* 144, no. 5 (March 4, 2011): 646–674, https://doi.org/10.1016/j.cell.2011.02.013.

5. Nobel Prize, "James P. Allison: Facts," https://www.nobelprize.org /prizes/medicine/2018/allison/facts/, accessed January 12, 2020.

6. D. R. Leach, M. F. Krummel, and J. P. Allison, "Enhancement of Anti-tumor Immunity by CTLA-4 Blockade," *Science* 271, no. 5256 (March 22, 1996): 1734–1736, https://doi.org/10.1126/science.271.5256.1734.

7. J. Yuan, B. Ginsberg, D. Page, Y. Li, T. Rasalan, H. F. Gallardo, Y. Xu, et al., "CTLA-4 Blockade Increases Antigen-Specific CD8+ T Cells in Pre-vaccinated Patients with Melanoma: Three Cases," *Cancer Immunology, Immunotherapy* 60, no. 8 (August 2011): 1137–1146, https://doi.org/10 .1007/s00262-011-1011-9.

8. L. E. Fulbright, M. Ellermann, and J. C. Arthur, "The Microbiome and the Hallmarks of Cancer," *PLoS Pathogens* 13, no. 9 (September 21, 2017): e1006480, https://doi.org/10.1371/journal.ppat.1006480.

9. Nobel Prize, "The Nobel Prize in Physiology or Medicine 2005: Press Release," https://www.nobelprize.org/prizes/medicine/2005/press-release, accessed January 12, 2020.

10. A. Elkrief, L. Derosa, L. Zitvogel, G. Kroemer, and B. Routy, "The Intimate Relationship between Gut Microbiota and Cancer Immuno-therapy," *Gut Microbes* 10, no. 3 (October 19, 2019): 424–428, https://doi .org/10.1080/19490976.2018.1527167.

11. L. Zitvogel, oral interview with authors, February 1, 2019.

12. G. Kroemer and L. Zitvogel, "Cancer Immunotherapy in 2017: The Breakthrough of the Microbiota," *Nature Reviews Immunology* 18, no. 2 (February 2018): 87–88, https://doi.org/10.1038/nri.2108.4.

13. Zitvogel, oral interview.

14. Zitvogel, oral interview.

15. J. Ferlay, M. Colombet, I. Soerjomataram, T. Dyba, G. Randi, M. Bettio, A. Gavin, O. Visser, and F. Bray, "Cancer Incidence and Mortality Patterns in Europe: Estimates for 40 Countries and 25 Major Cancers in 2018," *European Journal of Cancer* 103 (November 1, 2018): 356–387, https://doi.org/10.1016/j.ejca.2018.07.005.

16. Skin Cancer Foundation, "Skin Cancer Facts and Statistics," https://www.skincancer.org/skin-cancer-information/skin-cancer-facts/, accessed January 11, 2020.

17. Global Burden of Disease Cancer Collaboration, "Global, Regional, and National Cancer Incidence, Mortality, Years of Life Lost, Years Lived with Disability, and Disability-Adjusted Life-Years for 32 Cancer Groups, 1990 to 2015: A Systematic Analysis for the Global Burden of Disease Study," *JAMA Oncology* 3, no. 4 (April 2017): 524–548, https://doi.org/10.1001/jamaoncol.2016.5688.

18. C. P. Wild, "Complementing the Genome with an 'Exposome': The Outstanding Challenge of Environmental Exposure Measurement in Molecular Epidemiology," *Cancer Epidemiology, Biomarkers and Prevention* 14, no. 8 (August 2005): 1847–1850, https://doi.org/10.1158/1055-9965.epi-05-0456.

19. Centers for Disease Control and Prevention, National Institute for Occupational Safety and Health, "Exposome and Exposomics Overview," https://www.cdc.gov/niosh/topics/exposome/default.html, accessed January 12, 2020.

20. Centers for Disease Control and Prevention, National Institute for Occupational Safety and Health, "Exposome and Exposomics Overview."

21. S. Senger, A. Sapone, M. R. Fiorentino, G. Mazzarella, G. Y. Lauwers, and A. Fasano, "Celiac Disease Histopathology Recapitulates Hedgehog Downregulation, Consistent with Wound Healing Processes Activation," *PLoS One* 10, no. 12 (December 9, 2015): e0144634, https://doi.org/10.1371/journal.pone.0144634.

22. Cancer Treatment Centers of America, "How Does the Immune System Work? When It Comes to Cancer, It's Complicated," October 19, 2017, https://www.cancercenter.com/community/blog/2017/10/how-does-the-immune-system-work-when-it-comes-to-cancer-its-complicated, accessed January 12, 2020.

23. Zitvogel, oral interview.

24. European Commission, Cordis, Horizon 2020, "Gut OncoMicrobiome Signatures (GOMS) Associated with Cancer Incidence, Prognosis and Prediction of Treatment Response," https://cordis.europa.eu/project /id/825410, accessed January 12, 2020.

25. Zitvogel, oral interview.

26. Zitvogel, oral interview.

27. Zitvogel, oral interview.

28. Zitvogel, oral interview.

29. Zitvogel, oral interview.

30. Zitvogel, oral interview.

13 From Association to Causation

1. M. I. McBurney, C. Davis, C. M. Fraser, B. O. Schneeman, C. Hutten-hower, K. Verbeke, J. Walter, and M. E. Latulippe, "Establishing What Constitutes a Healthy Human Gut Microbiome: State of the Science, Regulatory Considerations, and Future Directions," *Journal of Nutrition* 149, no. 11 (November 2019): 1882–1895, https://doi.org/10.1093/jn /nxz154.

2. Ley et al., "Obesity Alters Gut Microbial Ecology."

3. N. K. Surana and D. L. Kasper, "Moving beyond Microbiome-Wide Associations to Causal Microbe Identification," *Nature* 552, no. 7684 (December 14, 2017): 244–247, https://doi.org/10.1038/nature25019.

4. Surana and Kasper, "Moving beyond Microbiome-Wide Associations."

5. N. Geva-Zatorsky, E. Sefik, L. Kua, L. Pasman, T. G. Tan, A. Ortiz-Lopez, T. B. Yanortsang, et al., "Mining the Human Gut Microbiota for Immunomodulatory Organisms," *Cell* 168, no. 5 (February 23, 2017): 928–943, https://doi.org/10.1016/j.cell.2017.01.022.

6. J. L. Round, S. M. Lee, J. Li, G. Tran, B. Jabri, T. A. Chatila, and S. K. Mazmanian, "The Toll-Like Receptor 2 Pathway Establishes Colonization

by a Commensal of the Human Microbiota," *Science* 332, no. 6032 (May 20, 2011): 974–977, https://doi.org/10.1126/science.1206095.

7. Turnbaugh et al., "An Obesity-Associated Gut Microbiome."

8. P. J. Turnbaugh, V. K. Ridaura, J. J. Faith, F. E. Rey, R. Knight, and J. I. Gordon, "The Effect of Diet on the Human Gut Microbiome: A Metagenomic Analysis in Humanized Gnotobiotic Mice," *Science Translational Medicine* 1, no. 6 (November 11, 2009): 6ra14, https://doi.org/10.1126/scitranslmed.3000322.

9. N. W. Palm, M. R. de Zoete, T. W. Cullen, N. A. Barry, J. Stefanowski, L. Hao, P. H. Degnan, et al., "Immunoglobulin A Coating Identifies Colitogenic Bacteria in Inflammatory Bowel Disease," *Cell* 158, no. 5 (August 28, 2014): 1000–1010, https://doi.org/10.1016/j.cell.2014.08.006.

10. R. Salerno-Goncalves, F. Safavie, A. Fasano, and M. B. Sztein, "Free and Complexed-Secretory Immunoglobulin A Triggers Distinct Intestinal Epithelial Cell Responses," *Clinical and Experimental Immunology* 185, no. 3 (September 2016): 338–347, https://doi.org/10.1111/cei.12801.

11. J.-C. Lagier, S. Khelaifia, M. T. Alou, S. Ndongo, N. Dione, P. Hugon, A. Caputo, et al., "Culture of Previously Uncultured Members of the Human Gut Microbiota by Culturomics," *Nature Microbiology* 1 (November 7, 2016): 16203, https://doi.org/10.1038/nmicrobiol.2016.203.

12. J. T. Lau, F. J. Whelan, I. Herath, C. H. Lee, S. M. Collins, P. Bercik, and M. G. Surette, "Capturing the Diversity of the Human Gut Microbiota through Culture-Enriched Molecular Profiling," *Genome Medicine* 8, no. 1 (July 1, 2016): 72, https://doi.org/10.1186/s13073-016-0327-7.

13. K. M. Ellegaard and P. Engel, "Beyond 16S rRNA Community Profiling: Intra-Species Diversity in the Gut Microbiota," *Frontiers in Microbiology* 7 (September 2016): 1475, https://doi.org/10.3389/fmicb.2016.01475.

14. J. Mestecky, W. Strober, M. W. Russell, B. L. Kelsall, H. Cheroutre, and B. N. Lambrecht, *Mucosal Immunology,* 4th ed. (San Diego, CA: Elsevier, 2015).

15. K. P. Nickerson, S. Senger, Y. Zhang, R. Lima, S. Patel, L. Ingano, W. A. Flavahan, et al., "*Salmonella* Typhi Colonization Provokes Extensive Transcriptional Changes Aimed at Evading Host Mucosal Immune Defense during Early Infection of Human Intestinal Tissue," *EBioMedicine* 31 (May 1, 2018): 92–109, https://doi.org/10.1016/j.ebiom.2018.04.005.

16. Senger et al., "Human Fetal-Derived Enterospheres."

17. R. Freire, L. Ingano, G. Serena, M. Cetinbas, A. Anselmo, A. Sapone, R. I. Sadreyev, A. Fasano, and S. Senger, "Human Gut–Derived Organoids Provide Model to Study Gluten Response and Effects of Microbiota-Derived Molecules in Celiac Disease," *Scientific Reports* 9 (May 7, 2019): 7029, https://doi.org/10.1038/s41598-019-43426-w.

18. W. Gou, Y. Fu, L. Yue, G. Chen, X. Cai, M. Shuai, F. Xu, and J. S. Zheng, et al., "Gut Microbiota May Underlie the Predisposition of Healthy Individuals to COVID-19," *medRxiv*, preprint, posted April 25, 2020, https://doi.org/10.1101/2020.04.22.20076091.

14 Preventive Medicine

1. L. Fiechtner, oral interview with authors, October 25, 2019.

2. Fiechtner, oral interview.

3. Fiechtner, oral interview.

4. Office of the Chief Economist, U.S. Department of Agriculture, "U.S. Food Waste Challenge: FAQs," https://www.usda.gov/foodwaste/faqs, accessed February 3, 2020.

5. Centers for Disease Control and Prevention, "Childhood Obesity Facts: Prevalence of Childhood Obesity in the U.S.," https://www.cdc.gov/obesity/data/childhood.html, accessed January 31, 2020.

6. Fiechtner, oral interview.

7. L. Fadulu, "Trump Targets Michelle Obama's School Nutrition Guidelines on Her Birthday," *New York Times*, January 17, 2020, https://www.nytimes.com/2020/01/17/us/politics/michelle-obama-school-nutrition-trump.html.

8. Fiechtner, oral interview.

9. Fiechtner, oral interview.

10. K. Fox, "United Supermarkets Offering Weekend Health Screenings," News Channel 6, Wichita Falls, Texas, January 8, 2020, https://www.newschannel6now.com/2020/01/08/united-supermarkets-offering-weekend-health-screenings/, accessed January 31, 2020.

11. R. Aguilar-Santos, "Latino-Owned Grocery Store Uses Bilingual Marketing to Inspire Healthy Shopping," *Salud America!*, April 4, 2014, https://salud-america.org/latino-owned-grocery-store-chain-uses-bi-lingual-marketing-labeling-program-to-inspire-healthy-shopping/, accessed January 31, 2020.

12. Fiechtner, oral interview.

13. A. Robinson, L. Fiechtner, B. Roche, N. J. Ajami, J. F. Petrosino, C. A. Camargo Jr., E. M. Taveras, and K. Hasegawa, "Association of Maternal Gestational Weight Gain with the Infant Fetal Microbiota," *Journal of Pediatric Gastroenterology and Nutrition* 65, no. 5 (November 2017): 509–515, https://doi.org/10.1097/MPG.0000000000001566.

14. M. Simione, S. G. Harshman, I. Castro, R. Linnemann, B. Roche, N. J. Ajami, J. F. Petrosino, et al., "Maternal Fish Consumption in Pregnancy Is Associated with a *Bifidobacterium*-Dominant Microbiome Profile in Infants," *Current Developments in Nutrition* 4, no. 1 (January 2020): nzz133, https://doi.org/10.1093/cdn/nzz133.

15. V. M. Martin, oral interview with authors, November 1, 2019.

16. Martin, oral interview.

17. V. M. Martin, Y. V. Virkud, H. Seay, A. Hickey, R. Ndahayo, R. Rosow, C. Southwick, et al., "Prospective Assessment of Pediatrician-Diagnosed Food Protein–Induced Allergic Proctocolitis by Gross or Occult Blood," *Journal of Allergy and Clinical Immunology: In Practice* 8, no. 5 (May 2020): 1692–1699, https://doi.org/10.1016/j.jaip.2019.12.029.

18. Martin, oral interview.

19. Martin, oral interview.

20. Martin, oral interview.

21. D. Ierodiakonou, V. Garcia-Larsen, A. Logan, A. Groome, S. Cunha, J. Chivinge, Z. Robinson, et al., "Timing of Allergenic Food Introduction to the Infant Diet and Risk of Allergic or Autoimmune Disease: A Systematic Review and Meta-Analysis," *Journal of the American Medical Association* 316, no. 11 (September 20, 2016): 1181–1192, https://doi.org /10.1001/jama.2016.12623.

22. Martin, oral interview.

23. Martin, oral interview.

24. Martin, oral interview.

25. Martin, oral interview.

26. Sellitto et al., "Proof of Concept of Microbiome-Metabolome Analysis."

27. Sellitto et al., "Proof of Concept of Microbiome-Metabolome Analysis."

28. M. Olivares, A. Neef, G. Castillejo, G. De Palma, V. Varea, A. Capilla, F. Palau, et al., "The HLA-DQ2 Genotype Selects for Early Intestinal Microbiota Composition in Infants at High Risk of Developing Coeliac Disease," *Gut* 64, no. 3 (March 2015): 406–417, https://doi.org/10.1136 /gutjnl-2014-306931.

29. E. Lionetti, S. Castellaneta, R. Francavilla, A. Pulvirenti, E. Tonutti, S. Amarri, M. Barbato, et al., "Introduction of Gluten, HLA Status, and the Risk of Celiac Disease in Children," *New England Journal of Medicine* 371, no. 14 (October 2, 2014): 1295–1303, https://doi.org/10.1056 /nejmoa1400697; S. L. Vriezinga, R. Auricchio, E. Bravi, G. Castillejo, A. Chmielewska, P. C. Escobar, S. Kolaček, et al., "Randomized Feeding Intervention in Infants at High Risk for Celiac Disease," *New England Journal of Medicine* 371, no. 14 (October 2, 2014): 1304–1315, https://doi.org/10 .1056/nejmoa1404172.

30. M. Leonard, oral interview with authors, November 4, 2019.

31. Leonard, oral interview.

32. Leonard, oral interview.

33. Leonard, oral interview.

34. Leonard, oral interview.

35. S. J. Blumberg, M. D. Bramlett, M. D. Kogan, L. A. Schieve, J. R. Jones, and M. C. Lu, "Changes in Prevalence of Parent-Reported Autism Spectrum Disorder in School-Aged U.S. Children: 2007 to 2011–2012," *National Health Statistics Reports*, no. 65 (March 20, 2013), https://www.cdc.gov/nchs/data/nhsr/nhsr065.pdf.

36. B. B. Nankova, R. Agarwal, D. F. MacFabe, and E. F. La Gamma, "Enteric Bacterial Metabolites Propionic and Butyric Acid Modulate Gene Expression, Including CREB-Dependent Catecholaminergic Neurotransmission, in PC12 Cells—Possible Relevance to Autism Spectrum Disorders," *PloS One* 9, no. 8 (August 2014): e103740, https://doi.org/10.1371/journal.pone.0103740; D. MacFabe, "Autism: Metabolism, Mitochondria, and the Microbiome," *Global Advances in Health and Medicine* 2, no. 6 (November 1, 2013): 52–66, https://doi.org/10.7453/gahmj.2013.089.

37. G. A. Rook, C. L. Raison, and C. A. Lowry, "Microbiota, Immunoregulatory Old Friends and Psychiatric Disorders," in *Microbial Endocrinology: The Microbiota-Gut-Brain Axis in Health and Disease*, ed. M. Lyte and J. F. Cryan (New York: Springer, 2014), 319–356, https://doi.org/10.1007/978-1-4939-0897-4_15; Y. E. Borre, R. D. Moloney, G. Clarke, T. G. Dinan, and J. F. Cryan, "The Impact of Microbiota on Brain and Behaviour: Mechanisms and Therapeutic Potential," in *Microbial Endocrinology,* ed. Lyte and Cryan, 373–403, https://doi.org/10.1007/978-1-4939-0897-4_17.

38. B. O. McElhanon, C. McCracken, S. Karpen, and W. G. Sharp, "Gastrointestinal Symptoms in Autism Spectrum Disorder: A Meta-Analysis," *Pediatrics* 133, no. 5 (May 2014): 872–883, https://doi.org/10.1542/peds.2013-3995.

39. E. Y. Hsiao, "Gastrointestinal Issues in Autism Spectrum Disorder," *Harvard Review of Psychiatry* 22, no. 2 (March–April 2014): 104–111, https://doi.org/10.1097/hrp.0000000000000029; S. M. Finegold, J. Downes, and

P. H. Summanen, "Microbiology of Regressive Autism," *Anaerobe* 18, no. 2 (April 2012): 260–262, https://doi.org/10.1016/j.anaerobe.2011.12.018.

40. D. R. Rose, H. Yang, G. Serena, C. Sturgeon, B. Ma, M. Careaga, H. K. Hughes, and P. Ashwood, et al., "Differential Immune Responses and Microbiota Profiles in Children with Autism Spectrum Disorders and Co-morbid Gastrointestinal Symptoms," *Brain, Behavior, and Immunity* 70 (May 2018): 354–368, https://doi.org/10.1016/j.bbi.2018.03.025.

41. D.-W. Kang, J. B. Adams, A. C. Gregory, T. Borody, L. Chittick, A. Fasano, A. Khoruts, et al., "Microbiota Transfer Therapy Alters Gut Ecosystem and Improves Gastrointestinal and Autism Symptoms: An Open-Label Study," *Microbiome* 5, no. 1 (January 23, 2017): 10, https://doi.org/10.1186/s40168-016-0335-7.

42. B. Palsson and K. Zengler, "The Challenges of Integrating Multi-Omic Data Sets," *Nature Chemical Biology* 6, no. 11 (November 2010): 787–789, https://doi.org/10.1038/nchembio.462.

43. Global Burden of Disease 2017 Diet Collaborators, "Health Effects of Dietary Risks in 195 Countries, 1990–2017: A Systematic Analysis for the Global Burden of Disease Study 2017," *The Lancet* 393, no. 10184 (May 11, 2019): 1958–1972, https://doi.org/10.1016/s0140-6736(19)30041-8.

44. D. Spector, "An Evolutionary Explanation for Why We Crave Sugar," *Business Insider*, April 25, 2014, https://www.businessinsider.com/evolutionary-reason-we-love-sugar-2014-4, accessed January 31, 2020.

45. M. Singh, "Why a Sweet Tooth May Have Been an Evolutionary Advantage for Kids," *The Salt: What's on Your Plate, Eating and Health*, March 19, 2014, NPR, https://www.npr.org/sections/thesalt/2014/03/19/291406696/why-a-sweet-tooth-may-have-been-an-evolutionary-advantage-for-kids, accessed January 31, 2020.

15 Treatments for Disease

1. G. R. Gibson and M. B. Roberfroid, "Dietary Modulation of the Human Colonic Microbiota: Introducing the Concept of Prebiotics,"

Journal of Nutrition 125, no. 6 (June 1995): 1401–1412, https://doi.org/10.1093/jn/125.6.1401.

2. G. R. Gibson, R. Hutkins, M. E. Sanders, S. L. Prescott, R. A. Reimer, S. J. Salminen, K. Scott, et al., "Expert Consensus Document: The International Scientific Association for Probiotics and Prebiotics (ISAPP) Consensus Statement on the Definition and Scope of Prebiotics," *Nature Reviews Gastroenterology and Hepatology* 14, no. 8 (August 2017): 491–502, https://doi.org/10.1038/nrgastro.2017.75.

3. M. M. Kaczmarczyk, M. J. Miller, and G. G. Freund, "The Health Benefits of Dietary Fiber: Beyond the Usual Suspects of Type 2 Diabetes Mellitus, Cardiovascular Disease and Colon Cancer," *Metabolism* 61, no. 8 (August 1, 2012): 1058–1066, https://doi.org/10.1016/j.metabol.2012.01.017.

4. A. M. Doherty, C. J. Lodge, S. C. Dharmage, X. Dai, L. Bode, and A. J. Lowe, "Human Milk Oligosaccharides and Associations with Immune-Mediated Disease and Infection in Childhood: A Systematic Review," *Frontiers in Pediatrics* 6 (April 2018): 91, https://doi.org/10.3389/fped.2018.00091.

5. J. Van Loo, P. Coussement, L. De Leenheer, H. Hoebregs, and G. Smits, "On the Presence of Inulin and Oligofructose as Natural Ingredients in the Western Diet," *Critical Reviews in Food Science and Nutrition* 35, no. 6 (November 1995): 525–552, https://doi.org/10.1080/10408399509527714.

6. D. Vandeputte, G. Falony, S. Vieira-Silva, J. Wang, M. Sailer, S. Theis, K. Verbeke, and J. Raes, "Prebiotic Inulin-Type Fructans Induce Specific Changes in the Human Gut Microbiota," *Gut* 66, no. 11 (November 2017): 1968–1974, https://doi.org/10.1136/gutjnl-2016-313271.

7. S. Fukuda, H. Toh, K. Hase, K. Oshima, Y. Nakanishi, K. Yoshimura, T. Tobe, et al., "Bifidobacteria Can Protect from Enteropathogenic Infection through Production of Acetate," *Nature* 469, no. 7331 (January 27, 2011): 543–547, https://doi.org/10.1038/nature09646.

8. W. S. F. Chung, M. Meijerink, B. Zeuner, J. Holck, P. Louis, A. S. Meyer, J. M. Wells, H. J. Flint, and S. H. Duncan, "Prebiotic Potential of Pectin and Pectic Oligosaccharides to Promote Anti-inflammatory Commensal

Bacteria in the Human Colon," *FEMS Microbiology Ecology* 93, no. 11 (November 2017): fix127, https://doi.org/10.1093/femsec/fix127.

9. R. Corrêa-Oliveira, J. L. Fachi, A. Vieira, F. T. Sato, and M. A. R. Vinolo, "Regulation of Immune Cell Function by Short-Chain Fatty Acids," *Clinical and Translational Immunology* 5, no. 4 (April 22, 2016): e73, https://doi.org/10.1038/cti.2016.17.

10. G. den Besten, K. van Eunen, A. K. Groen, K. Venema, D.-J. Reijngoud, and B. M. Bakker, "The Role of Short-Chain Fatty Acids in the Interplay between Diet, Gut Microbiota, and Host Energy Metabolism," *Journal of Lipid Research* 54, no. 9 (September 2013): 2325–2340, https://doi.org/10.1194/jlr.r036012.

11. Y. Zhao, F. Chen, W. Wu, M. Sun, A. J. Bilotta, S. Yao, Y. Xiao, et al., "GPR43 Mediates Microbiota Metabolite SCFA Regulation of Antimicrobial Peptide Expression in Intestinal Epithelial Cells via Activation of mTOR and STAT3," *Mucosal Immunology* 11, no. 3 (May 2018): 752–762, https://doi.org/10.1038/mi.2017.118.

12. G. Serena, S. Yan, S. Camhi, S. Patel, R. S. Lima, A. Sapone, M. M. Leonard, et al., "Proinflammatory Cytokine Interferon-γ and Microbiome-Derived Metabolites Dictate Epigenetic Switch between Forkhead Box Protein 3 Isoforms in Coeliac Disease," *Clinical and Experimental Immunology* 187, no. 3 (March 2017): 490–506, https://doi.org/10.1111/cei.12911.

13. Freire et al., "Human Gut–Derived Organoids."

14. R. Y. Wu, M. Abdullah, P. Määttänen, A. V. C. Pilar, E. Scruten, K. C. Johnson-Henry, S. Napper, C. O'Brien, N. L. Jones, and P. M. Sherman, "Protein Kinase C δ Signaling Is Required for Dietary Prebiotic-Induced Strengthening of Intestinal Epithelial Barrier Function," *Scientific Reports* 7 (January 18, 2017): 40820, https://doi.org/10.1038/srep40820.

15. Nobel Prize, "The Nobel Prize in Physiology or Medicine 1908," https://www.nobelprize.org/prizes/medicine/1908/summary/, accessed January 25, 2020.

16. E. Metchnikoff, *The Prolongation of Life: Optimistic Studies*, English trans., ed. P. Chalmers Mitchell (New York: G. P. Putnam's Sons; London: Knickerbocker Press, 1908).

17. Food and Agriculture Organization of the United Nations/World Health Organization, "Probiotics in Food: Health and Nutritional Properties and Guidelines for Evaluation," FAO Food and Nutrition Paper 85, Rome, 2006, http://www.fao.org/3/a-a0512e.pdf, accessed January 26, 2020.

18. T. Wilkins and J. Sequoia, "Probiotics for Gastrointestinal Conditions: A Summary of the Evidence," *American Family Physician* 96, no. 3 (August 1, 2017): 170–178, https://www.aafp.org/afp/2017/0801/p170.html.

19. J. Vanderhoof, oral interview with authors, August 21, 2019.

20. Vanderhoof, oral interview.

21. Vanderhoof, oral interview.

22. Vanderhoof, oral interview.

23. E. Isolauri, S. Rautava, and S. Salminen, "Probiotics in the Development and Treatment of Allergic Disease," *Gastroenterology Clinics* 41, no. 4 (December 2012): 747–762, https://doi.org/10.1016/j.gtc.2012.08.007.

24. Vanderhoof, oral interview.

25. Vanderhoof, oral interview.

26. Vanderhoof, oral interview.

27. Vanderhoof, oral interview.

28. P. V. Kirjavainen, S. J. Salminen, and E. Isolauri, "Probiotic Bacteria in the Management of Atopic Disease: Underscoring the Importance of Viability," *Journal of Pediatric Gastroenterology and Nutrition* 36, no. 2 (February 2003): 223–227, https://doi.org/10.1097/00005176-200302000-00012.

29. Vanderhoof, oral interview.

30. Vanderhoof, oral interview.

31. Vanderhoof, oral interview.

32. Vanderhoof, oral interview.

33. C. de Simone, "The Unregulated Probiotic Market," *Clinical Gastroenterology and Hepatology* 17, no. 5 (April 2019): 809–817, https://doi.org/10.1016/j.cgh.2018.01.018.

34. de Simone, "The Unregulated Probiotic Market."

35. de Simone, "The Unregulated Probiotic Market."

36. de Simone, "The Unregulated Probiotic Market."

37. de Simone, "The Unregulated Probiotic Market."

38. U.S. Department of Health and Human Services, Food and Drug Administration, Center for Biologic Evaluation and Research, "Early Clinical Trials with Live Biotherapeutic Products: Chemistry, Manufacturing, and Control Information: Guidance for Industry," February 2012, updated June 2016, https://www.fda.gov/media/82945/download, accessed January 26, 2020.

39. U.S. Department of Health and Human Services, Food and Drug Administration, "Statement from FDA Commissioner Scott Gottlieb, M.D., on Advancing the Science and Regulation of Live Microbiome-Based Products Used to Prevent, Treat, or Cure Diseases in Humans," August 16, 2018, https://www.fda.gov/news-events/press-announcements/statement-fda-commissioner-scott-gottlieb-md-advancing-science-and-regulation-live-microbiome-based, accessed January 26, 2020.

40. National Institutes of Health, National Center for Complementary and Integrative Health, National Health Interview Survey 2012, "Probiotics, Prebiotics," https://nccih.nih.gov/research/statistics/NHIS/2012/natural-products/biotics#child-data, accessed January 26, 2020.

41. National Institutes of Health, National Center for Complementary and Integrative Health, National Health Interview Survey 2012, "Probiotics, Prebiotics."

42. S. Guandalini, L. Pensabene, M. A. Zikri, J. A. Dias, L. G. Casali, H. Hoekstra, S. Kolacek, et al., "*Lactobacillus* GG Administered in Oral Rehydration Solution to Children with Acute Diarrhea: A Multicenter European Trial," *Journal of Pediatric Gastroenterology and Nutrition* 30, no. 1

(January 2000): 54–60, https://doi.org/10.1097/00005176-200001000
-00018.

43. Guandalini et al., "*Lactobacillus* GG Administered in Oral Rehydration Solution."

44. S. J. Allen, B. Okoko, E. G. Martinez, G. V. Gregorio, and L. F. Dans, "Probiotics for Treating Infectious Diarrhoea," *Cochrane Database of Systematic Reviews* 2003, no. 4: cd003048, https://doi.org/10.1002/14651858 .cd003048.pub2.

45. Allen et al., "Probiotics for Treating Infectious Diarrhoea."

46. S. B. Freedman, S. Williamson-Urquhart, K. J. Farion, S. Gouin, A. R. Willan, N. Poonai, K. Hurley, et al., "Multicenter Trial of a Combination Probiotic for Children with Gastroenteritis," *New England Journal of Medicine* 379, no. 21 (November 22, 2018): 2015–2026, https://doi.org /10.1056/nejmoa1802597.

47. D. Schnadower, P. I. Tarr, T. C. Casper, M. H. Gorelick, J. Michael Dean, K. J. O'Connell, P. Mahajan, et al., "*Lactobacillus rhamnosus* GG versus Placebo for Acute Gastroenteritis in Children," *New England Journal of Medicine* 379, no. 21 (November 22, 2018): 2002–2014, https://doi .org/10.1056/nejmoa1802598.

48. S. Hempel, S. J. Newberry, A. R. Maher, Z. Wang, J. N. V. Miles, R. Shanman, B. Johnsen, and P. G. Shekelle, "Probiotics for the Prevention and Treatment of Antibiotic-Associated Diarrhea: A Systematic Review and Meta-Analysis," *Journal of the American Medical Association* 307, no. 18 (May 9, 2012): 1959–1969, https://doi.org/10.1001/jama .2012.3507.

49. J. Suez, N. Zmora, G. Zilberman-Schapira, U. Mor, M. Dori-Bachash, S. Bashiardes, M. Zur, et al., "Post-Antibiotic Gut Mucosal Microbiome Reconstitution Is Impaired by Probiotics and Improved by Autologous FMT," *Cell* 174, no. 6 (September 6, 2018): 1406–1423, https://doi.org /10.1016/j.cell.2018.08.047.

50. N. Zmora, G. Zilberman-Schapira, J. Suez, U. Mor, M. Dori-Bachash, S. Bashiardes, E. Kotler, et al., "Personalized Gut Mucosal Colonization

Resistance to Empiric Probiotics Is Associated with Unique Host and Microbiome Features," *Cell* 174, no. 6 (September 6, 2018): 1388–1405, https://doi.org/10.1016/j.cell.2018.08.041.

51. R. M. Thushara, S. Gangadaran, Z. Solati, and M. H. Moghadasian, "Cardiovascular Benefits of Probiotics: A Review of Experimental and Clinical Studies," *Food and Function* 7, no. 2 (February 2016): 632–642, https://doi.org/10.1039/c5fo01190f.

52. F. L. Collins, N. D. Rios-Arce, J. D. Schepper, N. Parameswaran, and L. R. McCabe, "The Potential of Probiotics as a Therapy for Osteoporosis," *Microbiology Spectrum* 5, no. 4 (August 2017): bad-0015-2016, https://doi.org/10.1128/microbiolspec.bad-0015-2016.

53. M. R. Roudsari, R. Karimi, S. Sohrabvandi, and A. M. Mortazavian, "Health Effects of Probiotics on the Skin," *Critical Reviews in Food Science and Nutrition* 55, no. 9 (March 3, 2015): 1219–1240, https://doi.org/10.1080/10408398.2012.680078.

54. C. Hill, F. Guarner, G. Reid, G. R. Gibson, D. J. Merenstein, B. Pot, L. Morelli, et al., "Expert Consensus Document: The International Scientific Association for Probiotics and Prebiotics Consensus Statement on the Scope and Appropriate Use of the Term Probiotic," *Nature Reviews Gastroenterology and Hepatology* 11, no. 8 (August 2014): 506–514, https://doi.org/10.1038/nrgastro.2014.66.

55. P. Langella, F. Guarner, and R. Martin, "Editorial: Next-Generation Probiotics: From Commensal Bacteria to Novel Drugs and Food Supplements," *Frontiers in Microbiology* 10 (August 2019): 1973, https://doi.org/10.3389/fmicb.2019.01973.

56. Food and Agriculture Organization of the United Nations/World Health Organization, "Probiotics in Food."

57. A. W. Burks, L. F. Harthoorn, M. T. J. Van Ampting, M. M. O. Nijhuis, J. E. Langford, H. Wopereis, S. B. Goldberg, et al., "Synbiotics-Supplemented Amino Acid–Based Formula Supports Adequate Growth in Cow's Milk Allergic Infants," *Pediatric Allergy and Immunology* 26, no. 4 (June 2015): 316–322, https://doi.org/10.1111/pai.12390; L. B. van Der Aa,

H. S. Heymans, W. M. Van Aalderen, J. H. Sillevis Smitt, J. Knol, K. Ben Amor, D. A. Goossens, A. B. Sprikkelman, and the Synbad Study Group, "Effect of a New Synbiotic Mixture on Atopic Dermatitis in Infants: A Randomized-Controlled Trial," *Clinical and Experimental Allergy* 40, no. 5 (May 2010):795–804, https://doi.org/10.1111/j.1365-2222.2010.0346.x; L. B. van der Aa, W. M. C. van Aalderen, H. S. A. Heymans, J. Henk Sillevis Smitt, A. J. Nauta, L. M. J. Knippels, K. Ben Amor, A. B. Sprikkelman, and the Synbad Study Group, "Synbiotics Prevent Asthma-Like Symptoms in Infants with Atopic Dermatitis," *Allergy* 66, no. 2 (February 2011): 170–177, https://doi.org/10.1111/j.1398-9995.2010.02416.x.

58. E. Nikbakht, S. Khalesi, I. Singh, L. T. Williams, N. P. West, and N. Colson, "Effect of Probiotics and Synbiotics on Blood Glucose: A Systematic Review and Meta-Analysis of Controlled Trials," *European Journal of Nutrition* 57, no. 1 (February 2018): 95–106, https://doi.org/10.1007/s00394-016-1300-3; L. E. Miller, A. C. Ouwehand, and A. Ibarra, "Effects of Probiotic-Containing Products on Stool Frequency and Intestinal Transit in Constipated Adults: Systematic Review and Meta-Analysis of Randomized Controlled Trials," *Annals of Gastroenterology* 30, no. 6 (September 21, 2017): 629–639, https://doi.org/10.20524/aog.2017.0192; S. Arumugam, C. S. M. Lau, and R. S. Chamberlain, "Probiotics and Synbiotics Decrease Postoperative Sepsis in Elective Gastrointestinal Surgical Patients: A Meta-Analysis," *Journal of Gastrointestinal Surgery* 20, no. 6 (June 2016): 1123–1131, https://doi.org/10.1007/s11605-016-3142-y.

59. K. R. Pandey, S. R. Naik, and B. V. Vakil, "Probiotics, Prebiotics and Synbiotics: A Review," *Journal of Food Science and Technology* 52, no. 12 (December 2015): 7577–7587, https://doi.org/10.1007/s13197-015-1921-1.

60. J. E. Aguilar-Toalá, R. Garcia-Varela, H. S. Garcia, V. Mata-Haro, A. F. González-Córdova, B. Vallejo-Cordoba, and A. Hernández-Mendoza, "Postbiotics: An Evolving Term within the Functional Foods Field," *Trends in Food Science and Technology* 75 (May 2018): 105–114, https://doi.org/10.1016/j.tifs.2018.03.009; H. L. Foo, T. C. Loh, N. E. A. Mutalib, and R. A. Rahim, "The Myth and Therapeutic Potentials of Postbiotics," in *Microbiome and Metabolome in Diagnosis, Therapy, and Other Strategic*

Applications, ed. J. Faintuch and S. Faintuch (Amsterdam: Elsevier, 2019), 201–211, https://doi.org/10.1016/b978-0-12-815249-2.00021-x.

16 Microbiome Research in Gut-Brain Axis Diseases

1. T. G. Dinan, C. Stanton, and J. F. Cryan, "Psychobiotics: A Novel Class of Psychotropic," *Biological Psychiatry* 74, no. 10 (November 15, 2013): 720–726, https://doi.org/10.1016/j.biopsych.2013.05.001.

2. L.-H. Cheng, Y.-W. Liu, C.-C. Wu, S. Wang, and Y.-C. Tsai, "Psychobiotics in Mental Health, Neurodegenerative and Neurodevelopmental Disorders," *Journal of Food and Drug Analysis* 27, no. 3 (July 2019): 632–648, https://doi.org/10.1016/j.jfda.2019.01.002.

3. A. Slyepchenko, A. F. Carvalho, D. S. Cha, S. Kasper, and R. S. McIntyre, "Gut Emotions: Mechanisms of Action of Probiotics as Novel Therapeutic Targets for Depression and Anxiety Disorders," *CNS and Neurological Disorders—Drug Targets* 13, no. 10 (2014): 1770–1786, https://doi.org/10.2174/1871527313666141130205242.

4. S. Westfall, N. Lomis, I. Kahouli, S. Y. Dia, S. P. Singh, and S. Prakash, "Microbiome, Probiotics and Neurodegenerative Diseases: Deciphering the Gut-Brain Axis," *Cellular and Molecular Life Sciences* 74, no. 20 (October 2017): 3769–3787, https://doi.org/10.1007/s00018-017-2550-9.

5. A. P. Allen, W. Hutch, Y. E. Borre, P. J. Kennedy, A. Temko, G. Boylan, E. Murphy, J. F. Cryan, T. G. Dinan, and G. Clarke, "*Bifidobacterium longum* 1714 as a Translational Psychobiotic: Modulation of Stress, Electrophysiology and Neurocognition in Healthy Volunteers," *Translational Psychiatry* 6, no. 11 (November 2016): e939, https://doi.org/10.1038/tp.2016.191.

6. A. A. Mohammadi, S. Jazayeri, K. Khosravi-Darani, Z. Solati, N. Mohammadpour, Z. Asemi, Z. Adab, et al., "The Effects of Probiotics on Mental Health and Hypothalamic-Pituitary-Adrenal Axis: A Randomized, Double-Blind, Placebo-Controlled Trial in Petrochemical Workers," *Nutritional Neuroscience* 19, no. 9 (November 2016): 387–395, https://doi.org/10.1179/1476830515y.0000000023.

7. M. Messaoudi, R. Lalonde, N. Violle, H. Javelot, D. Desor, A. Nejdi, J.-F. Bisson, et al., "Assessment of Psychotropic-Like Properties of a Probiotic Formulation (*Lactobacillus helveticus* R0052 and *Bifidobacterium longum* R0175) in Rats and Human Subjects," *British Journal of Nutrition* 105, no. 5 (March 14, 2011): 755–764, https://doi.org/10.1017/s0007114510004319.

8. National Institutes of Health, U.S. National Library of Medicine, "Lactobacillus Plantarum PS128 in Patients with Major Depressive Disorder and High Level of Inflammation," ClinicalTrials.gov, last updated July 2, 2018, https://clinicaltrials.gov/show/nct03237078, accessed January 19, 2020; National Institutes of Health, U.S. National Library of Medicine, "Effects of Probiotics on Mood," ClinicalTrials.gov, May 29, 2018, https:// clinicaltrials.gov/ct2/show/nct03539263, accessed January 19, 2020; National Institutes of Health, U.S. National Library of Medicine, "Effect of Lactobacillus Plantarum 299v Supplementation on Major Depression Treatment," ClinicalTrials.gov, last updated September 7, 2018, https:// clinicaltrials.gov/show/nct02469545, accessed January 19, 2020; National Institutes of Health, U.S. National Library of Medicine, "The Probiotic Study: Using Bacteria to Calm Your Mind," ClinicalTrials.gov, January 3, 2019, https://clinicaltrials.gov/show/nct02711800, accessed January 19, 2020; National Institutes of Health, U.S. National Library of Medicine, "Probiotics and Examination-Related Stress in Healthy Medical Students," ClinicalTrials.gov, February 9, 2018, https://clinicaltrials.gov/ct2/show /nct03427515, accessed January 19, 2020; National Institutes of Health, U.S. National Library of Medicine, "Effects of Probiotics on Symptoms of Depression (ESPD)," ClinicalTrials.gov, last updated June 14, 2019, https:// clinicaltrials.gov/ct2/show/nct03277586, accessed January 19, 2020.

9. C. Jiang, G. Li, P. Huang, Z. Liu, and B. Zhao, "The Gut Microbiota and Alzheimer's Disease," *Journal of Alzheimer's Disease* 58, no. 1 (May 3, 2017): 1–15, https://doi.org/10.3233/jad-161141.

10. S. A. N. Azm, A. Djazayeri, M. Safa, K. Azami, B. Ahmadvand, F. Sabbaghziarani, M. Sharifzadeh, and M. Vafa, "Lactobacilli and Bifidobacteria Ameliorate Memory and Learning Deficits and Oxidative Stress in β-Amyloid (1–42) Injected Rats," *Applied Physiology, Nutrition, and Metabolism* 43, no. 7 (July 2018): 718–726, https://doi.org/10.1139/apnm-2017

-0648; N. H. Musa, V. Mani, S. M. Lim, S. Vidyadaran, A. B. A. Majeed, and K. Ramasamy, "Lactobacilli-Fermented Cow's Milk Attenuated Lipopolysaccharide-Induced Neuroinflammation and Memory Impairment In Vitro and In Vivo," *Journal of Dairy Research* 84, no. 4 (November 2017): 488–495, https://doi.org/10.1017/s0022029917000620; M. Nimgampalle and Y. Kuna, "Anti-Alzheimer Properties of Probiotic, *Lactobacillus plantarum* MTCC 1325 in Alzheimer's Disease Induced Albino Rats," *Journal of Clinical and Diagnostic Research* 11, no. 8 (August 2017): KC01-KC05, https://doi.org/10.7860/jcdr/2017/26106.10428.

11. A. Agahi, G. A. Hamidi, R. Daneshvar, M. Hamdieh, M. Soheili, A. Alinaghipour, S. M. E. Taba, and M. Salami, "Does Severity of Alzheimer's Disease Contribute to Its Responsiveness to Modifying Gut Microbiota? A Double Blind Clinical Trial," *Frontiers in Neurology* 9 (August 2018): 662, https://doi.org/10.3389/fneur.2018.00662.

12. F. Leblhuber, K. Steiner, B. Schuetz, D. Fuchs, and J. M. Gostner, "Probiotic Supplementation in Patients with Alzheimer's Dementia—An Explorative Intervention Study," *Current Alzheimer Research* 15, no. 12 (2018): 1106–1113, https://doi.org/10.2174/1389200219666180813144834.

13. E. Akbari, Z. Asemi, R. D. Kakhaki, F. Bahmani, E. Kouchaki, O. R. Tamtaji, G. A. Hamidi, and M. Salami, "Effect of Probiotic Supplementation on Cognitive Function and Metabolic Status in Alzheimer's Disease: A Randomized, Double-Blind and Controlled Trial," *Frontiers in Aging Neuroscience* 8 (November 2016): 256, https://doi.org/10.3389/fnagi.2016.00256.

14. P. Damier, E. C. Hirsch, P. Zhang, Y. Agid, and F. Javoy-Agid, "Glutathione Peroxidase, Glial Cells and Parkinson's Disease," *Neuroscience* 52, no. 1 (January 1993): 1–6, https://doi.org/10.1016/0306-4522(93)90175-f.

15. A. Fasano, N. P. Visanji, L. W. C. Liu, A. E. Lang, and R. F. Pfeiffer, "Gastrointestinal Dysfunction in Parkinson's Disease," *The Lancet Neurology* 14, no. 6 (June 1, 2015): 625–639, https://doi.org/10.1016/s1474-4422(15)00007-1.

16. M. G. Gareau, M. A. Silva, and M. H. Perdue, "Pathophysiological Mechanisms of Stress-Induced Intestinal Damage," *Current*

Molecular Medicine 8, no. 4 (June 2008): 274–281, https://doi.org/10
.2174/156652408784533760; M. Maes, "The Cytokine Hypothesis of
Depression: Inflammation, Oxidative and Nitrosative Stress (IO&NS)
and Leaky Gut as New Targets for Adjunctive Treatments in Depression," *Neuroendocrinology Letters* 29, no. 3 (June 2008): 287–291, http://
www.nel.edu/userfiles/articlesnew/NEL290308R02.pdf.

17. L. P. Kelly, P. M. Carvey, A. Keshavarzian, K. M. Shannon, M. Shaikh,
R. A. E. Bakay, and J. H. Kordower, "Progression of Intestinal Permeability Changes and Alpha-Synuclein Expression in a Mouse Model of
Parkinson's Disease," *Movement Disorders* 29, no. 8 (July 2014): 999–1009,
https://doi.org/10.1002/mds.25736.

18. O. R. Tamtaji, M. Taghizadeh, R. D. Kakhaki, E. Kouchaki, F. Bahmani, S. Borzabadi, S. Oryan, A. Mafi, and Z. Asemi, "Clinical and Metabolic Response to Probiotic Administration in People with Parkinson's
Disease: A Randomized, Double-Blind, Placebo-Controlled Trial," *Clinical
Nutrition* 38, no. 3 (June 1, 2019): 1031–1035, https://doi.org/10.1016/j
.clnu.2018.05.018.

19. S. Borzabadi, S. Oryan, A. Eidi, E. Aghadavod, R. D. Kakhaki, O. R.
Tamtaji, M. Taghizadeh, and Z. Asemi, "The Effects of Probiotic Supplementation on Gene Expression Related to Inflammation, Insulin and
Lipid in Patients with Parkinson's Disease: A Randomized, Double-Blind,
Placebo-Controlled Trial," *Archives of Iranian Medicine* 21, no. 7 (July
2018): 289–295, http://www.ams.ac.ir/AIM/NEWPUB/18/21/7/S1029-2977
-21(07)289-0.pdf.

20. M. Barichella, C. Pacchetti, C. Bolliri, E. Cassani, L. Iorio, C. Pusani, G.
Pinelli, et al., "Probiotics and Prebiotic Fiber for Constipation Associated
with Parkinson's Disease: An RCT," *Neurology* 87, no. 12 (September 20,
2016): 1274–1280, https://doi.org/10.1212/wnl.0000000000003127.

21. D. Georgescu, O. E. Ancusa, L. A. Georgescu, I. Ionita, and D. Reisz,
"Nonmotor Gastrointestinal Disorders in Older Patients with Parkinson's Disease: Is There Hope?," *Clinical Interventions in Aging* 11 (2016):
1601–1608, https://doi.org/10.2147/cia.s106284.

22. E. Cassani, G. Privitera, G. Pezzoli, C. Pusani, C. Madio, L. Iorio, and
M. Barichella, "Use of Probiotics for the Treatment of Constipation in

Parkinson's Disease Patients," *Minerva Gastroenterologica e Dietologica* 57, no. 2 (June 2011): 117–121, https://www.minervamedica.it/en/journals /gastroenterologica-dietologica/article.php?cod=R08Y2011N02A0117.

23. Tamtaji et al., "Clinical and Metabolic Response."

24. National Institutes of Health, U.S. National Library of Medicine, search results for "probiotics | autism," ClinicalTrials.gov, https://www .clinicaltrials.gov/ct2/results?cond=autism&term=probiotics&cntry= &state=&city=&dist=, accessed January 19, 2020.

25. S. Y. Shaaban, Y. G. El Gendy, N. S. Mehanna, W. M. El-Senousy, H. S. A. El-Feki, K. Saad, and O. M. El-Asheer, "The Role of Probiotics in Children with Autism Spectrum Disorder: A Prospective, Open-Label Study," *Nutritional Neuroscience* 21, no. 9 (November 2018): 676–681, https://doi.org/10.1080/1028415x.2017.1347746.

26. A. Fattorusso, L. Di Genova, G. B. Dell'Isola, E. Mencaroni, and S. Esposito, "Autism Spectrum Disorders and the Gut Microbiota," *Nutrients* 11, no. 3 (February 28, 2019): 521, https://doi.org/10.3390/nu11030521.

17 Artificial Intelligence, Synthetic Biology, and the Microbiome

1. Dartmouth Highlights, "Artificial Intelligence Coined at Dartmouth," https://250.dartmouth.edu/highlights/artificial-intelligence-ai-coined -dartmouth, accessed February 2, 2020.

2. A. Zomorrodi, oral interview with authors, August 13, 2019.

3. Zomorrodi, oral interview.

4. Zomorrodi, oral interview.

5. C. Ross, "In Hunt for New Drugs, Amazon and Other Tech Giants Are Using AI to Find Protein Structures," *Stat+*, September 24, 2019, https:// www.statnews.com/2019/09/24/amazon-google-facebook-ai-protein -structures/, accessed January 21, 2020.

6. HealthCatalyst, healthcare.ai, "Machine Learning versus Statistics: When to Use Each," *Data Science Blog*, https://healthcare.ai/machine -learning-versus-statistics-use/, accessed January 21, 2020.

7. Zomorrodi, oral interview.

8. Zomorrodi, oral interview.

9. R. A. Naqvi, M. Arsalan, G. Batchuluun, H. S. Yoon, and K. R. Park, "Deep Learning–Based Gaze Detection System for Automobile Drivers Using a NIR Camera Sensor," *Sensors* 18, no. 2 (February 2018): 456, https://doi.org/10.3390/s18020456.

10. A. Baldominos, Y. Saez, and P. Isasi, "Evolutionary Design of Convolutional Neural Networks for Human Activity Recognition in Sensor-Rich Environments," *Sensors* 18, no. 4 (April 2018): 1288, https://doi.org/10.3390/s18041288; A. Fernández, R. Usamentiaga, J. L. Carús, and R. Casado, "Driver Distraction Using Visual-Based Sensors and Algorithms," *Sensors* 16, no. 11 (November 2016): 1805, https://doi.org/10.3390/s16111805.

11. Zomorrodi, oral interview.

12. Z. Obermeyer and E. J. Emanuel, "Predicting the Future—Big Data, Machine Learning, and Clinical Medicine," *New England Journal of Medicine* 375, no. 13 (September 29, 2016): 1216–1219, https://doi.org/10.1056/nejmp1606181.

13. P. C. Konturek, J. Koziel, W. Dieterich, D. Haziri, S. Wirtz, I. Glowczyk, K. Konturek, M. F. Neurath, and Y. Zopf, "Successful Therapy of *Clostridium difficile* Infection with Fecal Microbiota Transplantation," *Journal of Physiology and Pharmacology* 67, no. 6 (December 2016): 859–866, http://www.jpp.krakow.pl/journal/archive/12_16/pdf/859_12_16_article.pdf.

14. T. Lu, oral interview with authors, August 23, 2019.

15. Lu, oral interview.

16. Lu, oral interview.

17. Lu, oral interview.

18. Lu, oral interview.

19. Lu, oral interview.

20. Lu, oral interview.

21. Lu, oral interview.

22. Lu, oral interview.

23. Lu, oral interview.

24. Lu, oral interview.

25. Lu, oral interview.

26. T. Ching, D. S. Himmelstein, B. K. Beaulieu-Jones, A. A. Kalinin, B. T. Do, G. P. Way, E. Ferrero, et al., "Opportunities and Obstacles for Deep Learning in Biology and Medicine," *Journal of the Royal Society Interface* 15, no. 141 (April 1, 2018): 20170387, https://doi.org/10.1098/rsif.2017.0387.

27. Lu, oral interview.

28. Lu, oral interview.

29. Lu, oral interview.

18   Maintaining a Resilient Microbiome through Old Age

1. E. Biagi, L. Nylund, M. Candela, R. Ostan, L. Bucci, E. Pini, J. Nikkïla, et al., "Through Ageing, and Beyond: Gut Microbiota and Inflammatory Status in Seniors and Centenarians," *PLoS One* 5, no. 5 (May 2010): e10667, https://doi.org/10.1371/journal.pone.0010667.

2. T. C. Cullender, B. Chassaing, A. Janzon, K. Kumar, C. E. Muller, J. J. Werner, L. T. Angenent, et al., "Innate and Adaptive Immunity Interact to Quench Microbiome Flagellar Motility in the Gut," *Cell Host and Microbe* 14, no. 5 (November 13, 2013): 571–581, https://doi.org/10.1016/j.chom.2013.10.009.

3. A. C. Hearps, G. E. Martin, T. A. Angelovich, W.-J. Cheng, A. Maisa, A. L. Landay, A. Jaworowski, and S. M. Crowe, "Aging Is Associated with Chronic Innate Immune Activation and Dysregulation of Monocyte Phenotype and Function," *Aging Cell* 11, no. 5 (October 2012): 867–875, https://doi.org/10.1111/j.1474-9726.2012.00851.x.

4. Metchnikoff, *The Prolongation of Life.*

5. C. Franceschi, M. Bonafè, S. Valensin, F. Olivieri, M. De Luca, E. Otta-viani, and G. De Benedictis, "Inflamm-Aging: An Evolutionary Perspective on Immunosenescence," *Annals of the New York Academy of Sciences* 908, no. 1 (June 2000): 244–254, https://doi.org/10.1111/j.1749-6632.2000.tb06651.x.

6. R. I. Clark, A. Salazar, R. Yamada, S. Fitz-Gibbon, M. Morselli, J. Alcaraz, A. Rana, et al., "Distinct Shifts in Microbiota Composition during *Drosophila* Aging Impair Intestinal Function and Drive Mortality," *Cell Reports* 12, no. 10 (September 8, 2015): 1656–1667, https://doi.org/10.1016/j.celrep.2015.08.004.

7. M. Rera, R. I. Clark, and D. W. Walker, "Intestinal Barrier Dysfunction Links Metabolic and Inflammatory Markers of Aging to Death in *Drosophila*," *Proceedings of the National Academies of Sciences of the United States of America* 109, no. 52 (December 26, 2012): 21528–21533, https://doi.org/10.1073/pnas.1215849110.

8. S. Oliviero and R. Cortese, "The Human Haptoglobin Gene Promoter: Interleukin-6-Responsive Elements Interact with a DNA-Binding Protein Induced by Interleukin-6," *The EMBO Journal* 8, no. 4 (April 1, 1989): 1145–1151, https://doi.org/10.1002/j.1460-2075.1989.tb03485.x.

9. Y. Qi, R. Goel, S. Kim, E. M. Richards, C. S. Carter, C. J. Pepine, M. K. Raizada, and T. W. Buford, "Intestinal Permeability Biomarker Zonulin Is Elevated in Healthy Aging," *Journal of the American Medical Directors Association* 8, no. 9 (September 1, 2017): 810.e1–810.e4, https://doi.org/10.1016/j.jamda.2017.05.018.

10. Qi et al., "Intestinal Permeability Biomarker Zonulin."

11. P. Carrera-Bastos, O. Picazo, M. Fontes-Villalba, H. Pareja-Galeano, S. Lindeberg, M. Martínez-Selles, A. Lucia, and E. Emanuele, "Serum Zonulin and Endotoxin Levels in Exceptional Longevity versus Precocious Myocardial Infarction," *Aging and Disease* 9, no. 2 (April 2018): 317–321, https://doi.org/10.14336/ad.2017.0630.

12. P. L. Minciullo, A. Catalano, G. Mandraffino, M. Casciaro, A. Crucitti, G. Maltese, N. Morabito, A. Lasco, S. Gangemi, and G. Basile, "Inflammaging and Anti-inflammaging: The Role of Cytokines in Extreme

Longevity," *Archivum Immunologiae et Therapiae Experimentalis* 64, no. 2 (April 2016): 111–126, https://doi.org/10.1007/s00005-015-0377-3.

13. C. Franceschi, M. Capri, D. Monti, S. Giunta, F. Olivieri, F. Sevini, M. P. Panourgia, et al., "Inflammaging and Anti-inflammaging: A Systemic Perspective on Aging and Longevity Emerged from Studies in Humans," *Mechanisms of Ageing and Development* 128, no. 1 (January 2007): 92–105, https://doi.org/10.1016/j.mad.2006.11.016.

14. N. Thevaranjan, A. Puchta, C. Schulz, A. Naidoo, J. C. Szamosi, C. P. Verschoor, D. Loukov, et al., "Age-Associated Microbial Dysbiosis Promotes Intestinal Permeability, Systemic Inflammation, and Macrophage Dysfunction," *Cell Host and Microbe* 21, no. 4 (April 12, 2017): 455–466, https://doi.org/10.1016/j.chom.2017.03.002.

15. Biagi et al., "Through Ageing."

16. Biagi et al., "Through Ageing."

17. E. Biagi, C. Franceschi, S. Rampelli, M. Severgnini, R. Ostan, S. Turroni, C. Consolandi, et al., "Gut Microbiota and Extreme Longevity," *Current Biology* 26, no. 11 (June 6, 2016): 1480–1485, https://doi.org/10.1016/j.cub.2016.04.016.

18. Biagi et al., "Gut Microbiota."

19. M. J. Claesson, I. B. Jeffery, S. Conde, S. E. Power, E. M. O'Connor, S. Cusack, H. M. B. Harris, et al., "Gut Microbiota Composition Correlates with Diet and Health in the Elderly," *Nature* 488, no. 7410 (August 9, 2012): 178–184, https://doi.org/10.1038/nature11319.

Epilogue

1. N. Pariente, "Milestones in Human Microbiota Research," *Milestones*, June 18, 2019, https://www.nature.com/immersive/d42859-019-00041-z/index.html, accessed February 2, 2020.

2. J. Leidy, *A Flora and Fauna within Living Animals* (New York: Putnam and Company, 1853).

3. P. J. Turnbaugh et al., "An Obesity-Associated Gut Microbiome."

Index